"A compelling and thoroughly researched narrative history of a seminal lawsuit aimed not only at a provision of U.S. patent law but at the ethical question of whether anyone should be allowed to own human genes."
—*The Wall Street Journal*

"In his eye-opening debut, law professor Contreras vividly tells the inside story of *AMP v. Myriad* . . . Contreras brings the large cast of case participants to life with vivid prose, and the exciting final spectacle before the Supreme Court is heart-pumping . . . The result is a thorough page-turner." —*Publishers Weekly* (starred review)

"Jorge L. Contreras, a law professor at the University of Utah, interviewed nearly 100 lawyers, patients, scientists and policymakers in this behind-the-scenes history of *Molecular Pathology vs. Myriad Genetics*, a long-shot lawsuit that culminated in a landmark 2013 Supreme Court decision that opened the human genome to the benefit of researchers, cancer patients and everyday Americans." —*The New York Times*

"Remarkable. Contreras manages to make a book about the lawsuit that ended gene patenting in America read like a thriller. This book will not only inform you and stir your moral outrage, it will keep you on the edge of your seat." —Ayelet Waldman, author of *A Really Good Day*

"Ably and affectingly detailed . . . This story stands as a guide to the forces that shape an increasingly important industry—and to the vexed influence of patents." —*Nature*

"*The Genome Defense* provides an incomparable perspective on a landmark Supreme Court case, and it is a testament to the importance of intellectual property law to humanity's prosperity."
—Jessica Silbey, author of *The Eureka Myth: Creators, Innovators, and Everyday Intellectual Property*

"An unflinching critique of the biotech industry's business practices . . . Contreras never presumes that his readers can't keep up, until suddenly a lay reader can likely understand the fundamentals of patent eligibility. That mix makes the book an incredibly accessible and engaging read."

—*Law360*

"Fascinating . . . Contreras goes behind the scenes with many of the key participants, offering an inside look at the legal strategy and the potential consequences on either side of the final decision."

—*Salt Lake City Weekly*

"Superb . . . Contreras assembles a large cast of lawyers, judges, activists, scientists, and patients and engagingly describes four years of tortuous legal action that saw victory in federal court, reversal on appeal, and a final triumph in the Supreme Court. A detailed account of patent law that, against all odds, turns out to be fascinating."

—*Kirkus Reviews* (starred review)

"Both a page-turner full of colorful characters and a profound commentary about how corporate giants use the law to monopolize knowledge and what we can do about it."

—Orly Lobel, author of *You Don't Own Me: The Court Battles That Exposed Barbie's Dark Side*

"A remarkable, fast-paced read. Contreras tells the behind-the-scenes story of how the Supreme Court stopped the patenting of the human genome. He does it in such an engaging style that it's almost like reading a legal thriller."

—Professor Mark A. Lemley, director of the Stanford Program in Law, Science, and Technology

"Masterfully written." —*Neo.Life*

"A gripping and important tale of how corporations were patenting our own genes and selling them back to us. Contreras gives us front-row-seat access, deft character sketches and crystal-clear explanations of law and science." —Jordan Fisher Smith, author of *Engineering Eden*

THE
GENOME
DEFENSE

**INSIDE THE EPIC LEGAL BATTLE TO
DETERMINE WHO OWNS YOUR DNA**

JORGE L. CONTRERAS

ALGONQUIN BOOKS OF CHAPEL HILL 2022

Published by
Algonquin Books of Chapel Hill
Post Office Box 2225
Chapel Hill, North Carolina 27515-2225

a division of
Workman Publishing
225 Varick Street
New York, New York 10014

First paperback edition, Algonquin Books of Chapel Hill, October 2022.
Originally published in hardcover by Algonquin Books of Chapel Hill
in October 2021.
Printed in the United States of America.
Design by Steve Godwin.

Appendix adapted from Jorge L. Contreras, *Association for Molecular
Pathology v. Myriad Genetics: A Critical Reassessment*, 27 Mich. Tech. L.
Rev. 1 (2020)

Library of Congress Cataloging-in-Publication Data
Names: Contreras, Jorge L., author.
Title: The genome defense : inside the epic legal battle to determine who owns
your DNA / Jorge L. Contreras.
Description: Chapel Hill, North Carolina : Algonquin Books of Chapel Hill,
[2021] | Includes bibliographical references and index. | Summary: "The
gripping true story of a Supreme Court civil rights battle to prevent biotech
companies from owning the very thing that makes us who we are—our
DNA"— Provided by publisher.
Identifiers: LCCN 2021021980 | ISBN 9781616209681 (hardcover) | ISBN
9781643752150 (ebook)
Subjects: LCSH: Association for Molecular Pathology,—Trials, litigation, etc. |
Myriad Genetics, Inc.,—Trials, litigation, etc. | Patent suits—United States. |
Human gene mapping—United States—Patents. | Human chromosome
abnormalities—Diagnosis—United States—Patents. | Human gene mapping—
Law and legislation—United States. | Human chromosome
abnormalities—Diagnosis—Law and legislation—United States.
Classification: LCC KF228.A8436 C66 2021 | DDC 346.7304/86—dc23
LC record available at https://lccn.loc.gov/2021021980

ISBN 978-1-64375-324-9 (PB)

10 9 8 7 6 5 4 3 2 1
First Paperback Edition

For my mother, who gave me a microscope
and showed me its magic, and my father,
who let me launch rockets.

NEARLY TWO CENTURIES AGO . . . Thomas Jefferson and a trusted aide spread out a magnificent map . . . The aide was Meriwether Lewis, and the map was the product of his courageous expedition . . . to the Pacific. It was a map that . . . forever expanded the frontiers of our continent and our imagination. Today the world is joining us . . . to behold a map of even greater significance. We are here to celebrate the completion of the first survey of the entire human genome.

—U.S. PRESIDENT BILL CLINTON (2000)

[I]F ANYTHING IS literally a common birthright of human beings, it is the human genome. It would thus seem that if anything should be avoided in the genomic political economy, it is a war of patents and commerce over the operational elements of that birthright.

—DANIEL J. KEVLES AND LEROY HOOD, THE CODE OF CODES (1992)

CONTENTS

Rally outside the U.S. Supreme Court
prior to oral arguments in AMP v. Myriad.

PREFACE

"FREE OUR GENES!" The chants began just after dawn on April 15, 2013, reverberating defiantly across the empty Supreme Court plaza. As the gray morning advanced, demonstrators—mostly women wearing ribbons and hoodies—posed for television crews and shouted into megaphones. They brandished signs and placards hand-printed with messages like OUTLAW HUMAN GENE PATENTS and HUMAN GENES BELONG TO HUMAN BEINGS.

The case being argued that morning concerned a technical question—the proper interpretation of Section 101 of the U.S. Patent Act. The legal briefs were laden with references to nineteenth-century precedents and details of molecular biology. Typically, cases like this attracted the attention of a few biotech lobbyists, science journalists, and curious law students. Yet on that day, the scene outside the court resembled a major civil rights demonstration. What was going on?

IN 2013, the United States Supreme Court was presented with a seemingly simple question: are human genes patentable? Its answer would have profound implications for the law, for science, for business, and for humankind in general. What exactly does it mean?

For the most part, the question is whether a single person or, more to the point, a single company can legally own the right to replicate the DNA that exists inside every man, woman, and child on the planet. And why would a company want to own this right and spend millions of dollars litigating over it? Because reproducing DNA in the laboratory is often the key to diagnosing deadly genetic conditions, assessing vulnerability to disease, and developing new drugs. In the United States alone, this is a half-trillion dollar market.

I spent much of one day in May 2009 in a large, artificially lit conference room at the Bethesda, Maryland, headquarters of the U.S.

National Institutes of Health (NIH). NIH, established in 1887, is the nation's biomedical research agency; it funds close to $40 billion in scientific research each year. Modern chemotherapy was invented there. Its researchers developed the vaccines for yellow fever and typhus and made significant contributions to those for COVID-19. And in the late 1990s NIH led the massive multinational effort to sequence the human genome.

I was there to attend a meeting of the National Advisory Council for Human Genome Research, the body that had overseen the Human Genome Project and now serves as the advisory group to NIH's Genome Institute. In 2000, the Human Genome Project announced the first draft map of the genome of the human species, followed three years later by its complete DNA sequence. Now, the Genome Institute was leading even more ambitious projects aimed at mapping the invisible "microbiome" of organisms inhabiting the human body, discovering the association between genetic mutations and a host of human diseases, and uncovering the genetic code of cancer.

I was the only practicing lawyer on the eighteen-member council; most of my colleagues were geneticists, data scientists, and clinicians from prestigious research institutions. I was there because I knew something about patents, specifically about patents covering human genes.

In the late 1990s, when I was a junior partner at a large Boston law firm, I advised a consortium of pharmaceutical companies that funded research alongside the Human Genome Project. The consortium released its results into the public domain, so that all could use them freely. Though I can't take credit for it, the SNP Consortium, as it was known, was phenomenally successful and became famous throughout the genomics research community. As a result I got to know many of the leaders of the Genome Institute, and was invited by its then-director, Francis Collins, to sit on its advisory council. There, I heard reports on the most cutting-edge genomics research being conducted around the country and got to help, in a small way, to chart the future of that research.

I clearly remember the council meeting in May 2009. Alan Guttmacher, the acting director of the Institute, had just delivered a report updating us on the many different projects under way. We were taking our morning coffee break, and I checked email on my Blackberry. An interesting news item caught my eye. The American Civil Liberties Union (ACLU) had just filed a lawsuit against a Utah-based biotech company called Myriad Genetics. The ACLU, in its first-ever patent suit, sought to invalidate several of Myriad's patents covering two human genes known as *BRCA1* and *BRCA2*.

The genes, and Myriad's patents, were well known in the field. Myriad had used them to build a lucrative business testing women for genetic variants that pointed to a dramatically increased risk of breast and ovarian cancer. I had discussed these very patents with Dr. Collins before I joined the Institute's council and shared his general unease with the practice of patenting complete human genes. Suffice it to say that many in the scientific community regarded Myriad's patents, and the business strategies that it used to exploit them, with discomfort. In the NIH conference room that morning, the buzz of conversation swelled as more council members viewed the news reports on their devices. The scientific director of the Institute, Eric Green, was standing. "The ACLU sued Myriad," he muttered to no one in particular. "Wow." Murmurs of agreement echoed around the room.

From that moment, I was hooked by the case, which was known officially as *Association for Molecular Pathology (AMP) v. Myriad Genetics*. Gene patents had been around for more than twenty-five years and were an accepted, if controversial, part of the legal landscape; could this unorthodox attack really work? At the end of 2009, for unrelated reasons, I left my law firm to pursue a career in academia. I gave talks about the case and wrote articles about it in law journals. But as the legal battle between the ACLU and Myriad progressed through the court system, from a district court in New York to the circuit court of appeals in Washington, DC, and eventually to the Supreme Court of the United States (twice), I became convinced that more was at stake than an obscure doctrine of patent law. The case involved a broad array of personalities, from scientists and doctors to lawyers, patients, and advocacy groups. It also attracted the attention of people at the highest levels of the Obama administration, which intervened in the case in unprecedented ways. And it garnered more coverage from the popular media than any other patent case in recent memory, including a segment on *60 Minutes*, not to mention op-eds by celebrities like Michael Crichton and Angelina Jolie. Something about the case had captured the public imagination like no other patent case had before, or has since.

In 2013, I began to interview everyone who would speak with me about *AMP v. Myriad*. In the midst of this work, I left American University in Washington, DC, to join the faculty at the University of Utah, the very place where Myriad was born. Over the course of the next seven years I interviewed nearly a hundred people involved in all aspects of the case: lawyers, judges, patients, scientists, doctors, genetic counselors, policy makers, and academics. In this book, I have tried to weave together their intertwined stories to show how they intersected at a unique moment in history.[1] More than simply deciding a narrow point of patent law, *AMP v. Myriad* illustrates how the law is made in this country, and how politics, science, and litigation can sometimes converge in unexpected ways. In short, it is about much more than one company, two genes, or seven patents; it is about how the law struggles to keep pace with scientific advances, and how it can be moved in one direction or another by those who seize the opportunity.

But this is also a story about people and the unlikely series of events that led them to the Supreme Court on a drizzly day in April 2013. In fact, despite the thousands of hours of hard work that so many people put into *AMP v. Myriad*, pure chance played a surprisingly large role in how the case developed and, some would argue, how it turned out. From the very outset, this unlikeliest of patent cases was filed by a civil rights lawyer who knew nothing about biology and a young science policy analyst who didn't have a law degree. The organization that employed them, the ACLU, had never filed a patent case in over a century of distinguished legal advocacy. Many of the experts they consulted said it was a lost cause. Yet this band of passionate advocates set out to overturn well-established policies that had been in place for more than three decades, taking on the biotech industry, the Department of Commerce, and the vociferous patent bar. If it sounds like the odds were against them, they were. This is their story.

---

1 The story of Myriad Genetics and its *BRCA* patents is not limited to the United States, and it has led to important legal developments in Canada, Europe, Australia, and elsewhere. But for the sake of the narrative, this book focuses largely on the debate and the case in the United States. Several excellent sources listed in the bibliography offer greater insight into the history of Myriad's international patenting program.

PART I

BUILDING THE CASE

THE HUMAN GENOME underlies the fundamental unity of all members of the human family, as well as the recognition of their inherent dignity and diversity. In a symbolic sense, it is the heritage of humanity.

—UNITED NATIONS UNIVERSAL DECLARATION ON THE
   HUMAN GENOME AND HUMAN RIGHTS (1997)

CHAPTER 1

Who Can We Sue?

THE NATIONAL HEADQUARTERS of the American Civil Liberties Union occupies three floors of 125 Broad Street, a forty-story office tower near the tip of lower Manhattan. The building's thin vertical ribs and smoky glass panes evoke the modernist 1970s design ethos that once pervaded the city, most memorably in the fallen twin towers that stood less than a mile away. From many of the ACLU's floor-to-ceiling windows, one has an unobstructed view of the East River, where on a sunny morning the endless procession of taxicabs, limos, and delivery trucks crawling across the upper deck of the Brooklyn Bridge sparkles like a string of glass beads. Occasionally a blue, red, or yellow helicopter floats into view, shuttling Wall Street tycoons to their mansions on Long Island or circling the Statue of Liberty for tourists.

The other tenants of 125 Broad Street include an enviable roster of blue-chip ad agencies, financial consultants, and law firms. The building is owned by its principal tenant, Sullivan & Cromwell, a venerable legal juggernaut founded in the decade after the Civil War. But stepping out of the elevator into the ACLU's offices, it is immediately apparent that one has not entered a high-priced legal boutique. There is no marble reception foyer, no chrome-and-leather seating, no oil paintings or Warhol lithographs—just a bustling bullpen of a workspace, teeming with office workers, photocopy machines, and student interns. The seventeenth floor is occupied by legal staff. Its internal corridors, designed by the building's Park Avenue architects to invoke air and spaciousness, are instead crammed with dark metal filing cabinets topped by teetering piles of storage boxes, three-ring binders, and miscellaneous office equipment. There is barely room to walk without brushing up against these steel cavern walls.

One Friday morning in early 2005, Tania Simoncelli threaded her way through this cluttered warren. In her early thirties with thick brown hair, dark brown eyes, and a friendly, easygoing manner, she walked briskly, smiling at everyone she passed.

As the ACLU's first and only scientific advisor, Simoncelli's role was unique. Two years earlier she had been handpicked for the job by the organization's executive director, Anthony Romero. Romero, a scrappy Puerto Rican wunderkind, had climbed his way from a Bronx housing project to Princeton and Stanford Law to become one of the youngest directors at the Ford Foundation. He took the reins at the ACLU, its first Latino and openly gay chief, six days before the 9/11 attacks. On September 14, with New York still blanketed by ash, the Reverend Jerry Falwell appeared on national television to blame the ACLU—with its support of abortion, gay rights, and secular schools—for turning God against the United States. But Romero didn't blink. During his first month on the job, he was thrust into the maelstrom, and it invigorated him and the organization that he led. The ACLU's campaigns against mass surveillance, the USA PATRIOT Act, and the incarceration of anyone with even the most tenuous link to Al-Qaeda gave the aging organization a vitality it hadn't enjoyed since the civil rights battles of the 1960s. Donations poured in and Romero quickly doubled the size of his staff.

To Romero, the ACLU was the last bulwark of American civil liberties. Throughout its long history, the organization had battled the government's use of technology in violating individual rights: wiretaps, hidden microphones, drug-sniffing dogs. Now, a whole new set of tools for abuse was emerging from the Wild West of science and technology. Things like internet surveillance, predictive profiling, and facial recognition had moved from the realm of science fiction to the nightly news. Romero wanted the ACLU to remain not only relevant, but indispensable, in the civil rights struggles of the future.

With the ACLU's formidable reputation, Romero could have tapped anyone to be his first science advisor: an Ivy League professor, a Silicon Valley pundit, a Nobel laureate. But he didn't want someone whose reputation was already cemented, who would mold the ACLU to his or her own views of the world. Romero and the ACLU's leadership wanted someone young, ambitious, and intellectually curious. Someone who could work with them to shape the ACLU's vision of science together and who had, in Romero's words "a very high tolerance for the unknown."

Tolerating the Unknown

THAT PERSON WAS TANIA SIMONCELLI. The youngest of four children, Simoncelli was raised outside of Philadelphia, the precocious daughter of an architect and an artist. As a student Simoncelli's interests were wide-ranging, encompassing music, biology, ethics, philosophy, government, and environmental policy, to name just a few. While an undergraduate at Cornell University, Simoncelli chanced upon a unique interdisciplinary program in Biology and Society, which combined many of her diverse interests. She enrolled and soon became a research assistant to Professor Sheila Jasanoff, a key figure in the emerging field of science and technology studies (STS).

After graduation, Simoncelli moved to DC, where she worked as a policy and legal analyst. But she soon realized that she had more to learn. So after three years, Simoncelli left DC to pursue a master's degree in international environmental policy at the University of California Berkeley. She wrote her thesis on conservation and bioprospecting in Costa Rica, but after getting her degree in 1999, she was not sure what to do next. So she turned back to one of her earliest passions, music, and devoted serious time to the cello—playing in ensembles, the occasional orchestral gig, and giving private lessons.

The San Francisco Bay Area at the turn of the millennium was a hive of intellectual activity. In nearby Silicon Valley companies were leading the global dot-com revolution, and biotech firms like Genentech were promising miracle cures based on the latest genetic discoveries.

In addition to a growing business landscape around human genetics, the 1990s saw increasing public awareness of the risks of genetic science. The film adaptation of Michael Crichton's novel *Jurassic Park* was released in 1993, introducing moviegoers to the frightening possibilities of genetic engineering, and the 1997 blockbuster *Gattaca* portrayed a not-too-distant world of genetic perfectionism that darkly echoed the eugenics movements of the past. In real life, genetic technology was in the news as well, with the highly publicized race to sequence the human genome in full swing, and reports from Scotland of a cloned ewe named Dolly.

In some neighborhoods, you couldn't enter a bar or café without overhearing at least one heated debate about the social and ethical implications of genetic technology. It was in this milieu that Simoncelli was approached by another former graduate student, Richard Hayes, who was putting together a group—part think tank, part agitprop—to focus on the social impact of genetic advances. He called it the Center for Genetics and Society, and he asked Simoncelli to join him. She was reluctant at first—Hayes had no money, he seemed more interested in banning farfetched research like human cloning than exploring more pressing issues, and her scientific interests were still focused on climate policy. But, eventually, she agreed, and they officially launched the Center in 2001 with a conference provocatively titled "Beyond Cloning: Protecting Humanity from Species-Altering Procedures."

Simoncelli helped run the Center on a part-time basis for a couple of years. Operating on a shoestring budget, the Center served as a venue for local intellectuals, policy wonks, and graduate students to debate the ethical and social implications of new genetic technologies . One observer aptly called them "genome huggers."

But after four years as a musician and part-time Center organizer, Simoncelli decided that it was time to move on with her career and get her PhD. With her credentials from Cornell and Berkeley, as well as her time at the Center, her record in science policy was strong. She applied to five different graduate programs and was accepted by all of them.

After some soul searching, she decided on the prestigious Program in Science, Technology and Society at the Massachusetts Institute of Technology. MIT's program accepted only four candidates that year, and they needed to know whether Simoncelli was coming. She had two weeks to accept.

Simoncelli was just finalizing her plans to move to Boston when a friend emailed her a job ad. It was for a new advisory position at the ACLU. Simoncelli said the job sounded interesting, but she was about to accept her offer from MIT. Her friend responded with one word: "Why?"

Soon, Simoncelli flew to New York for an interview. The rest happened quickly. Everyone at the ACLU who met Simoncelli wanted to hire her. For them, she offered the perfect blend of intelligence, savvy, and passion. The fact that she didn't have a PhD didn't bother them—none of them did either. For Simoncelli it was a once-in-a-lifetime opportunity to help shape an important institution's policies and perspectives on emerging questions at the intersection of science, law, and policy. She politely deferred her admission to MIT and packed her bags for New York.

A Science Angle

BY EARLY 2005 Simoncelli had spent nearly two years at the ACLU. Her role was to advise the organization about emerging scientific issues that had a civil liberties angle, then to help find ways to address them. The job was demanding. Simoncelli was a non-lawyer surrounded by a hundred of the most dedicated and hard-charging lawyers in the country. Getting them to pay attention to highly technical issues in the midst of a host of other responsibilities and crises was not easy, but it was just the kind of challenge that Simoncelli had been looking for.

At Cornell, Simoncelli's advisor Sheila Jasanoff had taught her how to think critically about emerging issues at the intersection of science, technology, law, and ethics. Simoncelli took these lessons to heart. One area that concerned her was criminal law. For years, MRI machines had

Tania Simoncelli, the ACLU's science advisor, who first raised the idea of a lawsuit challenging gene patents.

been used to scan patients' brains for tumors and blood clots. Now, a handful of researchers were proposing that "functional" MRI scans (fMRI) could reveal whether a person was *lying*. Was fMRI the Holy Grail of the criminal justice system, a scientifically validated lie detector? Some people, especially those in law enforcement, desperately wanted to believe so. But Simoncelli believed otherwise. Some red and green splotches on a computer monitor could not bare the soul of someone under interrogation. She organized a series of briefings for the ACLU on these issues, hoping that the organization could intervene before flimsy brain scan evidence was used to convict innocent people.

Then there was genetics. Scientists were increasingly linking a person's genetic code not only to physical characteristics and illnesses, but to behavioral traits—intelligence, musical talent, even the propensity to believe in God. In 2000, while Simoncelli was at Berkeley, the draft map of the human genome was released to international acclaim. But nagging concerns were emerging about the ways that this powerful new knowledge could be used. Could the police arrest someone with a genetic predisposition toward violence? Could an employer fire someone with a predilection for alcoholism?

This set of issues was connected to DNA warehouses. Since the 1990s, first the FBI, then local police departments, routinely collected DNA from criminal suspects. So-called DNA fingerprinting was the gold standard for placing individuals at crime scenes. Television shows like *CSI* and *Law & Order* convinced the public that a tiny speck of DNA could crack any case. But what happened to all of that DNA once an investigation ended? Could the police keep it on file, just in case? Should the government keep a perpetual DNA record of everyone in the country? The ACLU had long opposed such measures, and when Simoncelli arrived, these issues quickly landed on her plate.

The post-9/11 ACLU was a large and sprawling organization still coming to grips with its renewed stature. Not only was there a national office in New York, but local offices existed in nearly every state in the country. Groups within the ACLU focused on particular subject areas: racial discrimination, freedom of the press, women's rights, sexual identity issues. A big part of Simoncelli's job was engaging busy legal staff in an exploration of emerging scientific questions that had implications for civil liberties. Some topics, like DNA fingerprinting, tied naturally to the ACLU's existing programs and priorities. But other issues were harder to pigeonhole.

From graduate school, Simoncelli knew that one of the best ways to attract harried and overworked people was by offering them food. So she organized a series of informal lunchtime sessions with scientists and other experts who would talk with ACLU staff about emerging scientific issues over boxed salads and sandwiches. Sometimes the sessions were more extensive, and included one- or two-day offsite retreats. All of these events enabled Simoncelli to raise awareness about scientific issues that impacted civil rights and to build coalitions of interest within the ACLU.

But even when Simoncelli did manage to kindle interest, at the end of the day what the ACLU did best was litigation. To be most effective, it needed a litigation angle on an issue: someone to sue, someone to represent. Simoncelli lacked the hands-on legal experience to develop a case strategy for her science issues. To get this perspective, she sought out a veteran ACLU litigator named Chris Hansen.

Chris Hansen, a senior attorney on the ACLU National Legal Staff, who spearheaded the gene patenting lawsuit.

The Jedi Master

HANSEN WAS A member of the ACLU's National Legal Staff, one of a handful of civil rights defenders–at–large—free-ranging Jedi masters with the authority and financial backing to pursue whatever causes concerned them the most. As Hansen explained, "It was my job to look for an injustice, somewhere, anywhere in the country. I was unlimited by subject-matter area. I was unlimited by geography. All that I had to find was an injustice that was civil liberties–related, and it was my job to figure out a way to fix it. That was a pretty good job."

Tall and outdoorsy with salt-and-pepper hair, Hansen was seated behind a modest wooden desk, its scratched surface remarkably clear of papers, files, and other detritus. In keeping with the ACLU's egalitarian ethos, his office wasn't large, and its view was mostly obstructed by a gray office building across the street. One had to lean over the credenza to catch a glimpse of the sparkling river to the east.

A copy of the *New York Times* was often unfolded on Hansen's desk. For a civil rights advocate, early 2005 was not an encouraging time. George W. Bush and Dick Cheney had just been sworn in to a second term. Hopes for a reversal of the administration's controversial policies on surveillance, detention, and torture were all but dead. The crippling wars in Iraq and Afghanistan staggered on; the military officially ended its fruitless search for Saddam Hussein's weapons of mass destruction; countless detainees continued to languish at Guantanamo and other black sites; and extremists on three continents ramped up their targeting of major cities, usually followed by predictable law enforcement crackdowns. Overall, the news was not good.

Nevertheless, Hansen smiled when Simoncelli poked her head into his office. He liked her and, as he had told her more than once, thought she had the coolest job at the ACLU.

"What's up?" he asked, pivoting toward her in his high-backed leather chair, the only concession to traditional lawyerly accoutrements that Hansen made.

Simoncelli took a seat across the desk from Hansen. She glanced at the only decoration in his office—a disorienting surrealist print by the French painter Yves Tanguy, Hansen's favorite artist. She began by explaining that she was trying to find litigation angles for some of the science issues she had been thinking about.

"Like what?" Hansen asked. His voice was flat and unaccented, betraying his Midwestern roots. "Of all the issues you've been looking at, which are the most important?"

Simoncelli, becoming enthusiastic, sped up, speaking and gesturing with her hands while her words flew. She began with brain imaging—fMRI. If this technology were admitted as evidence in court, someone could be convicted of a crime based on nothing more than electrical activity in her brain. Was there a way to challenge the use of this technology in the courts?

Hansen thought for a moment. fMRI sounded like a new gloss on the old polygraph lie-detecting machine. They were banned from court in most states now, after the evidence proved them unreliable. Simoncelli nodded. Hansen suggested that she might want to review the cases rejecting lie detector evidence. Some of them might be broad enough to encompass fMRI results.

"Great." Simoncelli nodded, scribbling notes on a legal pad.

She moved on to genetic discrimination. Could they bring a lawsuit if an insurance company discriminated against someone who had the "wrong" DNA profile? Insurers already refused coverage to people infected with HIV—could a genetic propensity for heart disease or cancer be next? And what about employers? Could job interviews routinely require DNA tests, resulting in what one commentator called "cadres of the genetically unemployable"? Hansen considered the question. It was a good issue, he said, but without laws on the books, it was hard to see a litigation strategy. As Simoncelli knew, the ACLU was already supporting legislation to address the issue, and Hansen recommended that she try to engage further on the legislative front.

She nodded, again noting Hansen's suggestion. "Then there's gene patenting," she said.

Hansen furrowed his brow. "*Gene* patenting," he said. "What's that?"

"We discussed it at the meeting in Boston," she said. "Gene patents."

Hansen shrugged. He hadn't attended that meeting.

"OK," said Simoncelli. Last year she had convened a two-day symposium of top academics and policy makers in Boston to help the ACLU think through possible objections to gene patenting. She quickly summarized what had been discussed.

When she finished, Hansen shook his head. "That can't be right," he said.

"Chris, it's right," she said. "The Patent Office has been issuing patents on human genes for more than twenty years!"

Hansen shook his head again. He was no expert in patent law, but surely Simoncelli, who, after all, wasn't a lawyer, had to be misinterpreting something. "They must be patenting the method for getting the DNA, or the function of the gene, or something like that," he said.

"That, too," Simoncelli answered. "But that's not what I'm telling you. They have patents on the DNA. *The genes themselves are patented.*"

Hansen exhaled. What Simoncelli was saying made no sense. How could you patent a part of the human body? "I don't believe it," he said. "Show me."

"OK," Simoncelli said, rising to the challenge. She left Hansen's office and threaded her way through the corridors and filing cabinets back to her own. She turned on her old PC and once the computer came to life, searched through her folders, pulling up three short articles that explained what was happening with gene patents. She typed Hansen's email address and hit SEND. Then she waited.

Twenty minutes later, an agitated Hansen burst into her office. "My God," he exclaimed. "You're right!"

Simoncelli smiled. "Told you so," she said, only half joking.

Hansen shook his head in disbelief. "But that's just . . . *wrong*," he said. How could a human gene be patented? How could the genetic code inside a person be owned by a company, or parceled out to different buyers like lots in a housing development? The very notion was absurd, it was offensive. If this were true, and he still had his doubts, Simoncelli was on to something. Something the ACLU should get involved in. Something that the ACLU should *stop*.

Simoncelli was elated by Hansen's sudden enthusiasm. This was why she had joined the ACLU. "So what do we do now?" she asked.

Hansen folded his arms and smiled. The answer was obvious. "Who can we sue?" he said.

CHAPTER 2

The World in the Helix

THE NEXT MONDAY, Hansen dropped by Simoncelli's office. He was dressed in his typical attire: jeans, a collared shirt, comfortable loafers. The ACLU was casual, but not Greenpeace or Berkeley. People didn't come to work in flip-flops or protest T-shirts.

Before Simoncelli could even ask him how he was doing, Hansen began. He had been thinking about the gene patenting issue over the weekend and wanted to explore it further. "Where do we start?" he asked.

The question caught Simoncelli off guard. She had temporarily put aside their discussion of gene patenting while she turned to other work. She was thrilled but a little surprised, not only that Hansen remembered their conversation, but that he was so animated about it.

"I'm not sure," she said.

Hansen pressed her. "I'm serious."

Simoncelli took a deep breath. There was so much to cover—the gene patenting debate had been going on for years in academic circles. They would need to dive into a huge body of literature, cases, and the patents themselves.

Hansen, hooked on the idea, was drawing Simoncelli into a litigation battle plan. He'd read the articles that Simoncelli had sent, plus a few more.

"I cannot believe the U.S. government is issuing patents on these things," Hansen continued, pacing and waving a printout of one of the articles. "That is just totally offensive. We need to do something about this."

Simoncelli nodded. They would figure it out.

Meet Your Genome

THE HUMAN GENOME is embodied in an infinitesimally tiny, threadlike bundle in the nucleus of each of the hundred trillion cells in our bodies. It is a blueprint for everything that makes us human, from the biochemical inner workings of our bodies, to the structure of our organs, to our outward physical features. Genetic variations that have accumulated over the eons explain why the average Dutch man is six feet tall and the average Guatemalan woman stands only four feet eleven inches. Whether we have freckles or dimples or double-jointed thumbs, whether our hair recedes from the front or from the sides or not at all, and whether we are more or less susceptible to certain diseases are all determined by minute variations in an otherwise uniform genetic code.

A great deal of the genome is not uniquely human, but exists within every living organism, from blue-green algae to Siberian tigers. School children learn that we share 40 percent of our DNA with the humble banana plant, *musa acuminata*. But the commonalities increase as we move "up" the evolutionary ladder. We share about 99 percent of our DNA code with chimpanzees and 99.5 percent with every other human being. In that final half-percent is everything that makes us genetically unique from everyone else (except our identical twins).

Despite its power to define our physical characteristics, the genome is more than a recipe book. It is a living history of everything that came before us, as if every architect carries in his pocket the plans for the Great Pyramid, the Hoover Dam, and the Tokyo subway system. Long after the grass has swallowed their graves and their names are lost to time, there are still living vestiges of forgotten ancestors in our DNA. It records their migration over endless Asian steppes and dark Polynesian seas, along the Silk Road and across the Bering Strait. It hints at where they settled down, where they died, and with whom they bore children.

But the genome is also more than a record of human peregrination. It is a great chain of being linking us all the way back to the dawn of life on earth. Each of our cells dutifully replicates code from ancient viruses transmitted by insects to the earliest mammals. We bear vestiges of the first lungfish that crawled out of the ocean to breathe and live on dry land, and the genes of long-vanished hominid ancestors identified only by fossilized bone fragments—all still present within our vast bio-cellular archive.

The genome is made principally of a chemical called DNA, a long, spiraling molecule composed of four different nucleotides, or "bases"— adenine, guanine, thymine, and cytosine, usually abbreviated as A, G, T, and C. These bases occur in pairs that are strung together like the rungs of a ladder. The ladder, in turn, is twisted like a corkscrew, forming the famous "double helix" structure that James Watson and Francis Crick, with the help of images taken by Rosalind Franklin, discovered in 1953.

In total, each of us has about 3.2 billion of these ladder rungs, which are organized into twenty-three long, coiled strands called chromosomes. If all of the DNA contained in a single cell were unbundled and stretched out like a giant ticker-tape, with each base just an eighth of an inch wide, the whole thing would be about 6,300-miles long—the distance from Chicago to Tokyo. Along that 6,300-mile strip, the entire range of human diversity—every difference among us, good, bad, and irrelevant—would stretch just 31.5 miles (one half of one percent of the sequence), landing in the nearby Chicago suburb of Schaumburg, Illinois. Everything from Schaumburg to Tokyo is a compelling argument for brotherhood and sisterhood, not just with other human beings, but with all life.

The size of the genome may be impressive, but it is the *order* of its individual building blocks that is all-important. Determining the exact sequence of the 3.2 billion DNA base pairs in the human genome was a monumental scientific undertaking. When the first gene sequencing technology was developed in the 1970s, it took researchers years of pains-taking work, and more than a small degree of luck, to piece together the DNA segments linked to distinctive biological effects, usually debilitating hereditary diseases. Using these tools, early gene hunters discovered tiny irregularities in the genome—often a single switched pair of DNA bases somewhere among the 3.2 billion—that led to genetic conditions like Huntington's disease and cystic fibrosis.

How can such tiny genetic variations cause such grievous and far-reaching effects? Scattered throughout the genome are stretches of

DNA called genes. Genes encode more complex molecules called proteins, which carry out many of the important cellular functions in the body—breaking down sugars, repairing cellular damage, suppressing tumor growth. In an automobile, each discrete component—the brakes, the crankshaft, the hood ornament—is made according to a detailed mechanical plan. If the plan for a particular component contains even a small error, then that component may not work properly, or may not be included in the car at all. If the hood ornament is defective or missing, we don't care much—the car will still drive. But if the crankshaft is missing a bearing, or the brake system has a crossed pair of wires, then the car could break down and people could die. Likewise, when the gene responsible for making a particular protein contains a flaw—usually called a variant or mutation—the protein might not be made, or it might not function properly. Errors like this often have serious effects: diseases, birth defects, and even death.

Back in the 1980s, researchers estimated that each human has about one hundred thousand genes. Today, that estimate has dropped to around twenty thousand, with genes ranging in size from a few hundred to more than two million base pairs. Despite their importance, genes only represent about two percent of the DNA in the human genome. The rest of our DNA carries a wealth of historical and ancestry information, but its biological function, if any, remains murky.

By the mid-1980s, technological advances began to enable the large-scale, rapid determination of DNA sequences. For the first time, scientists were able to dream about decoding not only individual genes, but the entire human genome. Having a baseline human genome, they thought, would create a research tool of unparalleled potency. If we knew the "normal" genetic code for humans, it would be that much easier to identify the variants that cause genetic diseases and abnormalities. Thus, in 1988, the Human Genome Project (HGP) was born.

The HGP was one of the largest and most expensive scientific enterprises ever undertaken, rivaling the Manhattan Project and the Apollo moon program in both scope and expense. Working together, and racing with competing corporate efforts, government and academic labs around the world completed a rough map of the human genome in 2000. This monumental accomplishment was jointly announced by British prime minister Tony Blair and U.S. president Bill Clinton, who proclaimed that, "Today, we are learning the language in which God created life." The full sequence of the human genome was released in 2003, though refinements continue to be made today.

But even with its basic sequence decoded, the genome, stretching before us in all its microscopic immensity, remains one of the great mysteries of science. Like the deepest trenches of the ocean and the farthest reaches of space, what we have learned about it is dwarfed by what we still don't know. Though it is as close to us as it can be: coiled and working relentlessly within each of our trillions of cells, exactly how it makes us who we are still defies understanding. It is biomedical science's last frontier, a sprawling dark continent, unspoiled and unknown. To Simoncelli and Hansen, the very notion of laying claim to this common domain of humankind offended the conscience. It shouldn't be done, and it had to be stopped.

CHAPTER 3

The Gene Queen

DESPITE HIS INITIAL ENTHUSIASM, Chris Hansen knew better than anybody that the ACLU was a long way from challenging gene patents in court. The first obstacle was the simple fact that the ACLU had never brought a patent case. If Hansen were a gambling man, he would have bet that not a single lawyer at the ACLU had ever taken a law school class on patents.

Still, he thought, that deficiency could be overcome. ACLU attorneys are generalists. Unlike their counterparts at big law firms, they don't specialize in narrow fields like international debt restructuring or hospital regulation. While ACLU lawyers have a general focus on civil rights and know the Constitution backward and forward, the cases that they bring, whether dealing with school desegregation, prisoners' rights, or flag burning, all involve different sets of legal rules, customs, and factual circumstances that they learn as needed on the long road to trial. That was one of the best parts of the job. So with a little help, Hansen was confident that he could learn not only molecular biology and genetics, but patent law. How hard could it be?

One of the three articles that Simoncelli sent Hansen after their first conversation was by a law professor named Lori Andrews. Andrews taught at the Illinois Institute of Technology in Chicago, and had first met Simoncelli as an invited guest speaker at the Center for Genetics and Society at Berkeley.

To say that Andrews was an unconventional academic would be an understatement. Blonde and fashionable, with the penetrating gaze of a television detective, Andrews had a flair for publicity. She was a popular commentator on cable news shows, where she discussed

Lori Andrews, a law professor and advocate, was one of the first people to write about legal issues in genetics.

attention-grabbing issues like human cloning and designer babies. In profiling Andrews, the glossy women's magazine *MORE* aptly dubbed her "the Gene Queen."

Andrews, a Yale law graduate, started her legal career in the early 1980s focusing on cutting edge reproductive technologies like *in vitro* fertilization and embryo selection. This soon drew her into the world of genetics. In 1987 she published a book, *Medical Genetics: A Legal Frontier*, one of the first serious attempts to address the legal and ethical issues in the burgeoning field.

Andrews's timing couldn't have been better. The year after her book was released, the federal government kicked off the Human Genome Project. Its first director was James Watson, who shared the 1962 Nobel prize with Francis Crick and Maurice Wilkins for discovering the double helix structure of the DNA molecule. Conscious of the implications that mapping the human genetic code might have for humanity, Watson insisted that Congress earmark 3 percent of the massive project's budget for the study of its attendant ethical, legal, and social issues (ELSI). Arguably the nation's most revered scientist since Albert Einstein, Watson usually got what he asked for. Congress apportioned not three, but 5 percent of the HGP's budget for ELSI research and formed a permanent working group to advise the government on these issues. In 1995, Lori Andrews became its chair.

Thus, by the late 1990s the Gene Queen was well-known among scientists, lawyers, and policy makers within the closely knit genetics policy community. She was doing work that no one had even dreamed of a few years earlier.

So it was not surprising when, in early 2000, Andrews received a visit from a suburban Chicago couple, Dan and Debbie Greenberg, who were looking for an expert on genetics law. The Greenbergs' two children had died from a rare neurodegenerative disorder known as Canavan disease. The couple explained that in 1987 they approached a Chicago physician, Dr. Reuben Matalon, for help in solving the puzzle of this rare and deadly disease. Based on the recurrence of Canavan disease in certain families and populations (notably those of Ashkenazi descent), Matalon suspected that its roots were hereditary. This meant that if he could identify the specific genetic defect that caused the lethal condition—the slipped crankshaft or the unresponsive brake pedal—he could develop a prenatal screening test to tell parents whether their unborn children would be affected. This is what the Greenbergs were after.

The Greenbergs worked tirelessly to persuade other Canavan families to contribute their children's blood, urine, tissue samples, and even autopsy results to Matalon's research. Those who could afford it also gave money. Thanks to their support, in 1993 Dr. Matalon successfully identified the gene (known as *ASPA*) responsible for Canavan disease. He published his findings in the prestigious journal *Nature Genetics*, and soon clinics around the country began to offer families free screenings, much as they were doing for another deadly hereditary illness, Tay–Sachs disease.

But in 1997, Matalon's new employer, Miami Children's Hospital, sent shockwaves through the Canavan community. The hospital, with Matalon's cooperation, had obtained a patent covering the Canavan gene. Armed with the patent, Miami Children's Hospital began to demand payment from clinics that offered testing for Canavan disease. This was a problem, since most clinics tested patients for free. So when the hospital demanded payment, these clinics stopped offering the test. The Greenbergs and other Canavan parents felt betrayed. They had donated their own children's DNA to the research effort so that others could be tested. But now, the hospital claimed that it owned the results of that research. Dan Greenberg, a graduate of DePaul law school, decided that they should sue Miami Children's Hospital to prevent it from enforcing the patent. Luckily for the Greenbergs, Andrews, by then one of the nation's best-known experts in genetics law, was just a few miles away.

At the time she was approached by the Greenbergs, Andrews had not spent much time considering gene patents. Her intellectual calling cards were the ownership of body parts and new reproductive technology. Patents were an obscure sub-specialty of the law that primarily interested lawyers who had once been engineers and scientists. In her latest book, *Future Perfect: Confronting Decisions about Genetics*, Andrews included all of one sentence about patents. How, she wondered, could Miami Children's Hospital patent the Canavan gene? Wasn't that like owning something inside the Canavan patients' bodies, like their kidneys or their spleen?

After hearing the Greenbergs' story, Andrews offered to represent the family *pro bono*—without charge. Working with a law student, a clinical instructor, and volunteers from a large Chicago law firm, Andrews outlined a case against Miami Children's Hospital. They accused the hospital of violating its ethical duties to the Canavan parents, of using donated blood and tissue without informed consent, of fraudulent concealment, and several other forms of misconduct. At first things seemed to be going well and Andrews succeeded in generating some publicity for the case. She flew to Washington to speak with congressional staffers. She wrote articles and opinion pieces. She and Dan Greenberg were featured on the prime-time TV news show *60 Minutes*, where host Morley Safer ominously intoned, "Chances are, your genetic structure and mine, our most private property, may well belong to someone else."

But things soon took a turn for the worse. On the day their brief was due in court, the Chicago law firm suddenly backed out. Its representative explained that taking a stand against the hospital and its patents could damage the firm's lucrative biotech practice. Andrews was furious, but the big firm lawyers would not budge. What's more, the hospital was seeking to move the case from Chicago to Miami, which would significantly drive up the costs for Andrews and the Canavan families.

Now, desperate for help in what was proving to be a complex and difficult case, Andrews turned to an organization that had made its name representing unpopular clients and hopeless causes: the ACLU. But when she met with lawyers at the ACLU's Chicago office, she was turned away. Though sympathetic, the ACLU attorneys explained that they only sued the government, not private companies or institutions. With no help forthcoming, Andrews and her small academic team did their best to litigate the case, first in Illinois, and then in Florida. Not surprisingly, in 2003, out-gunned and out-lawyered, they lost. The court dismissed five of their six claims outright, and they lacked both the money and the resources to continue to pursue the sixth.

The disappointing outcome taught Andrews some valuable lessons about litigation. For one thing, it was clear that a law professor, a student intern, and a handful of well-meaning volunteers could not mount a legal challenge against a well-funded institution. Any serious lawsuit would require a lot more litigation muscle. Second, Andrews began to appreciate the power that patents could give their owners. Canavan disease is extremely rare. But what about more prevalent conditions? Genes linked to diseases like cystic fibrosis, hereditary deafness, Alzheimer's disease, and, the biggest of them all, cancer, were being discovered with growing regularity. Were these genes being patented too? Andrews was not yet a patent expert, but she had spent years writing about issues like organ donation and tissue ownership. Patenting genes— parts of the body—was just a small step from laying claim to someone's spleen.

Then, as fortune would have it, the ACLU hired Tania Simoncelli, Andrews's old acquaintance from the Center for Genetics and Society. Now Andrews had not only an ally in the growing field of genetics and law, but one who might give her an entrée into America's great civil rights litigation machine, the ACLU.

Simoncelli called Andrews shortly after she met with Hansen in 2005 to toss around ideas about how the ACLU might mount a legal challenge to gene patenting, a challenge that wasn't just technical, but had a civil rights angle. Andrews was ecstatic. She and Simoncelli considered different possibilities, and with each passing minute Andrews became more convinced that the ACLU might actually be interested in bringing such a suit. The old acquaintances ended the call on a positive note. Simoncelli would keep talking with the attorneys at ACLU, and Andrews promised to do some research into the case law.

For a few months, as Chicago thawed out from a deep winter freeze, Andrews dove into the issue, poring over cases, slogging through impenetrable patent claims, talking with doctors and scientists. Nobody was paying her to do it, but Andrews's conviction that gene patenting was wrong fueled her through more than a few cold nights in the law library. Then, in May, Andrews sent Simoncelli the fruits of her labor: a ninety-six-page, heavily footnoted memorandum that outlined a dozen legal theories for bringing a suit to challenge gene patents. The Gene Queen had come through.

CHAPTER 4

Mr. Lincoln's Boat

DEEP WITHIN THE archives of the Smithsonian Institution, in a humidity-controlled vault worthy of Indiana Jones, there sits an indexed crate containing a miniature wooden boat. It is a flat-bottomed barge, lovingly fashioned from polished oak and measuring about two feet from bow to stern. The vessel is unremarkable, save for four curious bellows mounted below its gunwales. Carved into its base, in neat cursive script, appears a single name: *Abraham Lincoln*.

This little boat is a working model—built and submitted to the U.S. Patent Office by the Great Emancipator himself—demonstrating an ingenious inflatable device for traversing shoals and sandbars.[1] In 1849, Lincoln's invention was awarded U.S. Patent No. 6469, conferring on him the exclusive right to the invention that he dreamed up while working riverboats along the banks of the muddy Mississippi.

Abraham Lincoln was the only U.S. president ever to receive a patent. Thomas Jefferson, himself a prolific inventor, served as one of the early republic's first patent examiners. In 1807 he wrote, "Nobody wishes more than I do that ingenuity should receive a liberal encouragement." That encouragement was amply in evidence by the nineteenth century, when Thomas Edison received more than a thousand patents for inventions, including not only the electric light bulbs for which he became famous, but for hundreds of other devices including batteries, phonographs, film

---

1 Today, the federal agency responsible for the examination and issuance of patents also processes applications for trademarks and is formally known as the U.S. Patent and Trademark Office. For the sake of simplicity, I refer to this agency as the "Patent Office."

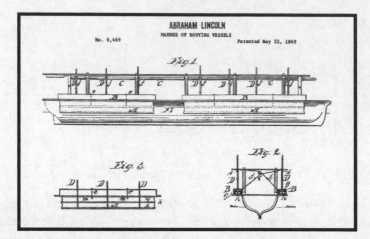

U.S. Patent No. 6,469 issued to Abraham Lincoln in 1849.

projectors, printing telegraphs, conveyor belts, cement kilns, and early electric cars.

Patents are part of America's legal fabric. They are enshrined in the Constitution, created to promote "the progress of science and useful arts" by giving inventors, for limited periods of time, the exclusive right to their discoveries. On the campaign trail in 1859, Lincoln said of patents that they add "the fuel of interest to the fire of genius, in the discovery and production of new and useful things."

The catalytic ingredient in this capitalistic fuel is exclusivity. Patents offer their owners—Lincoln, Edison, IBM (today's record-holder in terms of patents held), and tens of thousands of others—the *exclusive* right to exploit their inventions for limited periods of time. Thus, while a patent is in force, its owner is the only one legally authorized to make, use, or sell the patented invention in the United States (comparable rights exist in virtually every other country). In theory, this exclusive period, which today stands at twenty years, gives the inventor a chance to profit from the invention before it is copied by others.

This system makes sense, because if an inventor has no protection, and anyone who sees his or her invention can copy it, then the inventor may have little incentive to invent anything else, and, so the argument goes, science and technology will grind to a halt. In exchange for these twenty years of exclusivity, the inventor must disclose his or her invention to the public (today this is done only in writing—sadly, the working models of Lincoln's day are no longer required). Disclosure is necessary so that everyone will know how to make the invention once the patentee's exclusive period is over. This is why Edison's incandescent light bulbs, which made him a wealthy man, are today available for pennies and why, after a blockbuster drug goes "off patent," generic competitors often enter the market at a fraction of the cost.

But what kind of scientific and technological "discoveries" can be patented? Light bulbs, electric cars, and pharmaceuticals are one thing, but what about discoveries that govern the mechanisms of life itself? What about living organisms? Since 1930, horticulturalists could patent new plant varieties generated through cutting and grafting. But it was not until 1980 that the U.S. Supreme Court, in a sharply divided 5–4 decision, authorized the patenting of other living organisms. This landmark case, *Diamond v. Chakrabarty*, involved a genetically modified bacterium developed by a researcher at General Electric to help break down crude oil. It was here that the Supreme Court wrote the fateful words that would shape patent law for decades to come: "Anything under the sun made by man" is patentable.

GE's tiny petroleum-craving organism—*Pseudomonas putida*—began a trickle, then a flood, of patents covering biological "inventions." Armed with a patent, the discoverer of a new microorganism could lock up the rights to produce and commercialize it for years. And if a single-celled bacterium were patentable, why not a more complex organism? In 1988, Harvard scientists successfully patented a mouse that was genetically modified to develop cancer. The unfortunate creature was appropriately nicknamed "OncoMouse."

But the freight train that lurched into motion in 1980 did not stop with bacteria and mice. Fueled by the rapidly growing biotechnology industry, it took only a few years for the Patent Office to begin issuing patents on newly discovered sequences of *human* DNA. Though a few observers voiced ethical and religious concerns, the Patent Office and a brigade of patent attorneys convinced themselves, and the courts, that there was no problem. The patents didn't actually cover DNA *within* the human body, only DNA that had been *removed* from the body, which was then purified and processed like any other chemical compound.

Thus, despite the critics, DNA or "gene" patents were quietly issued throughout the 1990s and 2000s, expanding the notion of invention far from its roots in mechanical devices like Lincoln's boat pontoons. Research universities, biotech companies, and government labs all jumped on the gene patenting bandwagon, and by 2005, a pair of researchers at MIT estimated that 20 percent of known human genes were covered, to some degree, by patent claims. It seemed like a matter of time until the entire human genome was parceled out to private owners like lots in a suburban housing development. Who knew what lucrative therapies and medical procedures might emerge from our genes? One commentator called it a patent "gold rush." The commercial potential of human DNA seemed limitless.

Resistance to Gene Patenting

LIKE HANSEN, most Americans were unaware of gene patents. But the seeds of opposition to this practice had started to grow as early as the 1960s. It was then that scattered religious and activist groups began to express concern about genetic engineering and biotechnology in general. Large-scale genetic science arose, in no small part, from postwar studies of Hiroshima and Nagasaki survivors, linking the field to terrible, radiation-induced birth defects and diseases. Popular culture didn't help. Genetic mutants were depicted on film either as radiation-fueled monstrosities (Hollywood's giant ants in *Them!* and Tokyo's *Godzilla*, both released in 1954) or twisted versions of humanity (like the subterranean telepaths of *Beneath the Planet of the Apes* in 1970). This popular imagery, coupled with the looming specter of designer babies, racial eugenics, and human cloning, did little to help the reputation of genetic science in the public eye. By extension, patents on these technologies were viewed by some as encouraging companies to play God, a prospect both morally questionable and more than a little dangerous.

Responding to these fears, in the 1970s activist Jeremy Rifkin assembled a coalition of "Southern Baptists, mainline Protestants, Catholic bishops, Muslims, Hindus, and Buddhists" to oppose gene patenting. His group, obscurely named the People's Business Commission, submitted a brief to the Supreme Court in the *Chakrabarty* case. They argued that GE's patent on an oil-eating bacterium would open the door to a succession of moral depredations including "the manufacturing of . . . human beings, . . . the creation of super-intelligent beings; the asexual reproduction of organisms through cloning; . . . genetic surgery designed to alter the heredity of complex organisms" and so on. While taking note of this "gruesome parade of horribles," the Court nevertheless upheld GE's patent.

If fringe groups were up in arms over GE's patented bacterium, a broader segment of the public reacted when Harvard announced in 1988 that it had used genetic engineering to create, and patent, the ill-fated OncoMouse. Harvard licensed the genetically engineered creature to DuPont corporation, a move that embroiled the prestigious university in an unexpected public relations nightmare. Congress convened hearings and testimony opposing the mouse patents was heard from farmers, animal rights activists, environmentalists, and religious organizations, all of whom painted a grim picture of a future compromised by gene patents. Harvard and DuPont eventually bowed to public pressure and foreswore many of the most controversial aspects of their commercial pact.

But as the years passed and genetic doomsday failed to materialize, the dystopian visions that fueled this early opposition to gene patenting dwindled. Patents on human DNA were being used to protect things like artificially produced insulin and hemoglobin—beneficial inventions that sounded a lot like drugs. The courts considered DNA to be a chemical compound like any other, "albeit a complex one." Thus, throughout the 1980s, gene patents became the new normal. Biotech CEOs bragged about the size of their patent portfolios, venture capitalists built them into financial models, and patent attorneys developed a profitable niche in this hot new area.

Even the U.S. government got into the gene-patenting business. In the 1980s, NIH, like many government agencies, was trying to squeeze a financial return out of its research investments. One former NIH researcher recalls that, during this time, NIH's technology licensing officers would routinely visit labs, asking researchers whether they had come up with anything that might be patentable. In 1990, J. Craig Venter, a geneticist working at an NIH lab, announced that, with the agency's full support, he planned to file patent applications covering thousands of short DNA fragments known as expressed sequence tags or ESTs. Venter believed that ESTs could be used to locate genes more quickly and efficiently than sequencing the entire genome, and by 1991 he was using automated gene sequencing machines to identify 50 to 150 new ESTs every day.

It was Venter's announcement, perhaps more than any other event, that galvanized the scientific community against the patenting of human DNA. Unlike the religious objectors of the past, geneticists were outraged that someone could patent an EST, something that was so trivial to produce, something that was simply spit out of a sequencing machine with little or no human effort. James Watson, the outspoken Nobel laureate who led the Human Genome Project, called Venter's plan "sheer lunacy" and claimed that "virtually any monkey" could generate the EST sequences that Venter and NIH were patenting.

While Watson trivialized the scientific effort required to "invent" new DNA fragments, others saw these patents as potential roadblocks to scientific research. David Botstein, chair of the genetics department at Stanford, observed that "no one benefits from [these patents], not science, not the biotech industry, not American competitiveness." Watson and Botstein were soon joined by other influential voices, including the American Society of Human Genetics, the American College of Medical Genetics, and the Association for Molecular Pathology, in condemning Venter's plan and the patenting of human genes more generally. Even NIH's own internal advisory committees were "unanimous in deploring the decision to seek such patents."

Criticisms like these caused the agency to back down, and in 1994 NIH reversed course, effectively abandoning its EST patents. NIH's pivot away from DNA patents signaled a new period of openness in the world of genomics research. In early 1996, just as the Human Genome Project was beginning to gather steam, fifty leaders of the international scientific effort, including Francis Collins, the head of the U.S. program, and John Sulston, who led the UK effort, met on neutral ground at the coral pink Hamilton Princess Hotel in Bermuda. There, they hammered out a set of rules for publicly releasing the DNA sequences generated by the massive project. What they achieved over the course of a few rainy off-season days was nothing short of revolutionary.

The Bermuda Principles, as they became known, required that each lab contributing to the genome project release its DNA sequences to the public a mere twenty-four hours after generating them. In a world where scientific data was often hoarded for years and discoveries were jealously guarded until published in peer-reviewed journals, the Bermuda Principles were received with differing degrees of shock and acclaim around the world. They shaped attitudes and norms within the global research community and continue, even today, to define the standard for sharing scientific data. Another important purpose of the Bermuda Principles was to prevent others from filing patents on genetic sequences constructed by the international project.

But the private sector remained unmoved, and the patenting of human genes continued apace. By late 1996, the journal *Nature* reported that more than 350 new gene patent applications had been filed. Genes that could help to diagnose predispositions to more and more health conditions—Huntington's disease, diabetes, Alzheimer's, and several different cancers—became the subject of patents in the United States and elsewhere. Lori Andrews called it the "gene-of-the-week syndrome," and it was spreading.

CHAPTER 5

The ACLU Way

SINCE HER DISHEARTENING loss in the Canavan case, Lori Andrews had been spending more and more time thinking about gene patents. She had seen how powerful institutions like Miami Children's Hospital could crush resistance by poorly funded individuals, and had witnessed the hesitancy of public interest groups like the ACLU to get involved in the arcane field of patent law. So, even before Simoncelli called her in 2005, Andrews began to plan a campaign on two fronts: first was marshaling the legal theories that would be necessary to challenge gene patenting, and the other was turning public opinion against it.

The first prong of the attack was old hat to Andrews. Unlike the earlier religious and scientific opponents of gene patenting, Andrews focused on the law. While a court might be sympathetic to ethical and scientific considerations, it would not consider changing the law without solid legal arguments to back it up. And here, Andrews had plenty of help. Patent scholars had been writing about gene patenting for nearly two decades. Rebecca Eisenberg at the University of Michigan, one of the pioneers of biotech patent law, had thoroughly analyzed the validity of gene patents as early as 1990, and a steady trickle of academic writing had followed. The problem was that these legal theories were just that: theories. No one had yet sought to operationalize them in a real case argued before a real judge. That was what Andrews hoped to do.

This led to the second prong of Andrews's campaign: public relations. She might not be an elite ivory-tower law professor, but she knew how to get on the evening news. Andrews continued to produce a stream of magazine articles and TV interviews and even appeared in a documentary that premiered at Sundance. But these "talking head" gigs were just the beginning.

The turning point for Andrews was a grant from the Greenwall Foundation, a prominent bioethics think tank that gave her funding to create a series of "events" focusing on genetics policy. Over the years, Andrews had hosted her share of sparsely attended coffee-and-bagel symposia in dimly lit university lecture halls. But this time she wanted to aim higher. With her Greenwall funding, Andrews booked A-list venues like the Los Angeles County Museum of Art and the stunning oceanside headquarters of the Salk Institute in San Diego. And instead of the usual crew of academics, activists, and food-scrounging graduate students, she sought out the most prominent public intellectuals that she could find: artists, writers, Nobel laureates, MacArthur fellows. The level of discussion at these events was heady, and both Andrews and the attendees generally left pleased.

Perhaps the most significant thing that came out of the Greenwall events was the collaboration that Andrews developed with fellow Chicagoan Michael Crichton. Crichton, an imposing six-foot-nine Harvard-educated physician, single-handedly invented the techno-thriller genre in 1969 with his bestselling novel *The Andromeda Strain*. Then he redefined it again two decades later with his chart-topping tale of genetics-gone-mad, *Jurassic Park*. Now, as the producer of the hit television series *ER* and the driving force behind a velociraptor-fueled mega-franchise complete with movies, action figures, and theme parks, Crichton was casting about for a new cause. And Andrews was only too happy to oblige.

As soon as Andrews told him about gene patents and the Canavan case, Crichton was hooked. He abruptly set aside the draft of his new novel—a semi-historical yarn about Caribbean pirates—and began to work furiously on *Next*, a tale of corporate biotechnology gone terribly awry (think talking orangutans and fish that display bioluminescent corporate logos). Andrews, who at the time was working on her own genetics-based mystery novel, became Crichton's closest advisor on the project. She helped him to work a strong warning against gene patenting into the novel's convoluted plot. But Crichton did more than portray the evils of gene patenting in his fiction; he became its most outspoken public critic. He met with congressional aides and published an op-ed in the *New York Times* railing against gene patents. And when someone with the brand recognition of Michael Crichton spoke, people listened.

This gave Andrews cause for optimism. Thanks to people like Crichton, the public was starting to pay attention to gene patents. Tania Simoncelli at the ACLU seemed interested in challenging them in court, and Andrews had sent her a roadmap of the legal arguments that could be made. All she needed now was a good case, a second Canavan tragedy that could leverage the ACLU's litigation muscle to drive the final nail into the coffin of gene patenting.

And then, on Halloween of 2005, it materialized like a ghost: a case called *LabCorp v. Metabolite.* Unlike the Canavan case, in which the Greenbergs came to Andrews looking for help, this case was already in progress. It involved a complicated dispute between two large companies: Metabolite, a biotech company that had patented the correlation between low levels of an amino acid called homocysteine and vitamin B deficiency, and Laboratory Corporation of America, commonly known as LabCorp, a major U.S. diagnostic laboratory. Metabolite claimed that LabCorp was encouraging doctors to use homocysteine levels to diagnose vitamin B deficiency, a diagnosis that infringed Metabolite's patent. In 2001, a Colorado jury sided with Metabolite, finding that LabCorp had assisted doctors in infringing the patent (a theory called "inducing" infringement) and awarded Metabolite $4.6 million. LabCorp appealed, but in 2004 the appellate court affirmed the jury's verdict. So LabCorp threw the legal equivalent of a "Hail Mary" pass. It submitted a petition for writ of certiorari—commonly called a *cert.* petition—asking the U.S. Supreme Court to hear the case and overturn the appellate court's decision.

The Supreme Court grants only 1 to 2 percent of the seven to eight thousand *cert.* petitions that are filed each year, so getting any case heard by the highest court in the land is a long shot. To make matters worse, in those days patent cases weren't particularly interesting to the Supreme Court, which generally agrees to hear a case when there is an important constitutional issue at stake, or when the lower appellate courts have different interpretations of the same federal statute. Patent cases are highly technical in nature and seldom raise issues of constitutional law. What's more, because there is only one appellate court in the United States that hears patent cases (the Court of Appeals for the Federal Circuit), there are seldom circuit "splits" to resolve.

Nevertheless, to the surprise of many, the Supreme Court announced that it would hear LabCorp's appeal. And, even better, the high court said that it would consider only one issue in the case: whether Metabolite's patent was invalid because it attempted to claim a "law of nature."

For Andrews, the announcement was like winning the lottery. For nearly two centuries courts had agreed that one couldn't patent laws of nature like the Pythagorean theorem or $E = mc^2$. And Andrews had been writing for years that patenting genes was just like patenting laws of nature and, hence, just as impermissible. She had tried to make that argument in the Canavan case, but the litigation process defeated her before she hit her stride. But now, in a case already on its way to the Supreme Court, she could file a brief as an *amicus curiae* or "friend of the court." The Supreme Court regularly received *amicus* briefs from scholars, trade associations, and a range of other interested citizens. Though *amici* did not have a personal stake in the outcome of the case, the Court respected these opinions and sometimes gave them weight. And among the most respected voices at the Court was the ACLU's.

Will You Be My Amicus?

THE NINETY-SIX-PAGE MEMO from Andrews had been sitting on Hansen's desk for a few months and he was not sure what to make of it. Entitled "Human Gene Patents: The Challenge to Individual Rights, Public Health, and the Constitutional Incentive for Invention," the memo outlined a dozen legal theories relating to gene patenting, citing a vast array of authority from the Federalist Papers to the latest scientific journals. In it Andrews argued, with varying degrees of conviction, that gene patents violated Sections 101, 102, 103, and 112 of the Patent Act, and the First, Fourth, Fifth, Thirteenth, and Fourteenth Amendments of the

U.S. Constitution, as well as a handful of international treaties. The memorandum ended with a nine-page appendix analyzing a portfolio of patents covering genes that appeared to be linked, of all things, to asthma—a serious illness, but hardly the stuff of passionate advocacy. The memo, as well-intentioned as it was, was the legal equivalent of a blunderbuss. For Hansen to bring a case, he needed a high-caliber rifle.

To complicate matters further, Hansen was already starting to sense unease within the upper ranks of the ACLU. It wasn't exactly skepticism, but at least a desire to proceed with caution. Steve Shapiro, the ACLU's National Legal Director, was nervous about diving into an area in which the ACLU had no track record, and about which the organization knew next to nothing. Contacts on the outside were telling Shapiro that patents were essential for the health of the biotech economy, that without them new drugs wouldn't be produced and new cures for disease wouldn't be discovered. Back in 2000, when Bill Clinton and Tony Blair announced that the newly mapped human genome should be made freely available to scientists everywhere, biotechnology stocks plummeted, erasing tens of billions of dollars in market value. Is that what the ACLU wanted? If they won such a case and eliminated gene patents, would it be good for science or bad for science? Shapiro and Hansen had grown up together at the ACLU, each a twenty-five-plus-year veteran of the National Office, and trusted each other implicitly. After one of their many meetings on the topic, Shapiro pulled Hansen aside. "Just make sure that we're on the right side of this thing," he begged.

Now, sitting across from Lori Andrews in one of the ACLU's river-view conference rooms, Hansen felt like he was watching an episode of *Frontline*. Andrews was poised and intense, wearing a power suit and trying to convince him to file an *amicus* brief in the *Metabolite* case. Simoncelli sat beside Hansen, uncharacteristically quiet. If anything, the science advisor was slightly nervous. She hoped that her long-time ally would persuade Hansen to lend the ACLU's name to the fight, but she also sensed Hansen's hesitancy.

From the moment Andrews began talking, her iron-clad conviction shone through. Andrews was an advocate; she was trying to convince Hansen, to persuade him that the case could be made. Like Simoncelli, she was a true believer and to them, the answer was obvious. Gene patenting was unconstitutional, contrary to the law, and it never should have been allowed. It was a position to which Hansen was intuitively drawn, but he was not convinced that *Metabolite* was the right case to deal the fatal blow to gene patenting.

For one thing, *Metabolite* didn't actually involve genes or gene patents. Metabolite's patent covered a blood test for vitamin B deficiency. Andrews said that didn't matter. If they could establish that the "laws of nature" doctrine invalidated Metabolite's patent, it would be a substantial step toward invalidating other kinds of patents that covered natural phenomena, including genes. Hansen was skeptical. In his mind, there was a big conceptual gap between testing for vitamin B deficiency and owning human genes. No matter how flawed Metabolite's patent might be, the ACLU had no interest in waging a multi-front war on the patent system itself.

More importantly, Hansen couldn't see a civil rights angle to LabCorp's case. This was basically a commercial dispute between two large companies that would have little impact on the lives of patients. Why would the ACLU intervene to help the country's largest diagnostic laboratory avoid paying royalties to a biotech company? LabCorp was hardly a sympathetic plaintiff and, frankly, Hansen couldn't get very emotional about vitamin B.

Finally, from a procedural standpoint, *Metabolite* had some serious flaws. Foremost among these: LabCorp hadn't actually challenged Metabolite's patent under the "law of nature" doctrine. Courts weren't supposed to consider issues that weren't raised by the parties. Even the U.S. solicitor general was of the opinion that the "law of nature" question wasn't properly before the Court. Why the Supreme Court had agreed to hear LabCorp's appeal was a mystery to Hansen, but the entire foundation of the case seemed shaky to him. It would be much better to bring a case of their own, one that they built from the ground up, rather than jumping into a messy dispute between two companies that lacked any real moral salience.

The answer, Hansen told Andrews, was no. If he was going to take on gene patenting, they would do it his way, the ACLU's way.

The Road from Glen Ellyn

CHRISTOPHER ALAN HANSEN was born in the leafy Chicago suburb of Glen Ellyn during the post-war boom year of 1947. In the Hansen household, politics were a regular topic of dinner table conversation, and young Chris developed an early taste for the subject from his Goldwater Republican parents. But more than politics, the clever, argumentative boy gravitated toward the law—at the age of ten his best friend's mother predicted that he would become either a great con man or a great lawyer.

In high school Hansen joined the debate team and the student government, then went to Carleton College in rural Minnesota. At Carleton he was a DJ at the campus radio station during an era that witnessed the evolution of American music from Sinatra to Dylan, with all that transition implied. Nevertheless, the conservative campus was mostly insulated from the political turmoil surrounding the Vietnam War and the civil rights movement that roiled larger schools around the country. Over the summers Hansen mowed lawns and took long road trips with his friends. He was in Chicago during the 1968 Democratic National Convention, but kept his nose clean and didn't participate in the protests or the rioting.

True to his plan, Hansen entered the University of Chicago Law School in 1969 on a full scholarship. The venerable school was known for its focus on law and economics and its distinguished faculty included Ronald Coase, a future Nobel prize–winning economist, and Richard Posner, who would later become a prominent federal judge. Yet Hansen hated law school. To him, it was a boring waste of time, catering to students who aspired to work at big corporate law firms and little else. That life held no appeal to him.

Luckily, one on-campus interviewer did intrigue Hansen: the recruiter for the Legal Aid Society of New York. Legal Aid provided *pro bono* lawyers for indigent criminal defendants. Hansen applied and landed his first legal job as a public defender in Brooklyn.

Even as a new recruit Hansen appeared in court every day, representing defendants who ran the gamut from purse snatchers to hit men. It was invigorating: some days the newly minted lawyer handled up to a hundred different cases, all while his former classmates spent stultifying weeks and months buried in the labyrinthine bowels of giant business transactions and complex insurance settlements.

Despite the heavy workload, Hansen managed to indulge his love for outside reading. During the winter of 1973, he picked up a book by a civil rights attorney named Bruce J. Ennis. The book, *Prisoners of Psychiatry*, described the abuses that psychiatric patients suffered at institutions across the United States, a plight that had gained national attention after Ken Kesey's 1962 novel *One Flew Over the Cuckoo's Nest*. But Ennis surpassed the novelist, at least in Hansen's eyes, because he critiqued not only the deplorable conditions at state mental institutions, but the laws that permitted them to exist.

Hansen was hooked. Here was an emerging area of law that could be shaped to improve the lives of some of society's most vulnerable members. Ennis's wife was one of Hansen's coworkers at Legal Aid. A month after Hansen generously praised the book, she told him that Ennis was hoping to expand his group at the New York Civil Liberties Union, a local affiliate of the ACLU. Hansen applied and, a few weeks later, went to work for Ennis.

Ennis was in charge of a new group at NYCLU devoted to mental health law. In 1973 Hansen became its second member and was assigned to work on a massive lawsuit against the six-thousand-resident Willowbrook State School on Staten Island, the country's largest public institution for children and adults with developmental disabilities. Willowbrook had become infamous in the 1960s as the site of a series of involuntary hepatitis experiments performed on the inmates. In 1972, investigative journalist Geraldo Rivera released a further expose of the institution documenting its overcrowded, unsanitary, and abusive conditions. Senator Robert Kennedy called the institution a "snake pit."

Hansen spent Thanksgiving Day in 1973 conducting his first on-site inspection as a lawyer. What he saw at Willowbrook—naked inmates crowded together, feces smeared on the walls, developmentally disabled children piled in wooden "cripple carts"—horrified him. Hansen spent the next ten years pursuing the institution through the courts. That case,

NY State Association for Retarded Children v. Carey, became a landmark in the field, prompting Congress to pass multiple statutes protecting the rights of the mentally handicapped. And in 1983, the state of New York announced that it would permanently shutter the beleaguered institution.

But just as the Willowbrook litigation was winding down, NYCLU experienced a financial crisis and the Mental Health Law Project was defunded. Ennis, its founder, had previously moved to the national office of the ACLU and Hansen followed him there.

At the national office, another major lawsuit was just getting rolling: re-opening the landmark 1954 school desegregation case *Brown v. Board of Education of Topeka* to force the Topeka School Board to comply with the federal court's desegregation order. Hansen spent a year working exclusively on *Brown*, immersing himself in school desegregation law as he had earlier with mental disability law. But although *Brown* was a massive case that would last for decades, it did not require Hansen's full-time attention. So, in 1984, recalling his success at Willowbrook, he applied to become the deputy director of ACLU's Children's Rights Project, a kind of watchdog over state child welfare agencies.

Over the next ten years Hansen litigated cases against agencies that discriminated against minority children in New York, provided inadequate care in Kansas City, delayed adoptions in Kentucky, and committed a range of other legal and ethical violations against children across the country. In total, he litigated more than fifteen such cases, firmly securing his national reputation as a lawyer to be contended with. In 1994, he transferred to the ACLU's elite National Legal Staff.

To Hansen, the work was everything. He had no family or pets to distract him. He lived alone, occupying a small, neatly trimmed house in Mt. Vernon, New York, a working-class suburb in lower Westchester County. He rode a commuter train to Grand Central Station every day, then took the subway to the ACLU's lower Manhattan headquarters. Quiet, solitary, intense, a die-hard civil libertarian, Hansen had improved the lives of thousands upon thousands of people during his long tenure at the ACLU. He knew ACLU's internal politics, idiosyncrasies, and procedures better than anyone. He had an excellent instinct for what cases he could bring to trial and, more importantly, win. He always trusted his instinct, and history had shown that it was usually right.

Improvident Grant

WHEN HANSEN PASSED on the *Metabolite* case, Andrews was crestfallen. She thought that *Metabolite* presented a perfect opportunity to limit patents on biomedical technologies. So whether or not the ACLU was on board, she wasn't giving up.

As soon as Andrews got back to Chicago, she began working on her own *amicus* brief and looking for a client to file it on behalf of. Eventually she found a non-profit publisher, People's Medical Society, that had previously released information about vitamin B deficiency and was willing to sign its name to her brief. She filed it with the Court three days before Christmas.

Oral arguments in *Metabolite* were scheduled for March 2006. Being merely an *amicus* (one of twenty), Andrews would not argue before the Court and did not attend the hearing in person. But Michael Crichton, who was watching the case closely, flew to Washington on his private jet to listen to the arguments in person. Sadly, the performances were disappointing. Though both companies were represented by top-tier inside-the-Beltway law firms, the arguments were a mess; neither the Court nor the advocates really seemed to know what the issue was, whether the appellate court got it wrong, or how the "law of nature" doctrine was relevant to the case. Nobody left with a clear view of the Court's thinking.

Then in June, when the Court released its decision, it did something that not even the most skeptical observer had predicted. In a one-sentence order, the Court dismissed LabCorp's *cert.* petition as having been "improvidently granted." In effect, the Supreme Court admitted that it never should have agreed to hear the case. All of the briefs that were filed and the arguments that were made were for naught—legal nullities. The Federal Circuit's 2004 decision, and Metabolite's patent, would stand. The patent bar and the biotech industry collectively breathed a sigh of relief and scratched their heads.

But by far the most interesting, and encouraging, aspect of the *Metabolite* case was a lengthy dissenting opinion written by Justice Stephen Breyer. Breyer, a former Harvard professor, was the Court's science maven. Two other justices joined him in disagreeing with the Court's dismissal of the case, reasoning that the "law of nature" question was important enough to be considered, no matter how poorly the parties may have presented it. And, going further, Breyer felt that Metabolite's patent should *not* have been upheld because it embodied an unpatentable "law of nature." In making this argument, Breyer cited older Supreme Court cases, which held that fundamental scientific principles are "part of the storehouse of knowledge . . . free to all men and reserved exclusively to none." This, at least, was a sign of hope, and Hansen, as well as Andrews, took note.

CHAPTER 6

Product of Nature

THE ACLU TODAY has more than 1.75 million individual members, thousands of volunteers, and hundreds of employees. Unlike a typical non-profit organization, the ACLU is not a monolithic entity, but an unruly confederation of semi-autonomous affiliates in nearly every U.S. state, Puerto Rico, and the District of Columbia led, to some degree, by its New York–based National Office. While the National Office, overseen by the dynamic Anthony Romero and his team, spearheads fundraising and sets priorities for the conglomerate, it does not call all of the shots, and many decisions regarding which cases to bring and which causes to champion are made at the local level.

Sitting atop this sprawling organization is a National Board consisting of eighty delegates from across the country. Board seats are highly coveted and are generally doled out as rewards for years of dedicated service to the ACLU. The delegates, a graying assembly of die-hard civil rights warriors, many of whom knew and marched beside Dr. Martin Luther King Jr. and other civil rights leaders, relish the opportunity to demonstrate their commitment to the cause vocally, loudly, and at every given opportunity. Their meetings bear little resemblance to the staid board deliberations of other major charities, sometimes approaching the emotional pitch and volume of tent revivals.

The ACLU National Board meets twice per year, usually at a hotel ballroom in New York City fitted with oversized chandeliers and industrial carpet. Among the Board's many functions is setting ACLU policy. This is not a trivial task. The ACLU gives its local affiliates and National Office attorneys great latitude in bringing cases, with the sole caveat that these cases must be consistent with ACLU policy. And policy, at the ACLU, is a serious matter.

Over the years the ACLU has adopted hundreds of official policies. There are policies against racism, bigotry, inequality, and oppression of minorities, policies supporting freedom of speech and peaceful protest, a policy concerning the right to bear arms, a policy opposing the death penalty, and many, many more.

Thus, one of the biggest questions that Hansen and Simoncelli faced within the ACLU was whether an existing policy could be invoked to support a case against gene patents. The ACLU had dealt with new technologies before, but largely in the realms of copyright law and freedom of speech. In the 1990s, Hansen himself had litigated a landmark case striking down parts of the federal statute that tried to outlaw online pornography. ACLU's "Policy 10" now dealt specifically with online content, privacy, and copyright. Likewise, the ACLU had, since the 1925 "Scopes Monkey Trial," championed open scientific inquiry, and its "Policy 68" enshrined these values. Could these existing policies, or any others, be enlisted to support an attack on gene patenting?

It was an important question—an existential question. So, after discussing the matter with Anthony Romero, the ACLU's executive director, Hansen agreed that a committee of the Board should be formed to study how ACLU policy should be interpreted in the area of patents.

Because the ACLU's board didn't include any patent attorneys, Romero assembled the most plausible group that he could. In effect, he recruited any board member who had even the slightest connection to intellectual property or technology law. The ACLU's new Patent Committee consisted of Marjorie Esman, a soft-spoken copyright and trademark lawyer from New Orleans, Bert Foer, the head of a left-leaning antitrust think tank in DC, and Hamid Kashani, an Indianapolis-based computer consultant. Romero tapped Albert "Buzz" Scherr, a law professor at Franklin Pierce Law Center in New Hampshire, to be the committee's chair. Though Franklin Pierce had a national reputation in patent law, Scherr was a former public defender who now studied the use of forensic DNA evidence at trial—an outlier at the staunchly pro-patent school.

At Simoncelli's suggestion, Lori Andrews was also invited to join the

committee. And, to round out the group, Scherr enlisted an old acquaintance, Dan Burk, a law professor at the University of Minnesota with a
master's degree in molecular biology. Of the six committee members,
Burk had the deepest background in patent law. Often displaying a curt,
smartest-guy-in-the-room attitude, Burk was widely acknowledged to be
in the top echelon of intellectual property scholars. As such, he was
openly skeptical of newcomers like Andrews who approached patent law
not from a holistic standpoint, but to serve a particular policy goal.

For his part, Hansen was delighted. He needed to hear from more
than the true believers. He needed to hear from the doubters, the skeptics,
the people who thought that patenting human genes was a great idea.
So while Burk was certainly sympathetic to the ACLU's cause, he would
provide a healthy counterpoint to Andrews.

The Patent Committee began to meet in late 2006. They started by
considering whether existing ACLU policy would support a case challenging gene patents and, if not, whether the ACLU should adopt a new
policy specifically addressing patents. They reported on their progress
to the National Board in October 2007. According to one delegate who
attended the meeting, "Board members looked on with glazed expressions
as the Patent Committee began their report, but gradually moved to the
edge of [their] chairs hands waving for attention." Clearly, the patent
issue was attracting interest within the ACLU.

Hitting the Books

WITH THE ACLU PATENT COMMITTEE CONVENED, Hansen and
Simoncelli began the hard work of developing a case strategy. They had
read Andrews's memo and much of the literature on gene patenting.
They had spent months interviewing scientists, physicians, and academics
around the country to find out whether gene patenting was still perceived
as an issue (it was) and whether it was impacting access to genetic testing
(it was). Convinced that the issues were real, they now needed to develop
an airtight set of legal arguments that they could take to court. Hansen,
especially, needed to be sure the law was on his side. No matter what his

gut told him, or what the social considerations were, the case needed to be rock solid. So they began to test the waters.

Save for Washington, DC, there are probably more patent lawyers in New York City than any other metropolitan region in the world. New York was the birthplace of the most prestigious, blue-blooded patent boutiques: firms like Pennie & Edmonds, Kenyon & Kenyon, and Fish & Neave, whose mahogany furnishings and Persian carpets gave testament to their sterling credentials and distinguished lineage. As expected, practicing patent attorneys were not eager to talk about challenges to gene patents. Many of these lawyers had clients in the biotechnology industry who might be upset by such a challenge and, like the Chicago firm that had left Andrews at the altar in the Canavan case, they were not interested in getting on the wrong side of a potentially important legal issue.

Law professors were not much better. Rebecca Eisenberg from Michigan didn't understand why the ACLU wanted to pursue such a case. Weren't there more important civil rights issues to tackle? Chris Holman at the University of Missouri, a PhD biochemist himself, came out largely in favor of gene patents. Lawrence Sung at the University of Maryland thought the solution lay not in eliminating gene patents, but in immunizing academic researchers from patent infringement claims. And at Duke, in leafy Durham, North Carolina, Simoncelli met with Arti Rai and Robert Cook-Deegan, who had close ties to the genomics program at NIH. Though Rai and Cook-Deegan believed that individual gene patents might have some vulnerabilities, launching a frontal assault on gene patenting itself, they advised, seemed both quixotic and inadvisable. Cook-Deegan suggested they narrow their sights, challenging vulnerabilities of selected gene patents. Couldn't a broad, frontal attack on all gene patents backfire and eliminate the financial incentives needed to develop useful biological agents like antibodies and artificial hemoglobin?

One law professor who was more sympathetic to their cause was Joshua Sarnoff, a former EPA lawyer who had turned to patents later in life. With the scruffy beard and wire-rimmed glasses of a die-hard environmentalist, Sarnoff taught at American University in Washington, DC. He took the train to New York to meet with Hansen and Simoncelli on several occasions, always impressing them with his thoughtful analysis. Sarnoff worked with Lori Andrews and the ACLU team to study the convoluted legal precedents that established exceptions to the rules of patentability. There they found what would become the basis for their case.

Compositions of Matter

WHEN CONGRESS ENACTED the first U.S. Patent Act in 1790, it authorized the issuance of patents on any "useful art, manufacture, engine, machine, or device." Section 101 of today's Patent Act, which was enacted in 1952, modifies these eighteenth-century concepts only slightly, providing that "whoever invents or discovers any new and useful process, machine, manufacture, or composition of matter . . . may obtain a patent therefor."

Of course, there are other requirements for obtaining a patent. For example, the invention has to be *novel*—it can't have been invented before, and it can't be *obvious* in view of what already exists (the so-called "prior art"). But before these tests are applied, the invention must first fall into one of four categories of things that are eligible for patent protection: it must be "patentable subject matter."

All of this is simply to say that determining whether a patent should be granted is not always a straightforward task. Today, the Patent Office, based in Alexandria, Virginia, employs nearly nine thousand examiners who review patent applications in a wide range of technical specialty areas, from food additives to automotive electronics. A patent examiner evaluates whether or not the invention meets the different criteria for patentability, and usually concludes that the applicant is asking for too much. When that happens, the application is returned to the applicant, who may then amend and narrow it to address the objections raised by the examiner. This back-and-forth process goes on, often for years, until the examiner is either satisfied that the patent claims only new and inventive technology, or rejects the application. If the application is rejected, the applicant can appeal through various levels of the Patent Office and the courts, a process that can take decades.

At first glance, having a test for patentable subject matter may seem unnecessary: what's not a *process, machine, article of manufacture,* or *composition of matter*? Many things, it turns out. A new poem, for example, can't be patented. Nor can a television show, or a decorative sculpture, or a corporate slogan. Even though these things may be new and innovative, they are not patentable (though they are often protectable by other forms of intellectual property such as copyrights and trademarks).

So what *does* fall into the categories of patentable subject matter? Many of these are easy to understand. Machines, for example, include all kinds of mechanical devices. Historically, they were things like new plow blades, coffin lids, locomotive engines, and better mousetraps. Abraham Lincoln's barge pontoons fall squarely within this category. Today, patented machines include smart watches, power tools, self-driving cars, and anything else that is "a concrete device or thing consisting of parts."

Articles of manufacture, likewise, are *made* from other materials— nonstick skillets, rubber galoshes, ceramic dental implants. Things get a bit more complicated when we come to processes. The Patent Office helpfully defines a process as "an act, or series of acts or steps." So, for example, a new method of fermenting beer or an improved way of curing rubber might qualify for patent protection.

But the trickiest types of inventions are compositions of matter. As originally understood, compositions were new substances or compounds: things like chemical mixtures and metallic alloys. Most patented drugs fall into this category, as do a wide range of other products from gasoline additives to synthetic fibers to food colorings to pesticides.

But with compositions of matter, an issue arises that doesn't usually come up with machines or articles of manufacture. That is, many compositions of matter simply exist in nature without any human intervention at all. Plants, minerals, and animal products may all have useful properties, but these naturally occurring substances are not "inventions" that are eligible for patent protection. For example, if a botanist discovers a previously unknown mushroom deep in the jungle, he cannot patent it. He did not *invent* the mushroom, he was simply the first to *find* it.

Of course, new *uses* for natural substances can be patented. So a method of treating burns with the mushroom's pulp could conceivably be patented. But the fungus itself, to the extent that it already exists in nature, does not qualify as a patentable composition of matter. While this difference may seem like legal hair-splitting, it is critical. If someone patents the use of the pulp to treat burns, then he can prevent others from using the pulp for burn treatment. But if he patents the mushroom *itself*, he can prevent others from using it for *any* purpose: as a burn treatment, as an adhesive, or even as a salad topping. With compositions of matter, the substance itself is the patented invention, which gives very broad rights to anyone who invents a new substance.

The patentability of natural products has always been a bit metaphysical. At some level, everything on earth is composed of naturally occurring substances: the basic elements hydrogen, oxygen, carbon, and so forth. These can't be patented. It takes some human intervention to move from a naturally occurring substance to a patentable composition of matter. The courts, which fill the gaps in our statutory law, have wrestled with the distinction between natural and man-made substances for some time.

But what about *purifying* a substance found in nature? Is the purified form of the substance patentable by the first one to purify it? For example, suppose that a prospector panning for gold observes a few gold flakes floating in a pan filled with silt and mud. Suppose that he then invents an ingenious method of separating the gold flakes from the other contents of the pan. Clearly, that innovative separation method will be patent-eligible. But what about the concentrated gold flakes that result? One would never find a pile of gold flakes in nature—the pile exists only because of the prospector's ingenuity. But since the nineteenth century, courts have held that the flakes are, by and large, not patentable. They are still naturally occurring substances, and merely isolating, purifying, and collecting them does not make them into patentable compositions of matter.

With these principles well-established, lawyers for the biotech firm Genentech knew that they would face the product of nature doctrine when they tried to patent synthetic insulin. Insulin, the hormone that enables humans to process certain sugars, is naturally created by the pancreas. When Genentech perfected a method for synthetically producing insulin molecules using recombinant DNA technology in the 1970s, its lawyers worried that the new molecules would not qualify as new compositions of matter, given that they also occurred naturally in the human body. As a result, they applied for, and obtained, patents covering the novel method that Genentech used to create synthetic insulin and insert it into a bacterial cell rather than the synthetic molecules themselves.

But Genentech's strategy would not work for genes, as the methods that scientists used to discover new genes were well-known in the field. And the genes themselves were nothing more than "products of nature." By granting companies exclusive rights over bits and pieces of the human genome, the Patent Office was allowing companies, to borrow Jonas Salk's famous phrase, to "patent the sun." The vast human genome, virtually uncharted until 2000, was rightfully part of the public domain, belonging to everyone, like the immense wilderness of a national park, or the oceans, or the sky. Together, we, everyone on earth, were its conservators, its citizens, its beneficiaries, but not its owners.

But despite the rhetorical and intuitive appeal of this argument, the product of nature doctrine, it turns out, had some warts.

Learned Hand's Legacy

BILLINGS LEARNED HAND was born in 1872 to an old-line Albany family, the son and grandson of distinguished judges. He studied philosophy and law at Harvard, then began a successful legal career on Wall Street. But by the age of thirty-five, Hand had tired of legal practice. Letting this become generally known, he was offered an appointment to the

*Judge Learned Hand's 1910
ruling in* Parke-Davis v.
Mulford *was a precedent
that the ACLU would have
to overcome.*

federal bench in Manhattan, and gladly accepted. He served as a judge
in the Southern District of New York for fifteen years, deciding many
landmark cases and developing a reputation as a leading legal mind. At
that point, President Coolidge promoted Hand to the United States Court
of Appeals for the Second Circuit. His thoughtful and incisive opinions
there eventually earned him the office of Chief Judge. When Learned
Hand died in 1961, after more than fifty years on the bench, he was
widely hailed as "the greatest jurist of his time."

But back in 1910, just a year after Hand took the oath of federal office,
a complex case brought by pharmaceutical giant Parke-Davis & Co. came
before him. Parke-Davis held the rights to a 1903 patent claiming an
isolated form of the human hormone adrenaline. Parke-Davis's Adrenalin
became one of the early twentieth century's first blockbuster drugs.
Touted as both a cardiac stimulant and a blood loss preventative,
Adrenalin sold for nearly $7,000 per pound (an astronomical sum in
those days). Not surprisingly, competitors including Eli Lilly and Wyeth
soon began to offer their own versions of Adrenalin. Parke-Davis, eager
to eliminate competition in this lucrative market, began to assert its
patents, selecting as its first target H.K. Mulford Company, a small
Philadelphia-based manufacturer of an adrenaline-based concoction

known as Adrin. In addition to competing with Parke-Davis in the market for adrenaline-based medicines, H.K. Mulford had also acquired a reputation as an anti-patent advocate, viewing patents as contrary to the ethical standards of the medical profession.

The young judge Hand approached the case with a degree of trepidation, but also a feeling of empowerment. He recognized the ability of the law to address even the most complex scientific disputes, and wrote, "I cannot stop without calling attention to the extraordinary condition of the law which makes it possible for a man without any knowledge of even the rudiments of chemistry to pass upon such questions as these."

With this eloquent introduction, however, Hand went on to make a crucial mistake. When Mulford, in its defense, challenged Parke-Davis's patent, Hand upheld the patent, reasoning that "even if [Adrenalin] were merely an extracted product without change, *there is no rule that such products are not patentable.*"

Despite his intellectual brilliance, the young Judge Hand was wrong about the absence of such a rule. It was clearly laid out in earlier cases. But this was long before the days of electronic databases and armies of legal researchers. Mulford did a poor job presenting its case, and that was that. Hand's decision in *Parke-Davis v. Mulford* was affirmed in glowing terms by the court of appeals. Over the years it became a key precedent, supporting the view that isolated and purified natural substances are, indeed, eligible for patent protection. This early opinion by one of the giants of American jurisprudence, and the many subsequent cases that cite it, threw the "product of nature" doctrine into disarray. Somehow, Hansen and Simoncelli would have to stave off the ghost of Learned Hand if they wished to eliminate patents on human genes.

Specific, Substantial, and Credible

LEARNED HAND WASN'T the only obstacle they would have to overcome in seeking to overturn gene patents. Their largest opponent was likely to be the Patent Office itself. In 1990, when Craig Venter at NIH filed his first patent applications on the short DNA fragments known as ESTs, it was far from assured that the Patent Office would ever grant those patents. One question raised by the examiners who evaluated Venter's applications was whether ESTs had any known use. If not, then they would fail to meet the Patent Act's requirement that patented inventions had to be both novel and "useful." Though Venter claimed that ESTs could be used as DNA probes and for research purposes, the argument was never carried to its logical end. NIH, at the urging of Watson and others, abandoned Venter's patent applications in 1994 and he left the agency to pursue even bigger projects, like sequencing the human genome.

Nevertheless, the EST episode did cause the Patent Office to think hard about what types of genetic "inventions" deserved patent protection. As law professor Rebecca Eisenberg has pointed out, the first human DNA patents covered the production of artificial proteins like insulin and erythropoietin, which resembled drugs more than anything else. Because traditional, small-molecule drugs were a mainstay of chemical patent practice, it was only a small conceptual step to accept the same framework for protecting biological compounds like synthetic insulin. Then when institutions like the University of Michigan and University of Utah began to patent newly sequenced human disease genes in the late 1980s and early 1990s, the same framework remained in place. They were all patentable.

But objections began to arise, and so the Patent Office began to review its "utility" rules for patenting genes. In 1999, it published a new set of examination guidelines for public comment (known as the Utility Guidelines). These Utility Guidelines required that, in order for an invention to be patentable, it must have "specific, substantial, and credible" utility. Thus, short DNA fragments with no known use would not meet the utility threshold for patentability, but full genes linked to specific diseases might.

The draft Utility Guidelines attracted significant attention in the genetics community. Thirty-five individuals and seventeen organizations submitted written comments to the Patent Office. This is when the modern debate over gene patenting effectively began. The comments were submitted by organizations on both sides. In favor of allowing patents on human genes with demonstrated utility were groups like the American Intellectual Property Law Association, the Biotechnology Industry Organization (BIO), and several companies. The opposing camp consisted primarily of medical and scientific research organizations including the Association for Molecular Pathology (AMP), the American College of Medical Genetics, and the American Society of Human Genetics.

In the end, the Patent Office ignored most of these objections and made its new Utility Guidelines effective in January 2001, just as Simoncelli and Rich Hayes were launching the Center for Genetics and Society in Berkeley. This was the last official word on gene patenting from the Patent Office, which continued to allow gene patents at a rapid clip. But while the collective objections of the scientific community failed to alter Patent Office policy toward gene patenting, it did serve to raise public awareness of the issue. In fact, the Patent Office's failure to address these objections convinced Simoncelli, observing the ensuing debate from the Center, that something was very wrong with gene patenting. It was an issue that she would not forget.

CHAPTER 7

On the Hill

WHILE THE ACLU mulled the implications of challenging gene patents in court, others were moving ahead on a different front: legislation. Chief among these was Lori Andrews's protégé and convert to the cause, Michael Crichton. Frustrated by the Supreme Court's abrupt reversal in *Metabolite*, the writer-producer quickly channeled his substantial energy away from the courts and toward Congress. His latest techno-thriller *Next*, which had just appeared in bookstores across the country, included an unusual author's note that explicitly urged Congress to outlaw gene patents. Now it was time to take the fight directly into the lion's den.

This was not the first time that legislation seeking to limit gene patenting had come before Congress. Back in the early 1990s, during the EST patenting debate, Senator Mark Hatfield (R-Ore.) sought to amend an NIH funding reauthorization bill by placing a three-year moratorium on patenting genetic material. In theory, this moratorium would have given Congress time to analyze the issue more fully and develop a well-considered solution. But others saw no need to delay. Orrin Hatch, the long-serving Republican senator from Utah, warned that "a freeze of any duration on the issuance of patents for biotechnology will retard the continued growth of this vital industry." Hatch's subcommittee held a day of non-committal hearings on gene patenting, Hatfield withdrew his amendment, and the matter was put to rest for a decade.

Then in 2002, while the Canavan case was ongoing, Andrews helped Representative Lynn Rivers (D-Mich.) introduce legislation that would have immunized physicians, hospitals, and scientists from being sued for infringing gene patents. This approach was modeled on a similar immunity that had been extended a few years earlier to physicians who performed patented surgical techniques and other medical procedures. But the gene patent immunity bill gained no traction and died quietly in committee.

This time around, Crichton and Andrews would be more direct. Instead of trying to limit who could be sued for infringing gene patents, they would try to eliminate the offending patents altogether. Crichton enlisted the aid of Congressman Xavier Becerra (D-Cal.), whose district straddled West Los Angeles and Hollywood. Becerra liked the idea. Andrews worked with his staff to draft a bill titled the "Genomic Research and Accessibility Act." Its operative language consisted of one sentence: "no patent may be obtained for a nucleotide sequence, or its functions or correlations, or the naturally occurring products it specifies."

Becerra, joined by Republican Dave Weldon of Florida, himself a physician, introduced the bill in February 2007. On the House floor, Becerra explained that his bill was intended to fix a "regulatory mistake" that had allowed genes to be patented, a mistake that had "dramatic, costly, and harmful implications for every American."

Becerra and Weldon timed the introduction of their bill to coincide with Crichton's publication of a *New York Times* op-ed entitled "Patenting Life." Crichton began with the dramatic flair of a seasoned novelist: "You, or someone you love, may die because of a gene patent that should never have been granted in the first place. Sound far-fetched? Unfortunately, it's only too real." After discussing the many real-world issues raised by gene patenting, Crichton concluded with a ringing endorsement of the Becerra bill, stating that it would "fuel innovation, and return our common genetic heritage to us."

The Stroke of a Pen

BY EARLY 2007 it had been two years since Simoncelli first raised gene patenting with Hansen, and the ACLU's new Patent Committee had barely begun its deliberations. In contrast, Simoncelli was impressed by the legislative assault being orchestrated by Crichton and Andrews.

Maybe legislation was the best way to contend with gene patents. After all, if the stars aligned, Congress could end gene patenting with the stroke of a pen. And even if the legislation failed (which Simoncelli thought was likely), it would at least raise public awareness of the issue.

Simoncelli conferred with Hansen, who had no problem with her pursuing whatever strategy seemed most likely to advance their cause. That being said, he was a litigator, not a lobbyist. If Simoncelli wanted to go to Washington, she was on her own. Hansen had an allergic reaction to politics and no desire to spend the last years of his career drafting talking points and buttering up congressional staffers.

So Simoncelli began to spend more time in DC. She was no stranger to the nation's capital. Before heading to Berkeley, she had spent three years at the Environmental Defense Fund and then at a law firm run by a couple of ex-Carter administration officials. There, she had tracked legislation, done legal research, and interacted with clients. It was work that she enjoyed and was good at. Several of the firm's partners had encouraged her to apply to law school—something that her ACLU experience was making her consider again. But for now, she had all that she could handle.

Early 2007 was a busy time in DC. With President Bush's popularity at an all-time low, the Democrats had swept the 2006 mid-term elections and regained majorities in both the House and the Senate. It was unclear how the power shift would play out with the Bush White House, but things were clearly changing.

In March, Becerra's bill was referred to the House Subcommittee on Courts, the Internet, and Intellectual Property, chaired by Representative Howard Berman (D-Cal.). Gene patenting was now on the radar screen of the nation's capital, and dozens of lobbyists, policy wonks, and trade associations were trying to figure out whether the issue deserved their support, opposition, or indifference.

In Washington, Simoncelli found an ally in Christopher Mason, now a professor of genetics at Weill Cornell Medical College. Mason has the square jaw and clean-cut good looks of a college rugby captain. Today he is probably best known for an experiment in which he collected samples of grime from every subway station in New York City to develop an

*Christopher Mason advised
the ACLU litigation team on
scientific issues while a genetics
graduate student at Yale.*

eye-popping, stomach-turning bacterial map of the five boroughs. But
back in the mid-2000s, Mason was a graduate student at Yale developing
genetic analysis tools called microarrays, searching for new genes in the
fruit fly, and co-teaching a law course titled "Law & Order: Special
Genomics Unit."

Simoncelli recalls meeting the peripatetic Mason on a train in
Chicago, where they were both heading to a policy conference organized
by Lori Andrews. With a wealth of common interests, Simoncelli and
Mason hit it off. After returning to New York they kept in touch and
regularly saw each other at conferences and symposia. Like Simoncelli,
Mason was drawn to the policy debates that accompanied the introduction of Rep. Becerra's bill. Soon, they joined forces to advance the issue
on Capitol Hill.

Clearly, Becerra's bill would not go unopposed. In addition to the
powerful and well-funded biotech lobby, the patent bar seemed to be
uniting against any potential incursions onto their turf. One outspoken
champion of gene patenting was Kevin Noonan, a Chicago-based biochemist turned patent attorney who wrote the popular *Patent Docs* blog.
In a series of posts, Noonan referred to Becerra as "a wrong-thinking
political meddler," a "political Cassandra," and an "undistinguished"

member of Congress. Noonan and others in the patent community—many holding PhDs in one branch of the sciences or another—effectively told Congress to butt out of the patent world. They claimed that such interference, no matter how well-intentioned, would do more harm than good and damage the economy in ways that Congress did not—and could not possibly—understand.

Staffers

DURING HER YEARS in DC, Simoncelli learned that most members of Congress were clueless about science. They focused largely on raising funds for their re-election campaigns and seeking special favors to benefit their districts. In 1999, bioethicist Arthur Caplan had asked a group of Pennsylvania legislators if they knew where their genome was located. As reported by journalist James Shreeve, "roughly one third answered that it was in the brain, and another third thought it was in the gonads. The others weren't sure."

As a result, the key to any legislation were congressional aides—commonly known as staffers. These staffers are the central nervous system of Congress. They read the bills and proposals that would be considered in committee, summarized their main points, met with lobbyists, trade groups, and constituents, and developed positions that they could recommend to their bosses. Some long-time congressional staffers were the most knowledgeable people on the Hill. Getting to them was the way to get something done.

One of Mason's old friends from graduate school, Ian Simon, was now an intern in the office of Harry Reid, the new Senate Majority Leader from Nevada. Mason asked Simon if he could drop by the senator's office to chat with his staff about genetics. Simon agreed and set up a meeting between Mason and a small group of Reid's aides. The session went well. Though the staffers knew next to nothing about science, they were intrigued by Mason's claim that companies were patenting the human genome bit by bit. They left the meeting assuring him that they would look for opportunities to help.

That gathering energized Mason, and now, together with Simoncelli, he planned something more ambitious: an open briefing for House staffers to generate broad congressional support for Becerra's anti-gene-patenting bill. They coordinated with the congressman's office, and in June—as temperatures approached the ninety-degree mark in Washington—reserved a conference room in the Longworth House Office Building, a grand old neoclassical pile just a block from the Capitol.

Mason and Simoncelli invited their small but growing list of DC contacts to the event: congressional aides, policy wonks, law professors, healthcare advocates, and non-governmental organizations focusing on biomedical issues. They filled a windowless meeting room, again deploying the tried-and-true tactic of offering free lunch to the attendees. The faces watching them from a half-dozen rows of hotel ballroom chairs were mostly far-too-young interns, interspersed with a smattering of grizzled Beltway insiders. Observing quietly from the back, bumping awkwardly into the food table, was a small contingent from the National Academies of Science. Among these sages of science policy was Steve Merrill, a tired-looking six-foot-three program officer at the Academies who had just released two influential and damning reports on the state of patent law in the United States. He was curious to see what these Beltway newcomers had to say about the subject.

One last-minute addition to the invite list was Ed Ramos, a Puerto Rican-Peruvian twenty-something from Queens with the compact physique of a bantamweight prizefighter. Ramos, even in a suit, looked out of place among the sandy-haired prep school graduates that formed the core of the congressional staff. But with a PhD in molecular biotechnology and a prestigious fellowship from the American Society of Human Genetics, Ramos probably knew more about genetics than anyone else in the audience. He was there on behalf of his equally precocious boss, another outsider—the freshman senator from Illinois, Barack Obama.

The Becerra bill, with its focus on genetics, had caught Ramos's attention. He followed congressional activity on science and healthcare for Obama, who, despite his background as a community organizer and constitutional law professor, had a knack for science. In Obama's first year in the Senate, he had tackled tough science-related issues, including President Bush's executive order banning embryonic stem cell research, efforts to contain the spread of avian flu, and the increasing health disparities that affected poor communities. Which is why Ramos, who had also been courted by the office of Ted Kennedy, one of the Senate's most powerful members, chose to work for the junior legislator from Illinois.

As it turned out, Obama was sponsoring his own genetics bill in the Senate, though it was mostly administrative in nature and had little to do with patents. With Ramos's help, Obama introduced the Genomics and Personalized Medicine Act in 2006, impressing scientists around the country. And even without a focus on patents, Simoncelli and Mason were thrilled that the well-spoken and charismatic Obama, who had inspired much of the nation during the 2004 Democratic National Convention, was already focused on their issues. An ally like him could only help.

Mason kicked off the briefing with a lengthy PowerPoint presentation that explained why legislation was needed to limit patents on human genes. The slides went on and on with too many diagrams and too many words, and the audience's attention began to drift in the increasingly warm post-lunch atmosphere, already beginning to smell of mayonnaise and salad dressing. Simoncelli occasionally interjected a humorous quip, getting some laughs and keeping the atmosphere light. But the real momentum did not begin until the question-and-answer period. The hands went up, first a few, and then a lot. The reactions among the staffers were similar to Hansen's. How was it possible to patent a human gene? Was this really happening? Who was being harmed? And how would a patent ban affect the biotech industry, a darling of Silicon Valley and Senator Kennedy's hometown of Boston?

As they left the session, Mason and Simoncelli gave each other big smiles. They had reached their audience. Now, hopefully, these assorted aides and staffers would go back to their bosses and explain the issues raised by Becerra's bill. And even if the legislation didn't pass, some of Washington's most influential players would have heard about it.

Business as Usual in the Swamp

IN JULY, a hot, muggy pall fell over the capital. Then the House adjourned for its August recess, and when it returned in September, it considered not Becerra's anti-patenting bill, but a sweeping patent reform package sponsored by Representative Howard Berman, another Democrat from Southern California. Berman's Patent Reform Act of 2007 had been in the works for years and modified numerous procedural aspects of the patent system, none of which was very relevant to gene patenting. Berman chaired the House subcommittee that reviewed patent matters and ensured that his bill was quickly referred out of committee. It passed the full House by a vote of 220–175 soon thereafter. To Berman, patent reform, years in the making, had finally been accomplished. In his view, Xavier Becerra's one-sentence effort to limit gene patents was a mere oddity, an unwanted appendage that had not been part of the delicate compromise that was struck to get the real patent reform bill through the House.

In October, mostly as a courtesy to Becerra, a fellow Democrat from Southern California, Berman held a single day of hearings on gene patents. But the hearings were not specifically directed to Becerra's bill, nor did Berman invite Becerra, Mason, Simoncelli, Andrews, or Crichton to testify. No further action was taken on Becerra's bill and it died in committee that term.

Congress may have felt that it had already done its part with respect to genetics. Looming far larger than Becerra's modest bill was the Genetic Information Nondiscrimination Act (GINA), a landmark piece of legislation that prohibited health insurers and employers from discriminating against individuals using genetic information. GINA, backed by Francis Collins and other powerful allies at NIH, as well as the ACLU and other civil rights groups, had taken more than a decade to claw its way through Congress, and the House finally passed an acceptable version in April 2007. After that, Congress may have needed a break, both from patents and genetics. Senator Obama's bill also died in committee, lost in the shadow of GINA and perhaps falling victim to the larger events that would soon overtake the young senator.

CHAPTER 8

Speaking of Patents

CHRIS HANSEN WAS, by background and temperament, what civil rights lawyers call a "speech guy." Defending the right to free expression under the First Amendment was among the ACLU's oldest and most important charges. The organization grew up in the shadow of J. Edgar Hoover's no-holds-barred campaign against incipient Bolshevism in America, and some of its earliest victories were against governmental censorship and suppression of speech. Since then, the ACLU had gained notoriety for defending those who expressed unpopular beliefs—Klansmen marching in white bedsheets, hippies burning draft cards, protesters defiling the flag—breathing life into Voltaire's apocryphal proclamation, "I disapprove of what you say, but I will defend to the death your right to say it."

Hansen was born, it seemed, to defend the First Amendment. When he was choosing a college, he picked Carleton not for any of the usual reasons—reputation, scholarships, sports—but because he read a student newspaper story about a free speech dispute that was brewing at the school. The issue was whether a student should have been suspended for using a profanity at the school snack bar. Though it may have seemed trivial to others, Hansen, looking at the turbulent world of the mid-1960s, thought the case had big implications. He enrolled at Carleton and never looked back.

At the ACLU, Hansen was among the First Amendment's most ardent champions. In 1997, he led a team that struck down parts of the federal Communications Decency Act (CDA), the government's ham-fisted attempt to curb online porn by criminalizing anything online that might be offensive to anybody. In the case, *Reno v. ACLU*, the Supreme Court agreed with Hansen, ruling that the government may not limit adults to viewing only online content fit for children.

It was a landmark decision, and to Hansen it exemplified what was wrong with gene patenting. Just as the CDA attempted to eliminate everyone's access to adult-oriented material, essentially controlling the types of things they could see and think about, gene patents could prevent, or at least discourage, researchers from experimenting on human genes. It wasn't just a matter of owning something inside the human body. Gene patents prevented research and scientific inquiry, too. If a company owned a gene, other researchers couldn't study it, which meant they couldn't make hypotheses or draw conclusions about it, they couldn't publish articles about it, they couldn't talk about it at conferences. It was almost as if people couldn't even *think* about the gene without infringing the patent. Thus, even though patents didn't restrict speech or thought *per se*, they did limit some important precursors to expression. In the same way, the government's restriction of a journalist's right to investigate a story, or a voter's right to contribute to a political campaign, would invariably curtail their ability to express their views. To Hansen, restrictions like this were a problem.

The Indispensible Condition

THE IDEA THAT the First Amendment protects not only free speech but also thought had been floating around since the early twentieth century. Supreme Court Justice Benjamin Cardozo famously wrote in 1937—nearly a decade before Orwell—that "freedom of thought . . . is the matrix, the indispensable condition, of nearly every other form of freedom." Extending this theory to patents was a natural step. In the late 1990s, Dan Burk, one of the two law professors on the ACLU's Patent Committee, had been arguing that patents on computer software could impair the free expression of ideas by software developers. And Rebecca Eisenberg, the Michigan law professor who was never optimistic about challenging gene patents, thought that if a challenge were to be brought, the preservation of free speech was among the more promising arguments that could be made.

But it was Lori Andrews who first tried to operationalize the free speech/free thought theory in a patent lawsuit. Her lengthy 2005 memo to Simoncelli contained an entire section devoted to potential First Amendment challenges to gene patenting, as did her *amicus* brief in the *Metabolite* case. *Metabolite* was especially germane, as the defendant, LabCorp, hadn't actually infringed Metabolite's patent, it had merely published articles telling doctors *how* to infringe the patent. LabCorp was nevertheless found liable for encouraging or *inducing* those doctors to infringe. To Andrews, and to Hansen, it seemed like LabCorp was being punished for speech, a theory that Andrews hit hard in a provocative article titled "The Patent Office as Thought Police." And though the Supreme Court refused to hear LabCorp's appeal, and even Justice Breyer didn't mention the First Amendment in his stinging dissent, Hansen was intrigued by this line of reasoning.

Not only was bringing a case involving free speech something that the ACLU was well-equipped to do, such a case would present other tactical advantages. First, the ACLU had longstanding and incontrovertible policies supporting free speech. The Patent Committee, dithering over whether, and how, ACLU policy might support a patent lawsuit, would have no problem finding support for a suit that was speech-based. Second, the Constitution guarantees certain freedoms, such as the freedom of expression, against interference by the government. The ACLU almost always sued the government. The fact that the government wasn't the defendant was the principal reason that the Chicago branch of the ACLU had declined to help Andrews in the Canavan case, and why it didn't intervene in *Metabolite*. With a First Amendment argument, though, the ACLU could sue the Patent Office— the federal agency that issued the offending patents in the first place. All of this would help Hansen build support for the case within the ACLU.

But beyond academic papers, First Amendment arguments had achieved little traction in the patent world, and Hansen wanted to proceed cautiously. What he needed was a test balloon—a case in which he could try out the theory without risking too much. And, as chance would have it, an opportunity to float such a balloon soon appeared.

The Energy Guys

THE PATENT THAT Bernard Bilski and Rand Warsaw applied for had nothing to do with genes. They were in energy trading—the business that made Enron one of the most valuable firms in the world before it imploded. Bilski and Warsaw developed a formula to help municipal utilities—local gas, electricity, and water companies—charge their customers a fixed monthly rate without going broke if a summer was particularly hot or a winter was particularly cold.

Not surprisingly, the Patent Office rejected Bilski and Warsaw's patent application, as did its appeals board. They each ruled that one can't patent a mathematical formula. But the energy traders would not give up. They appealed to the Federal Circuit, and the assigned three-judge panel of the court ruled against them in October 2007.

The case thrust Bilski and Warsaw into the midst of a raging controversy over patents on so-called "business methods." Even though these patents didn't claim any kind of mechanical, chemical, or electrical device, they did, so it was argued, cover new methods—methods of running a business. The Federal Circuit allowed the first of these business method patents back in 1998—for the "hub-and-spoke" method of organizing mutual funds—and the patent world had been in turmoil ever since.

After being rejected by the Federal Circuit panel, Bilski and Warsaw petitioned the Federal Circuit to re-hear their case *en banc*—with all twelve sitting judges, rather than three. Petitions for *en banc* rehearing are not uncommon, though the court rarely grants them. One Federal Circuit judge recently called these requests "a waste of time and a waste of money." Nevertheless, in February 2008, the Federal Circuit surprised everyone by announcing that it would re-hear Bilski and Warsaw's patent case. To gather as much information as possible before the re-hearing, the court issued a public call for supplemental briefing by the parties and interested *amicus curiae*.

This got Chris Hansen's attention. The Federal Circuit, which hears all patent appeals in the United States, was interested in what *amici* thought should and shouldn't be patented. This was an opportunity to test his First Amendment theory, but without the time and expense of filing his own case from scratch. Better still, *Bilski* involved a patent that even the Patent Office didn't want to grant. Unlike gene patents, which the Patent Office was granting with abandon, the Patent Office and the ACLU were on the same side when it came to Bilski's flimsy invention. So if Hansen filed an *amicus* brief advancing a First Amendment argument, there was a good possibility that the court, or even the Patent Office, might validate it in some way. And anything the court said here could later be used in an attack on gene patents, which would be much harder.

So Hansen got to work. Writing an *amicus* brief was an intellectual exercise, the purest form of legal analysis. It didn't require witnesses or depositions or fact-gathering or messy procedural disputes. An *amicus* brief simply needed to present the court with a legal argument in its essential form, as clearly and persuasively as possible. For Hansen, this was a one-man job. He occasionally bounced ideas off of Dan Burk, who had written about the topic, and Simoncelli, who was supportive. The three-year-old memo from Lori Andrews—a compendium of legal theories and arguments—proved helpful, but Andrews herself was busy with other projects. She was also anxious about her friend Michael Crichton, who had recently been diagnosed with cancer. For Steve Shapiro, the ACLU's Litigation Director, the prospect of a full-scale patent lawsuit was still worrisome, but he didn't object when Hansen suggested that they test the waters with an *amicus* brief in the *Bilski* case.

Enter the Amici

THE COURT'S CALL for *amicus* briefs was issued on February 15, 2008. Hansen worked fast. He filed his brief, a concise and razor-sharp fifteen pages, on April 3. He mentioned Justice Breyer's dissent in *Metabolite* several times, reminding the Federal Circuit that the highest court in the land had also taken an interest in these issues.

In addition to the ACLU, nearly thirty other parties filed *amicus* briefs with the Federal Circuit in *Bilski*. Many of them questioned the wisdom of granting patents on business methods and computer software, but a few touched on biotech issues.

As expected, after the hearing not one of the twelve Federal Circuit judges voted to award Bilski and Warsaw a patent on their energy-pricing formula. But, as often happens in complex cases, the circuit judges did not agree on *why* the energy traders did not deserve a patent. Nine of them concurred with the basic holding that such a patent must claim a method of using a machine or a method that *transforms* some physical thing (a substance, a chemical, a device) into some other state—what came to be known as the "'machine or transformation test." Yet some of the judges wrote separate opinions finding that the fundamental weakness in Bilski's and Warsaw's patent application was their attempt to patent an abstract idea. Others criticized the very notion of "business method patents"—suggesting that the court reverse its ten-year-old precedent allowing such patents.

But one judge, Haldane Robert Mayer—a gruff West Pointer who, after a tour of duty in Vietnam, went on to graduate first in his class from William & Mary Law School—took Hansen's bait. Though he did not cite Hansen's brief, Judge Mayer dissented from the majority's "machine or transformation" test, reasoning, instead, that patents on methods of conducting business "raise significant First Amendment concerns by imposing broad restrictions on speech and the free flow of ideas."

Bingo. It was the free speech argument. Though it wasn't part of the majority's opinion, Hansen's free speech patent theory was now validated by at least one appellate judge in a written opinion. Hansen was pretty sure that *Bilski* would be appealed to the Supreme Court, so the argument would be aired there as well. The groundwork was laid for the ACLU to bring its own case, this time challenging patents on human genes.

CHAPTER 9

The Power of Pink

BY LATE 2008, nearly four years after Simoncelli's first ACLU symposium on gene patenting, the stars were finally beginning to align in favor of a lawsuit. Anthony Romero, the ACLU's executive director, was increasingly enthusiastic about bringing a case. But Hansen and Simoncelli still needed to win over Steve Shapiro, who oversaw all litigation brought by the ACLU's National Office. Despite approving the ACLU's *amicus* brief in *Bilski,* Shapiro, to Hansen's annoyance, harbored lingering doubts about a gene patenting case. First, Shapiro wanted to be absolutely sure they weren't missing anything. Despite the team's strong convictions, Shapiro couldn't believe that the answer was as simple as Hansen made it out to be. Hansen wasn't a patent lawyer, Simoncelli wasn't a scientist, and the ACLU had never brought a patent case. How did they know that their suit wouldn't inadvertently sabotage the biotech industry, harming not only the economy, but public health? And why, Shapiro wondered, hadn't gene patents been challenged during the twenty-five years that the Patent Office had been issuing them? Surely the ACLU, of all organizations, was not the first to notice this practice.

To convince Shapiro, they needed someone from the biotech sector to explain why their case wouldn't cripple the industry. That person, it turned out, was a Bay Area lawyer named Barbara Caulfield. Caulfield, well known in biotech circles, was the general counsel of a Santa Clara-based company called Affymetrix. Simoncelli had been introduced to Caufield by a common friend, a DC biotech consultant named Jennifer Lieb. Lieb, who began her career as an NIH genetic counselor, was now well-connected in the biotech policy world of DC. Lieb had met Caulfield when they both worked at Affymetrix. When Simoncelli told Lieb about the suit that the ACLU was considering, Lieb suspected that Caulfield might be able to help.

Caulfield's company Affymetrix made "gene chips"—devices that could sample hundreds of thousands of DNA fragments simultaneously. As such, Affymetrix and Caulfield were concerned about patents that companies had acquired on individual genes. With enough of these patents floating around, Affymetrix's gene chips could become unaffordable or, worse, useless. Caufield had made this point in an *amicus* brief that she filed in the *Metabolite* case, and had become something of a spokesperson for the wing of the biotech industry that opposed gene patenting.

With a quick phone call, Simoncelli persuaded Caulfield to speak with Shapiro. When she did, Caulfield explained that the industry wasn't monolithic in its support for gene patents, and that at least some companies, like hers, found them problematic. The call seemed to have tipped the balance in Shapiro's mind. The next time that he saw Hansen, Shapiro suggested, though not in so many words, that he wouldn't oppose a direct ACLU suit against the Patent Office.

With this tacit blessing, Hansen and Simoncelli could start to map out a case. But first, they needed a patent attorney. Patents existed in an arcane and little-known area of the law—to Hansen it was clear that they needed somebody who knew the field inside out and wouldn't be caught off-guard by some off-the-wall question from a judge or opposing counsel.

Finding such a person, though, was more difficult than Hansen had anticipated. Big law firms usually lined up to help the ACLU, whether to generate good PR, boost employee morale, or place themselves in a positive light with government regulators. But in this case, no firm that specialized in patents would pick up the phone, let alone set up a meeting, to talk about challenging gene patents. Which of them wanted to be branded as "anti-patent"? That was not a smart way to expand a firm's business in the booming field of biotech patenting.

In effect, the entire patent bar was against them. Those lawyers were perfectly happy to represent one side or the other in litigation about gene patents, and even to argue that a particular patent was invalid for some reason, but no self-interested law firm had the slightest interest in abolishing the practice of patenting genes altogether. It was like trying to persuade a professional jockey to argue that thoroughbred racing was cruel to animals.

And the law professors? Even the ones most sympathetic to their cause were, well, just professors. As Hansen knew very well, the bare-knuckled brawl of real world litigation was a far cry from analyzing cases in a lecture hall. So who could they recruit to their patent-slaying team?

Robin Hood

DURING THE FALL of 2002, Dan Ravicher spent a lot of time staring out of his office window. From the twenty-second floor of 1133 Avenue of the Americas, headquarters of the venerable law firm Patterson Belknap Webb & Tyler, Ravicher had a partial view of Bryant Park and the iconic New York Public Library. On paper, his job was to die for. Just two years out of law school, Ravicher, with a boyish face and mischievous grin, was making close to $175,000 per year. His firm had a blue-chip client list and alumni that included powerful judges, corporate titans, and ex-mayor Rudy Giuliani. As a junior patent attorney, Ravicher had a steady stream of work, he liked his co-workers, and even the partners at the firm seemed like decent people.

But helping big pharma get more patents, and then enforcing those patents against generic drug manufacturers, began to trouble Ravicher. Despite all the rhetoric about innovation and progress, patents today seemed to him a lawyers' game. Scientists reported genuine discoveries to their lawyers, who then spun them into legalistic flights of fancy— imagining scores of new applications and extensions of otherwise modest discoveries that probably never would have occurred to the inventors themselves and which they had no intention of pursuing. Why? Because it was better to claim the new application just in case somebody else chanced upon it in the real world. Then, the owner of the patent could shut them down, or hold them up for a royalty. One of the great things about patents—for a patent lawyer—was that somebody could infringe a patent without knowing it, or even being aware of the patent's existence. They could unwittingly sell an infringing product for years and then, wham! They're hit with a patent suit and could owe up to six years of retroactive damages, plus interest. And the lawyers were laughing all the way to the bank.

Ravicher, from the lowest rung of this vast money-printing machine, began to wonder why companies needed so many patents, and what would happen if there were fewer of them. Would the outrageous prices of prescription drugs go down, or would companies just give up and stop developing drugs, as the patent bar warned? Ravicher didn't know, and it depressed him. He began to dread Tuesday mornings, when the Patent Office announced the new crop of patents that had been granted during the prior week. They just kept coming—inexorably and mercilessly. Ravicher was a patent lawyer who hated a lot of patents, a position that was quickly becoming untenable.

One day, while flipping through a glossy business magazine instead of working on the patent application on his computer screen, an announcement caught Ravicher's eye. An organization called Echoing Green was offering grants to social entrepreneurs. There was money for anyone with an idea to change the world. As it turned out, Ravicher had just such an idea.

Earlier that year, Congress had passed a technical amendment to the Patent Act that Ravicher had researched. There was a seldom-used administrative procedure that allowed anyone who thought that a patent had been improperly granted to ask the Patent Office to reconsider it. The procedure—known as *inter partes* reexamination—was seldom used because more effective methods for challenging patents existed in the federal courts. But in 2002, Congress strengthened the *inter partes* reexamination procedure. While litigation in the courts might still be preferable for large companies seeking to challenge one another's patents, *inter partes* reexamination now gave members of the public an effective method for challenging patents without the expense of a full-on court battle.

The idea intrigued Ravicher. According to his calculations, for as little as $9,000 someone could challenge a patent potentially worth

Dan Ravicher was notorious in patent circles for challenging some of the pharmaceutical industry's most valuable patents.

billions and pursue the challenge all the way to the U.S. Supreme Court. He applied for an Echoing Green grant, proposing the formation of a non-profit organization that would "enable the public to access the patent system in order to prevent unsound patent policy, and provide them with representation, education, and advocacy." He called it the Public Patent Foundation—"PubPat." In 2003, Echoing Green awarded Ravicher and PubPat $60,000 to "protect freedom in the patent system."

Ravicher got to work quickly. He left his law firm and began to run PubPat out of his Manhattan apartment. He raised additional funding from George Soros's Open Society Institute, hired a couple of part-time contract attorneys, and began to hunt for vulnerable patents.

They weren't hard to find. PubPat's first target was a 2002 patent owned by Columbia University that covered a process for inserting foreign DNA into a host cell to mass-produce therapeutic proteins. The process, called cotransformation, was the basis for major drugs including Amgen's Epogen (anemia), Genentech's Activase (cardiovascular disease), Biogen's Avonex (multiple sclerosis), and Genetics Institute's Recombinate (hemophilia). PubPat argued that Columbia should not have received its latest patent on cotransformation because it had been awarded three prior patents for the same process in the 1980s and 1990s, all of which were now expired. Rather than defend the patent, Columbia agreed in 2004 not to assert the patent, effectively nullifying it. It was PubPat's first victory, and it was sweet.

Later that year, Ravicher and PubPat aimed even higher, this time at Pfizer's blockbuster cholesterol drug Lipitor. In 2003, Lipitor's first year on the market, the drug earned Pfizer over $9 billion. In 2004, that figure rose to $11 billion. Pfizer touted Lipitor as "the best-selling pharmaceutical product of any kind in the world." Ravicher submitted an *inter partes* reexamination request to the Patent Office in September 2004, arguing that the drug's active ingredient was not invented by Pfizer, but was actually based on work done by scientists at another company in the early 1990s. To the shock of the pharmaceutical world, the Patent Office agreed. In June 2005, it rejected all forty-four claims of Pfizer's Lipitor patent.[1] A glowing profile of the thirty-year-old Ravicher appeared in *Science*, wryly asking, "Did Pfizer get punked by a non-profit?" Long-time patent lawyer and blogger Hal Wegner dubbed Ravicher the Robin Hood of the patent world.

More success and notoriety followed. Researchers around the country began to call Ravicher, tipping him off to overly broad patents that rested on shaky ground. Through PubPat, Ravicher challenged three University of Wisconsin patents covering human embryonic stem cells, four Gilead patents relating to the AIDS/HIV drug TDF, and four Monsanto patents on genetically engineered seeds. He got a $300,000 grant from the Rockefeller Foundation. He hired a full-time attorney. He filed an *amicus* brief at the Supreme Court in the *Metabolite* case. And a leading trade magazine rated him one of the "Fifty Most Influential People in Intellectual Property," alongside luminaries like the Director of the Patent Office, the Chief Judge of the Federal Circuit, and the Chairman of the Senate Judiciary Committee.

But even Robin Hood occasionally needed help, and Ravicher was not a biochemist. Challenges to pharmaceutical patents were highly technical, requiring complex analyses of chemical formulae to show how existing drugs were based on earlier discoveries. To assist with this work, Ravicher enlisted Chris Mason, the young Yale geneticst, who was more than eager to help. Their relationship was symbiotic. Mason helped

1 Eventually Pfizer was able to convince the Patent Office to allow some of its Lipitor patent claims in a narrower form. Lipitor went on to earn Pfizer more than $100 billion before the patent expired in 2011.

Ravicher and his team—many of them students from nearby Cardozo School of Law—to understand the complex chemical structures described in pharmaceutical patents, and Ravicher gave Mason a hand with legal issues—like writing a letter to the New York City Transit Authority when it threatened to ban Mason's team from scraping DNA samples off the teeming floors of its subway stations.

Sometime in 2007, Mason invited Ravicher to a strategy session at Yale. He and other policy types would be discussing the anti-gene patenting bill introduced by Representative Becerra in the House. That's where Ravicher first met Tania Simoncelli, who already knew him by reputation. So when Hansen needed a bona fide patent lawyer to fill out his litigation team, Mason and Simoncelli knew who to call.

Pick a Gene, Any Gene

UNLIKE CONGRESS, courts in the United States don't make laws to solve social problems. Courts exist to resolve actual disputes among actual people—automobile accidents, divorce claims, inheritance disputes. To bring a lawsuit, at a minimum there must be a *plaintiff*—someone who claims to have been injured—and a *defendant*—someone who caused an injury.

The ACLU team knew that one defendant in their case would be the Patent Office—it was, after all, responsible for granting all those gene patents. But Hansen didn't want to base his case solely on constitutional arguments. They needed a story that was simple and compelling: Here is a particular gene, and here is a patent covering that gene. And here is how this "gene patent" is injuring real people.

But which patent? Which gene? There were plenty to choose from. In 2005, MIT researchers found that U.S. patents covered more than 4,300 of the 25,000 or so human genes. The number in 2008 was probably significantly higher. But there wasn't a relatable human story behind each of those patented genes.

The members of the ACLU team—Hansen, Simoncelli, and Ravicher—along with Chris Mason, the geneticist from Cornell Weill, were meeting in an ACLU conference room, their notes and laptops spread out across a long conference table. With them was Wendy Chung, who had taken the subway downtown from Columbia Medical School that morning.

Unlike Mason, who spent most of his time in the lab or the classroom, Chung was a medical geneticist—she saw patients. Winner of the national Westinghouse Science prize as a high school student, she held a joint PhD and MD and now, not quite forty, an endowed chair at Columbia. The press compared her to the television doctor House because clinicians around the world sent her their most mysterious cases to diagnose. In a growing number of instances, she had discovered entirely new diseases.

Simoncelli had first heard about Chung during the 2007 congressional hearings on Representative Becerra's gene patenting bill. Chung's testimony supported what many of the other speakers were saying: gene patents were impeding research and harming patients.

Even though the effects of some genetic defects—diseases, physical impairments, developmental abnormalities—were known, the biological mechanisms behind most of those effects were mysterious. And, surprisingly, one didn't need to know *why* something happened in order to patent it, one only needed to know that it *did* happen (or, stranger still, that it *might* happen).

But despite the thousands of gene patents on the books, only a few dozen really seemed to matter much. In Hansen's mind they were all equally unconscionable, but to try a case, they needed a patent that was actually affecting people. In the parlance of civil liberties suits, the case needed human appeal.

At first, when gene hunters in the 1980s started to find genes associated with certain diseases, and then in the 1990s, when the Human Genome Project got under way, people thought that discovering genes would rapidly lead to cures for a variety of hereditary disorders. Miracle drugs like Genentech's Herceptin (cancer) and Novartis's Gleevec (chronic myeloid leukemia) were developed in the 1990s by targeting specific genes or receptors in a patient's body. These drugs indisputably saved lives and also made billions for their manufacturers. Now, the goal of nearly every gene discovery program was to find new drug targets. The problem was that targeted drugs were exceedingly difficult to develop, and nothing in the past decade had even come close to the success of Herceptin or Gleevec.

Less interesting to the pharmaceutical world, and less profitable, were diagnostic tests. Generally speaking, a diagnostic test measures something in the body using a small quantity of saliva, blood, urine, spinal fluid, biopsied tissue, or other extracted sample. The results of such tests can indicate whether the individual has high blood sugar or low cholesterol, has been abusing alcohol or illicit drugs, is pregnant or menopausal, is infected with herpes, Ebola, rabies, or some other communicable disease, has a benign or malignant tumor, and a thousand other conditions known to medical science. Genetic diagnostic tests are the same, except that they involve testing a patient's DNA to determine whether he or she carries a genetic variant that is associated with a particular disease or condition. Before the early 1990s, most genetic tests were conducted for family-planning purposes—determining whether would-be parents carried genes that would make certain birth defects more likely, or testing the DNA of a developing embryo to determine whether it would be born with a genetic disease like Down syndrome.

But the discovery of gene-disease associations in the late 1980s opened the door to a whole range of new genetic diagnostic tests. For the first time, healthy (or seemingly healthy) individuals could be tested to see whether they carried genes that would increase their risk of disease later in life. Tests for Alzheimer's disease, cystic fibrosis, hearing loss, and various forms of cancer suddenly appeared on the market. In almost every case, the genes in question had been patented as soon as they were discovered.

Almost all disease-related genes were initially discovered by academic researchers whose universities then patented them. What the universities did with these patents fell into roughly three categories: (1) *open access* (non-enforcement or low-cost/no-cost licensing to anyone who wanted to offer a diagnostic test), (2) *paid access* (licensing to anyone who wanted to offer a diagnostic test, but at a cost), and (3) *exclusivity* (only one test provider is authorized by the patent holder to perform diagnostic testing, often at whatever price the market would bear).

Major Gene Patents in 2008

DISEASE	GENE(S)	PATENT OWNER(S)	EXCLUSIVE LICENSEE	MARKET STRATEGY
Alzheimer's	APOE, PSEN1, PSEN2, APP	Duke Univ.	Athena Diagnostics	Exclusive provider
Canavan	ASPA	Miami Children's Hosp.	None	Paid access
Cystic fibrosis	CFTR	Univ. Michigan, Hosp. Sick Children, Johns Hopkins	None	Open access (non-commercial) Paid access (commercial)
Breast cancer	BRCA1, BRCA2	Univ. Utah, NIH, Univ. Cal., CRCT (UK)	Myriad Genetics	Exclusive provider
FAP (colon cancer)	APC	Johns Hopkins, Univ. Utah	n/a	Paid access
Hearing loss	GJB2	Institut Pasteur	Athena Diagnostics	Paid access
Hemochromatosis	HFE	Bio-Rad Labs (by succession)	n/a	Paid access
Huntington's disease	IT-15	Mass. General Hosp.	none	Open access
Long-QT syndrome	LQTx	Univ. Utah	PGxHealth	Exclusive provider
Lynch syndrome	MLH1, MSH2	Oregon Health Sci. Univ., Dana Farber Cancer Inst., Johns Hopkins	None	Open access
Maturity-onset diabetes of the young (MODY)	GCK, HNF4A, 4 more	Univ. Chicago	Athena Diagnostics[2]	Exclusive provider

---

2 Ambry Genetics also offered MODY testing, though it is unclear whether it was authorized to do so by the patent holder (Pierce et al (2009, p. 204)).

DISEASE	GENE(S)	PATENT OWNER(S)	EXCLUSIVE LICENSEE	MARKET STRATEGY
Melanoma	MTSx	Myriad Genetics	Myriad Genetics	Exclusive provider
Ovarian cancer	BRCA2	Univ. Utah, CRCT (UK)	Myriad Genetics	Exclusive provider
Spinocerebellar ataxia	SCA	Baylor Coll. Med., Univ. Minn.	Athena Diagnostics	Exclusive provider
Tay-Sachs	HEXA	NIH	None	Open access

The team agreed that the gene patents in the open-access category were not causing much harm. For example, the *CFTR* gene associated with cystic fibrosis had been patented by the University of Michigan, but the university allowed any lab that wanted to test the gene to do so at no charge. A similar situation existed with the gene for Tay-Sachs disease, which was patented by NIH but also made available without charge. So while patents covered these genes, the relatively benign behavior of their owners made them poor choices for a suit attacking gene patents. With other disease genes, however, things quickly became more complicated.

In the paid-access category, multiple labs offered diagnostic testing using a patented gene even though they had to pay the patent holder to do so. In some cases, like Miami Children's Hospital's licensing of the Canavan gene, non-profit testing centers could not afford the fee and had to stop testing. Lori Andrews, still unhappy over the result in the Canavan case, might have favored a challenge to the Canavan gene patent. Even so, everyone agreed that there were better targets in the third category. In this group—exclusivity—the university holding the patent had granted a *single* company the exclusive right to perform testing on a gene, and that company would not allow others to compete.

A few players stood out in this category. One was Athena Diagnostics, a company that had made a business of obtaining exclusive rights to a whole range of gene patents and then setting itself up as the sole provider of diagnostic tests based on those genes. Myriad Genetics had done the same thing with the genes relating to breast and ovarian cancer. And a small company called PgxHealth had the rights to the gene associated with a condition called Long QT syndrome.

In her 2007 congressional testimony, Chung described how researchers at the University of Utah had discovered genetic variants linked with a rare form of cardiac arrythmia known as Long QT syndrome.[3] The disease was hereditary, but because sudden cardiac arrest and death were often the result, a person's first symptom was often her last. The onset of cardiac arrest was preventable, however, with beta-blocker drugs and implantable defibrillators. So knowing that one carried the genetic variants associated with the disease could enable an affected individual to take life-saving precautions.

The university had patented the genes associated with Long QT syndrome and licensed them exclusively to a small company called DNA Sciences. DNA Sciences, patent in hand, then proceeded to send cease-and-desist letters to every other lab that was performing genetic testing for Long QT. When one lab, GeneDx, did not cease testing, DNA Sciences sued it for patent infringement. By 2002, DNA Sciences had "cleared the market" of tests for Long QT syndrome, yet DNA Sciences did not offer a test of its own. Soon thereafter, DNA Sciences went into bankruptcy. It took two years for DNA Sciences' assets to be acquired out of bankruptcy by another company and for that company to begin to offer genetic testing for Long QT. During that two-year hiatus, when no other company was allowed to offer this testing, patients with a risk of hereditary Long QT syndrome could not obtain a genetic test for the condition anywhere in the United States.

Chung described the case of a ten-year-old girl named Abigail, whose parents brought her in for testing because the syndrome seemed to run in their family. If Abigail carried the Long QT mutation, they would start her on beta-blocker drugs. But Abigail couldn't be tested because nobody

---

3 The name "Long QT syndrome" comes from the designations for two of the five electrical pulses that characterize the human heartbeat (P, Q, R, S, and T). The time between the start of the Q pulse and the end of the T pulse (the QT interval) measures the time it takes for the heart to contract and then refill with blood before beginning the next contraction. Long QT syndrome is an abnormality in the heart's electrical system that creates a prolonged QT interval, which can lead to sudden cardiac arrest and death.

was offering the test. Abigail died while lawyers were picking at the bones of the defunct DNA Sciences company. The university did nothing to enable others to offer Long QT testing while the legal squabble dragged on. In Chung's mind, this set of patents stood out as the worst around.

Simoncelli agreed with Chung. Children had died because of this patent. How much more compelling could a case get? Chris Mason was also getting enthusiastic, looking up some details of the condition on his laptop.

Yet Hansen shook his head. "We can't use it," he told them.

"Why not?" somebody asked.

"Because," Hansen said softly, "no one will know what it is."

Hansen didn't want to downplay the seriousness of the disease or the suffering that it had caused. But going after a patent on a gene for a rare disease that nobody had ever heard of would condemn their case to the back pages of medical law journals, not the front page of the *Times*. The best choice, the only choice, on the long list of gene patents were the breast cancer genes: *BRCA1* and *BRCA2*.

The Problem with *BRCA*

AFTER A BRIEF pause, Chung nodded. "Myriad Genetics," she said.

"They're certainly the poster child for bad gene patents," Simoncelli added.

Breast cancer affects roughly one in eight women in the United States. In 2007 alone, about 240,000 new cases of breast cancer were diagnosed, and 40,000 women had died from it (the figures today are comparable). There is a set of well-known risk factors for breast cancer. But in some families, there is a clear genetic link to the disease. One of the key genes responsible for this link—*BRCA1*—was found and patented in 1994 by researchers at the University of Utah (just like the Long QT gene) working with a private company, Myriad Genetics. They patented a second gene, *BRCA2*, a year later.

The *BRCA* genes are tumor suppressors. If a woman has certain mutations in one of her *BRCA* genes, her risk of contracting breast cancer doesn't just double, it soars from about 12 percent to 72 percent, a six-fold increase. Compounding the threat, *BRCA* mutations also increase a woman's chance of contracting ovarian cancer—which is even deadlier and more difficult to detect—from about 1.3 percent to 44 percent over the span of her lifetime. The *BRCA* genes were stunning discoveries. But there was a catch. Only 5 to 10 percent of all breast cancer cases are hereditary. The rest arise from other causes—environment, substance use, diet, weight, and factors that we don't yet understand. So variants in the *BRCA1* genes have a devastating effect on a small group of women— many of whom are of Ashkenazi Jewish descent—but are not factors in the large majority of cancer cases in the United States.

This could be a problem. The *BRCA* genes were terrible news for the women (and some men) who carried mutations, but they formed only a small percentage of total breast and ovarian cancer cases. On the other hand, breast cancer was huge. Now that polio had been eradicated, breast cancer was probably the most dreaded disease in America. And the advocacy network for breast cancer was equally huge. There was no end to the high-profile women who could stand up and say, "I'm a breast cancer survivor"—Nancy Reagan, Betty Ford, Sandra Day O'Connor, the list went on and on. Hansen guessed that every person in America had been affected by breast cancer, either personally or through a close relative. It was the universal enemy. And it was patented.

"Let's think about Myriad," Hansen said. "Are they as bad as the QT people?"

"In some ways worse," Simoncelli said. She explained that people had been critical of Myriad's *BRCA* patents for more than a decade—even before the patents were issued. They were mentioned in academic articles, newspaper stories, even Lori Andrews and Dorothy Nelkin's popular book *Body Bazaar*. Researchers, physicians, health advocates, genetic counselors, and bioethicists all faulted the company for threatening a lawsuit against anyone who wanted to perform *BRCA* testing. And many women couldn't afford Myriad's test, which was priced at around $3,100. To make matters worse, insurance often didn't cover it.

Hansen nodded. This was the kind of story they needed to tell. Myriad Genetics would make a good defendant—many already believed it to be opportunistic, greedy, and unsympathetic. And breast cancer would resonate with the public. It was far more appealing than Long QT syndrome or any of the other obscure diseases they had discussed. The public would respond to a case that struck a blow against breast cancer, and a company that seemed to be withholding an important breast cancer test. It didn't matter that most cases of breast cancer were caused by something else. These *BRCA* genes were important. Their impact on women's lives was enormous. And all of this would feed into their legal argument that patents on those genes were simply wrong.

One-Trick Pony

DESPITE HANSEN'S INTUITION, not everyone they asked thought that Myriad would be the ideal defendant in a case challenging gene patents.

Bob Cook-Deegan, a researcher at Duke University, was in the process of coordinating the largest study of gene patents in history. His team was responsible for much of the data that the ACLU had reviewed on gene patents. And the number one player on that list? Athena. Myriad was a one-trick pony. All it had, practically speaking, was *BRCA*. Athena, on the other hand, had wrangled a whole basket of gene patents. It was becoming a kind of "patent troll" for biotechnology.[4] Wouldn't this better illustrate the problem of patent thickets, and how gene patents were quickly eating up the entire human genome?

At the same time, some thought that breast cancer was too politicized. The patent issues could be obscured by the other health and equity issues swirling around. Even Simoncelli viewed the complex politics of breast cancer with some trepidation. Her own mother had died from the disease when Simoncelli was thirteen; now she wasn't sure how she felt about wading back into the world of breast cancer advocacy all these years later. But political is exactly what Hansen wanted. He wasn't in this to

---

4 Patent "troll" is a pejorative term for a company that acquires patents for the sole purpose of monetizing and enforcing them without conducting any R&D of its own. The Obama administration took a strong public stance against patent trolls.

refine the patent law—for him this was a civil rights case. And civil rights were political. That was his element.

The Legacy Of RBG

HANSEN WAS PRETTY sure that they had their defendant. And if the disease was breast cancer, it opened a new door in the strategy. Because what they had now wasn't just a free speech case, and it certainly wasn't just a patent case. Now, they had a women's rights case.

To bring a women's rights case, Hansen needed to pay someone a visit. Back at the ACLU office in Lower Manhattan, he walked down the hall to visit Lenora Lapidus, head of the ACLU's Women's Rights Project. The project, founded by Ruth Bader Ginsburg in 1972, had grown into one of the ACLU's most influential and powerful units. It had achieved landmark legal victories on issues ranging from gender discrimination and sex trafficking to domestic violence and reproductive rights. Lapidus, in her early forties, ran the group with a potent combination of legal acumen and deep sympathy for women of all backgrounds.

Hansen knew that embarking on a case so closely tied to women's health would require political buy-in from the Women's Rights Project. Moving forward without its involvement could cause serious discord within the ACLU, something that Hansen had always been savvy enough to avoid.

Hansen found Lapidus in her office, cluttered with memorabilia from the struggle for gender equality. She smiled and greeted him, and he quickly explained the nature of the case they were considering—a company had patented genes for breast and ovarian cancer, lots of women couldn't afford it, nobody else could offer the test. He tentatively asked Lapidus whether her group would like to be involved in the case. Without hesitation, she responded, "Of course we want to be involved!" She smiled broadly. "We're in. We're completely in."

Lapidus assigned the case to Claudia Flores, a senior attorney at the Women's Rights Project, but Flores left the ACLU soon thereafter to help the government of Indonesia deal with its human trafficking problem.

At that point a young attorney named Sandra Park asked if she could step in. Park had joined the Women's Rights Project only a year earlier, in 2007. She was working with survivors of domestic violence, but had a longstanding interest in science issues.

Born in Seoul, Park had come to the United States at the age of three. She grew up in Wheaton, Illinois, just outside of Chicago and a couple of miles from Hansen's home town of Glen Ellyn. She attended the prestigious Illinois Mathematics & Science Academy, where she was a research assistant for a genetics professor, and completed the pre-med program at Harvard College before deciding to go to law school. Like Hansen, she enrolled at the University of Chicago Law School, but transferred to NYU a year later because of its generous loan forgiveness program for graduates who pursued public interest careers. And, on a personal note, Park's mother was a breast cancer survivor. Lapidus was convinced. She assigned Park to work on the case. And with that, the ACLU's legal team was more or less complete.[5] Now the real work would begin.

---

5 The ACLU legal team consisted of Hansen, Simoncelli, Lapidus, Park, and Aden Fine. At PubPat, Ravicher was joined by Sabrina Hassan.

CHAPTER 10

We've Got You Covered

LISBETH CERIANI WAS not intimidated by insurance. The petite native of Newton, Massachusetts, with sharp features and an acerbic wit, had spent a decade working at Blue Cross Blue Shield. She was a graduate of Boston College, the daughter of a civil engineer—educated, vegetarian, divorced.

Shortly after her daughter Isabella was born in 2001, Ceriani had to decide between putting her baby in daycare so that she could go back to work or figuring out something else. The advice she got from her friends was consistent: go back to work, don't sacrifice your career. But Ceriani was unconvinced. She wasn't married, she was pushing forty—what if this was her only shot at raising a child? In the immediate post-9/11 world, people were thinking more about families and last chances and cherishing the moments they had together. Ceriani decided to raise her daughter—she could always make money later.

Ceriani didn't have much savings, and neither her ex-husband nor her daughter's father (different people) would be of any use. But she was a planner; she could do it. She earned a little money doing childcare out of her home, and for insurance she relied on Medicaid. Her taxes had paid for the Medicaid system over the last decade, so why not put it to work for her? Living in Massachusetts, she was comfortable that her and her daughter's medical needs would be covered—Massachusetts under Governor Mitt Romney had the closest thing to socialized medical care that had ever existed in the United States.

Against the odds, Ceriani's plan worked. She had to economize, but at least she felt she could be there during her daughter's formative years.

When Isabella entered kindergarten, Ceriani took another part-time job—helping to coordinate a cross-cultural exchange program run by the State Department. And then, in 2008, at the age of forty-two, she was diagnosed with breast cancer.

Ceriani's oncologist at Massachusetts General Hospital—Mass General—suspected a genetic link. Ceriani was relatively young, with multiple tumors and a particularly aggressive form of cancer—all hallmarks of a hereditary disease. Ceriani wanted to know whether she had a *BRCA* mutation. The answer would affect the kind of breast surgery she had, whether she should consider having her ovaries removed, and whether Isabella would also have to contend with these issues. But the genetic counselor at Mass General politely explained that Medicaid didn't cover the test.

Ceriani couldn't believe it, so she went home and did some research. She found that MassHealth, the Medicaid program in Massachusetts, did, in fact, cover *BRCA* diagnostic testing. The catch was that it would only cover the test if it was performed by a contracted provider—a laboratory that had a contract with MassHealth. And MassHealth had decided that it would pay only $1,600 for the test—what it considered to be a "usual and customary" charge. But the only company that offered the test—Myriad Genetics—charged roughly $3,100. And under Massachusetts's progressive laws, a healthcare provider is not allowed to bill a patient for amounts in excess of what is reimbursed by her insurance carrier (a practice called "balance billing"). So if Myriad performed a *BRCA* test for a Medicaid patient in Massachusetts, it would receive only $1,600 from MassHealth, with no possibility of recovering the remaining $1,500 balance from anyone. This math apparently didn't work for Myriad, so it never entered into a contract with MassHealth. And because Myriad controlled the *BRCA* testing market in the United States, no other lab could perform *BRCA* testing. So even though MassHealth theoretically covered *BRCA* testing, there was no contracted provider to offer the test to Massachusetts Medicaid patients like Ceriani.

To Ceriani, this situation verged on the ridiculous. She argued with the genetic counselor at Mass General. Couldn't the hospital order the test from Myriad and then bill MassHealth? Mass General was a MassHealth contracted provider. The answer she got was no. If Mass General did that, who would cover the $1,500 that MassHealth didn't reimburse? She called Myriad, doggedly arguing her way up the telephone helpdesk chain until she reached somebody who actually knew something about contracts with insurance carriers. But when Ceriani asked the Myriad employee when the company intended to sign a contract with MassHealth, the employee told Ceriani that Myriad was not planning to contract with MassHealth. Ever. When Ceriani persisted, the employee got impatient and told her to stop calling.

Ceriani, who had to suspend her childcare job during chemo, couldn't afford the $3,100 on her own. Another representative at Myriad told her that it had a financial assistance program for low-income patients, but federal law prohibited them from offering financial assistance—known as inducements—to people on Medicaid.[1] Ceriani was incredulous. Assuming that Myriad was willing to reduce its fees for poor people who weren't on Medicaid, why wouldn't it sign a contract with MassHealth for the $1,600 that MassHealth was able to reimburse? Wouldn't the $1,500 that Myriad couldn't collect from MassHealth be the same as the financial assistance that Myriad wanted to offer, but couldn't? No, a polite Myriad representative informed her. It just wasn't the same. Why not? Ceriani asked. Well, it just wasn't.

In desperation, Ceriani's mother offered to pay for her BRCA test. "I can just put it on my MasterCard," she said. But Ceriani flatly refused. "It's extortion," she told her mother. It had become a matter of principle for Ceriani. Somebody—Myriad, Mass General, MassHealth, maybe even Mitt Romney—needed to figure it out, because she wasn't the only Massachusetts woman on Medicaid, and the others probably knew a lot less about insurance than she did.

---

1 Under the Social Security Act, a healthcare provider may not offer a Medicaid or Medicare patient any remuneration that is likely to affect her selection of services, other than inexpensive gifts such as toothbrushes, hospital slippers, and coffee.

The Dismal Science

NOTWITHSTANDING THE FRUSTRATION of patients like Lisbeth Ceriani, nobody wanted Myriad's *BRCA* test, which it marketed as BRACAnalysis, to be covered by insurance more than Myriad itself. The company's executives knew that Myriad couldn't achieve the kind of market penetration they wanted by selling only to customers who could afford the $3,100 price tag out of pocket. No, they needed every insurance carrier, health plan, and health maintenance organization (HMO) in America to cover BRACAnalysis. Their target market was nothing short of every woman in the country. But Myriad had not predicted how difficult or time-consuming it would be to get insurance carriers to pay for their test.

The health insurance landscape in the United States is a bewildering patchwork of private insurance plans, HMOs, and three massive government programs: Medicare, Medicaid, and the Veterans Health Administration (VA).[2] Medicare is a form of nationalized health insurance for individuals sixty-five and older; Medicaid covers the indigent. Both Medicare and Medicaid are administered by the Centers for Medicare & Medicaid Services (CMS), with individual coverage decisions made by fifty-seven different claims payment carriers in the case of Medicare, and fifty semi-autonomous state Medicaid plans. The VA covers approximately 9 million military veterans.

Given Myriad's target demographic—women between ages twenty-five and fifty-four, especially those of Ashkenazi Jewish heritage—neither Medicare, Medicaid, nor the VA would be the primary carriers paying for Myriad's tests. Myriad needed to focus on the private insurance market. Though there are only thirty or so major health insurance carriers in the United States, the market also includes about nine hundred smaller insurance carriers plus nearly five hundred HMOs. Each of these payers had the ability to decide whether or not to cover Myriad's genetic tests. Getting

---

2 This landscape is also in constant flux. The discussion in this chapter focuses on its characteristics during the period that Myriad struggled to get coverage for its tests—approximately 1997 to 2008.

them to agree to do so was a gargantuan task that would occupy Myriad's executives and legal counsel for years to come.

At first glance, the economics of getting a payer to cover the cost of BRACAnalysis might seem straightforward. After all, if a woman tests positive for a *BRCA* mutation, she has the option to undergo prophylactic surgery or preventative chemotherapy, greatly reducing her chance of contracting the disease later and eliminating the payer's need to cover her treatment after the disease is diagnosed. One might think that this alone would be enough to persuade payers to cover the test. But the math is a bit more complex.

First, the insurance industry is geared toward covering expenses for people who are sick. But diagnostic tests are not treatments. Unlike a cholesterol or blood sugar test, or a biopsy of a suspicious mole, BRACAnalysis isn't designed to detect the presence of an existing disease. Rather, it will detect a genetic marker that makes it more likely that an individual will, at some point in the future, develop a disease. But at least in the case of breast cancer (much less so for ovarian cancer) there are already a number of techniques—breast inspection, mammography, ultrasound, MRI—for detecting the disease at an early stage of development. And these are already covered by most health plans, are far less expensive than BRACAnalysis, and give a much more definitive result ("you have cancer" not "you might develop cancer"). So why test healthy people using a test that does not even detect the disease?

Second, if a woman develops breast or ovarian cancer, surgical and chemotherapy treatments will likely be prescribed. While on average the cost of these procedures is slightly higher when they are ordered to treat a known cancer than when they are only prophylactic, the price difference is not that great. So for a woman who is likely to get cancer, it could be cheaper for the payer to wait until she is diagnosed—when it has to pay only the cost of the treatment, as opposed to the cost of the test *plus* the treatment. Moreover, in some cases a disease diagnosis might come too late—the patient might die before an expensive procedure is performed or completed—again saving the payer some money.

Third, a positive *BRCA* test does not guarantee that a given woman is going to get cancer. While the odds are increased, they are not 100 percent, and in some cases not even 50 percent. Some women are just lucky, and some may die of other causes before their cancer would have appeared. Thus, in at least some cases, a woman who would not have developed cancer during her lifetime might be tested positive and get prophylactic surgery. The insurer in this case pays for both the test and the surgery. But if the woman was not tested, the payer would have paid nothing at all. So for these women, payers would strongly prefer not to pay for testing.

Fourth, payers play the long game—they consider their cost over an individual's lifetime (or at least until she goes onto Medicare). Suppose that a woman gets a *BRCA* test (at the payer's expense) and elects to undergo prophylactic surgery (another cost to the payer). She may now live longer, incurring additional costs over the duration of her life (hypertension and diabetes medications, hip replacements, etc.), and may eventually become afflicted by a disease that is even more expensive to treat than breast or ovarian cancer. Clearly, the payer's cost would have been less if the individual had never gotten the *BRCA* test and was simply allowed to develop terminal cancer.

Fifth, payers know that people change insurance plans over the course of their lives. Suppose that a *BRCA*-positive woman is tested at age twenty-five and elects to have a prophylactic mastectomy at age twenty-six. Her current payer is responsible for that bill. But then suppose that five years later she changes jobs, or moves to another state, or gets married and is covered under her spouse's health plan. Odds are that even without the prophylactic surgery, she wouldn't have developed cancer until her forties or later. This means that the payer who covered her at twenty-six would not have been the one to cover the cost of her treatment at forty. Which makes the first payer wonder why it should cover an expensive diagnostic test and an even more expensive prophylactic surgery when the cost being avoided—treatment for a future cancer—would in all likelihood be someone else's responsibility.

And, finally, even if a payer could be persuaded to cover Myriad's diagnostic tests, there remained the question of *how much* the payer would agree to reimburse for it. As experience in the pharmaceutical industry had shown, payers often drive a hard bargain, and they seldom reimburse 100 percent of a drug's list price. Working against Myriad was the expense of its test. Most lab tests—cholesterol, blood sugar, even prenatal genetic tests for diseases like Tay-Sachs—are fairly inexpensive, usually costing less than $100 each, and some less than $50. At first, Pete Meldrum, Myriad's CEO and co-founder, estimated that Myriad's *BRCA* test would fall into a similar price range. In 1993, he told potential investors that a single *BRCA* test should cost the company no more than $100 or so to run. But that was before *BRCA1* had been sequenced. And, as they discovered, *BRCA1* was a big, complicated gene.

Mark Skolnick, the University of Utah scientist who teamed up with Meldrum to found Myriad, insisted that their test involve full sequencing of the gene along both DNA strands, an expensive procedure. Even before Myriad launched BRACAnalysis, the University of Pennsylvania was charging around $1,600 for its *BRCA1* test, and it wasn't even as comprehensive as BRACAnalysis. Meldrum's challenge was convincing insurance carriers that Myriad's price was reasonable. It was a long, hard sell.

In addition to the significant economic and procedural hurdles, Myriad also faced legal challenges. Though the company did not expect a large number of its test subjects to be insured by Medicare, which provides healthcare coverage for individuals over the age of sixty-five, the massive federal program strongly influenced the coverage decisions of private insurers and HMOs. And persuading Medicare payers to cover BRACAnalysis was not easy.[3]

Under the Social Security Act, which created the Medicare program in 1965, no Medicare payment may be made for items or services that are not "reasonable and necessary for the diagnosis or treatment of illness or injury or to improve the functioning of a malformed body member." Thus, medical procedures, drugs, implants, mobility devices, and the like are covered, but things like genetic screening tests that bear on a woman's decision to become pregnant generally are not. However, if a genetic test is used to improve a patient's medical treatment (e.g., by

---

3 Though Medicare is centrally managed by CMS, the majority of coverage decisions for diagnostic tests during this period were made by local Medicare contractors. Lewin (2005, pp. 99–103).

adjusting a chemotherapy regimen), then Medicare will cover it. Myriad's *BRCA* tests fell somewhere in between these two poles.

As a result, no payer was willing to give Myriad a blanket pass. Rather, Myriad, or the patient or her physician or genetic counselor, would have to demonstrate medical necessity to the payer on a patient-by-patient basis. And to do this, they would have to satisfy certain criteria that Myriad negotiated with the payer. For example, in many cases BRACAnalysis would be covered only when administered to high-risk patients, meaning women within certain age brackets who had been diagnosed with breast cancer at an early age and who had two or more close relatives who had also been diagnosed at an early age.

Myriad's executives knew that overcoming this triple hurdle—economic, administrative, and legal—would not be easy. The company warned investors in 1995 (a year before the release of BRACAnalysis) that its success would depend on obtaining adequate reimbursement for its services, and that ongoing efforts to contain healthcare costs could hinder those efforts.

In addition to medical necessity, payers also required that, in order to be reimbursed, a diagnostic test must be generally accepted by the medical community and must be requested by physicians and patients. Myriad launched a series of advertising campaigns to address this need. Increasing public awareness of BRACAnalysis wouldn't just increase demand for Myriad's test (the primary goal of the ad campaigns), but it would also create consumer and political pressure on payers to *cover* the test. In this respect, Myriad's multimillion dollar print, radio, and television advertising campaigns in markets like Atlanta, Denver, and New York paid double dividends, both in terms of consumer demand and reimbursement coverage.

Three Cheers for GINA

THE EFFORT TOOK years, but little by little Myriad's *BRCA* tests began to be accepted by healthcare payers and HMOs. In its 1997 Annual Report to Stockholders, Myriad explained that "while reimbursement policies for BRACAnalysis are still under discussion with a number of insurance companies and managed care providers, several major insurance companies and HMOs have provided reimbursement for BRACAnalysis testing to their members." By 1998, Myriad had entered into coverage agreements with more than three hundred different payers and claimed that only 5.7 percent of reimbursement claims for BRACAnalysis had been denied.

In 1999, the company announced a major breakthrough. It signed a coverage agreement with Aetna U.S. Healthcare, one of the largest private health insurers in the country. Aetna would cover BRACAnalysis for each of its 23 million insureds who met the stringent qualification criteria negotiated with Myriad. Less than a year later, Myriad announced a similar agreement with Kaiser Permanente, covering another 8.6 million Americans, and then one with Empire BlueCross BlueShield, the largest health insurance carrier in New York. In 2002, Myriad told its stockholders that BRACAnalysis was "covered by *most* health maintenance organizations and health insurance providers in the United States," and by 2004, eight years after the test was introduced, Myriad's public filings claimed that BRACAnalysis was covered by *all* major U.S. HMOs and insurance providers. The head of its diagnostics business claimed that by 2009, Myriad had in place over four hundred contracts with private insurance payors covering over 130 million Americans.

For the company, this milestone was cause for celebration. But even while Myriad succeeded in securing some level of coverage from private insurers, not everyone who needed one was assured of receiving a *BRCA* test. For example, some patients with high deductibles and co-pays found themselves unable to pay the required balance for testing. And, most importantly, Myriad neglected patients on the margin of its target market—those covered by Medicaid. And this failure would cost it dearly.

In addition to the Medicaid coverage gap, there was another major problem that Myriad faced in terms of insurance coverage. No matter how many insurers and health plans Myriad signed up, there were individuals who were afraid to file claims with their insurance carriers. Why? Because they feared that those carriers could discriminate against them on the basis of their test results. For example, if a woman tests positive for a *BRCA* mutation, the predicted cost of that woman's healthcare over the remainder of her life suddenly skyrockets. Could an insurer drop her from coverage as a result? Could it increase her premiums? Could it use that information when making underwriting decisions about her family members? And what about other forms of coverage, like life and disability insurance? Could these carriers find ways to discriminate based on one's *BRCA* test results?

Researchers found that fears of genetic discrimination were real and pervasive. For Myriad, this was a big problem. If women were unwilling to submit *BRCA* testing claims to their insurance carriers for reimbursement, then many of them would not be tested at all. This was lost revenue. So Myriad worked from an early stage to allay fears of genetic discrimination. Most importantly, it adopted a policy of refusing insurers' requests for *BRCA* test results. That is, when Myriad ran a *BRCA* test and the patient submitted a claim to her insurer, Myriad would confirm to the insurer that the test had been provided, but would not disclose the results of that test to the insurer. While this policy caused some friction between Myriad and the carriers with which it had coverage agreements, the ability to assure the public that insurers would not get access to its test results was more important to the company's success.

The most important developments in the area of genetic discrimination, however, were not initiated by Myriad. Concerns about discrimination on the basis of one's genes had emerged in the early 1990s, years before *BRCA* testing became available. But despite a few anemic legislative Band-Aids, many felt that more comprehensive legislation was needed. The struggle to enact such legislation spanned thirteen difficult years. Finally, in 2008, President Bush signed the Genetic Information Nondiscrimination Act (GINA), which prohibits discrimination in both health insurance and employment on the basis of genetic information. Senator Ted Kennedy, a co-sponsor of the legislation, called it "the first major civil rights bill of the new century."

While critics claim that GINA did not go far enough—for example, life and disability insurance are still not covered—the enactment of a federal anti-discrimination law for genetic information helped Myriad immeasurably. Now, it could deflect the longstanding fears that genetic counselors and physicians had harbored about *BRCA* testing. Another barrier to widespread adoption of Myriad's test had fallen.

LISBETH CERIANI HAD plenty of time to think about *BRCA* testing, gene patents, and healthcare reimbursement while she was immobilized in the chemo room at Mass General during 2009. At least it had a nice view, she thought, gazing across the Charles River at the gleaming glass-and-steel biotech city that surrounded MIT. She had still not been tested for the *BRCA* mutations.

CHAPTER 11

BART

As an operations manager overseeing the installation of outdoor corporate signage across the country, Kathleen Maxian knew building codes and crane heights, labor contracts and completion schedules, and how to get things done in the male-dominated construction industry. Tall and assertive, with wavy black hair that fell below her shoulders, she spoke loudly and looked at ease in a hardhat. Nevertheless, she did cry a little when her parents called her on the Monday after Easter in 2007 to let her know that her sister, Eileen Kelly, had breast cancer.

Kathleen had just come home; the evening air was still crisp in the Buffalo suburb where she and her husband, Tom, lived. She was taking off her jacket when the phone rang. It was her parents—unexpected since they had just been together at the traditional Easter dinner that her mother cooked every year. Maybe Kathleen had left something at their house—a scarf or a pair of gloves. But no. It wasn't a scarf or gloves. Her mother's voice was unsteady. "Eileen . . . Eileen has breast cancer," she said. Kathleen's father, leaning close over the living room speakerphone and occasionally clearing his throat in the background, said nothing.

The news was the worst kind of surprise. Eileen was young—only forty. She lived in Atlanta and had just started a new business with her partner, Jean. They had gotten decent insurance for themselves, which prompted Eileen to schedule her first mammogram the previous Monday. They found a lump, and it was not small. She went back for more images on Friday, and this time there was no question what they were seeing—a large tumor with calcifications. Eileen was scheduled for a biopsy later that week.

Kathleen hadn't seen Eileen in a while. She and Tom lived in Buffalo, where Kathleen had been raised. They had met on the job about five years earlier at Niagara Falls State Park, where Tom was the park director and Kathleen worked in the engineering department. An argument over some damaged lampposts brought them together—first as adversaries and then as friends who could laugh about the overblown incident. Six months later they were married.

Kathleen was stunned as she hung up the phone that evening. She didn't speak with her sister much—the result of a long-ago sibling dispute—but still cared deeply about her health. Over the following weeks, Kathleen's parents kept her updated on Eileen's progress. The biopsy revealed an invasive grade 3 ductile tumor—not a good prognosis. Eileen's oncologist scheduled a lumpectomy in June, followed by a four-month course of chemo and radiation. Her parents went to Atlanta for the surgery, and afterward her mother stayed on to help.

Midway through chemo, Eileen's oncologist mentioned genetic testing. Because she was diagnosed young, if there was a family history of cancer, there could be a genetic root to her disease. But, as far as Eileen knew, she didn't have a family history of cancer. Her parents, both in their seventies, were relatively healthy. Her father, Arthur Kelly, had three brothers, two of whom had died from heart failure and one from complications from diabetes; her mother, Ann, the eldest in her family, had two sisters and a brother, all of whom were still living and in comparatively good health. Even so, Arthur, a professional engineer, wouldn't dismiss the genetic theory so quickly. He began to piece together a family tree.

Arthur was born in Yonkers, near New York City, but moved to western New York in the 1960s. His wife, Ann, immigrated to upstate New York from Newfoundland at the age of eighteen. Neither were close to their extended families. But Arthur was determined. He went through their Christmas card list and found phone numbers for many of those relatives that he remembered, some whom he hadn't spoken with in decades. He started making calls. It was strange at first, hearing a voice from a half-century ago, speaking with people with whom he shared blood but little else.

Arthur's cousin Gale was one of these people. Gale and her little sister, Moira, the daughters of Arthur's aunt Marie, had played with Arthur when they were all children in Yonkers. It was during the Depression, before the war—a long, long time ago. Arthur remembered that Aunt Marie had gotten sick and died. At the wake there was black crepe everywhere and the adults were either crying or trying not to. The girls' father remarried soon after that, and they moved away. Arthur hadn't seen them again. Now, when Gale answered the phone, she was not the playful, pigtailed girl whom he remembered, but a tired woman in her seventies. But it was her.

Gale confirmed the troubling memory that had been nagging at the back of Arthur's mind. His aunt Marie, Gale's mother, whom Arthur remembered as a fun-loving, irreverent woman in those days of Studebakers and console radios, had died of breast cancer. She had been in her early thirties. And Gale herself had been diagnosed with breast cancer, though much later and, thank God, she survived. Gale's sister, Moira, now deceased, had succumbed to leukemia just a few years earlier.

There was more. As Arthur continued to dig into the family's history, he found that his father's sister Loretta and her daughter Peggy, another cousin whom Arthur vaguely remembered, had both died of cancer.

That was it, then. They had a family history of cancer, even though few spoke about it. Arthur painstakingly drew their family tree on a big piece of folded-out paper. It listed names, ages, causes of death. He gave it to Eileen, who gave it to a genetic counselor at Piedmont Hospital, where she was being treated.

The counselor was impressed by the thoroughness of Arthur's research. She met with Eileen in her office and gave her a mini-lesson in molecular biology. With a diagnosis at a young age and a family history of breast cancer, Eileen would qualify for a genetic test to see if she had a mutation in *BRCA1* or *BRCA2*, the two major genes known to cause early-onset breast cancer. They would draw blood in Atlanta, but the sample would be sent to a special company in Utah—the only company that performed these tests. And it would all be covered by Eileen's Blue Cross Blue Shield policy, so there was nothing she would have to pay out of pocket. To Eileen, this all sounded fine. If their family carried a genetic mutation that could cause cancer, better to know about it than not. But, in the midst of a hard course of chemo, with cancer already contaminating her body, Eileen didn't really think much about it.

A few weeks later, Eileen got the test results back from Piedmont. They were negative for the known *BRCA* mutations. She was relieved, to a degree, but also confused. Now it was even less clear what had caused her breast cancer. What about Aunt Marie and the rest? Still in the midst of chemo, she was not inclined to dwell on it. Eileen told her mother the results, and then her mother called Kathleen. Kathleen was the more relieved of the sisters. She had read about what those genetic mutations meant. Women were getting mastectomies and hysterectomies—major surgeries—if their tests came up positive. Thank God at least she wouldn't have to make those choices.

The Diagnosis

IN THE FALL of 2008, the signage company that employed Kathleen closed its Buffalo office—another victim of the financial crisis that gripped the country. She received a generous severance package and decided to wait out the economic downturn. Tom's job with the state of New York seemed secure, and his health insurance would cover them both. So Kathleen turned her attention to some long-neglected household projects and, when spring finally came, to her garden.

But Kathleen soon began to feel unwell. She was bleeding between periods. She felt swollen, bloated. She was only forty-seven, but wondered if she could be experiencing the early signs of menopause. In July she visited her gynecologist, whom she had known for years. He was unsure what was going on. He ordered blood work, a uterine biopsy, and a transvaginal ultrasound. The ultrasound must have disturbed something deep within her, because Kathleen spent the weekend in pain, lying in bed and searching the internet for what it might be.

Her gynecologist called her on Monday morning with the bombshell

Kathleen Maxian (right) and her oncologist, Dr. Nefertiti duPont (left), who urged her to get BRCA testing.

that would change the course of Kathleen's life. She had ovarian cancer, and she needed surgery. Right away.

After that, things moved quickly. Kathleen, sick with dread, met with Nefertiti duPont, a gynecological oncologist at Buffalo's Roswell Park Cancer Institute. Kathleen was impressed by the dynamic young doctor with a string of university degrees and medical credentials. DuPont examined her, and when she insisted that Kathleen schedule surgery the next week, Kathleen knew her condition was grave.

The operation was devastating. Her cancer was Stage 3B—invasive and lurking throughout her abdomen. They scraped her out, ran her bowel (literally pulling it, foot by foot, through the surgeon's probing fingers to check for cancerous growths), then sewed her up again.

When the anesthesia wore off, Kathleen thought she had died. Her room was full of flowers, some from people she didn't even know. She heard voices—her husband, her mother, Dr. duPont, others—talking solemnly about her prognosis. There was a 20 percent survival rate for ovarian cancer this advanced—the odds were four to one that she would be dead within five years, probably sooner.

Kathleen went home and recovered over the next few weeks. During that time, when she couldn't get out of bed, her neighbor Mollie Hutton came to visit. Hutton was a genetic counselor at a local hospital. When she heard about Kathleen's family, she couldn't believe it. One sister with a breast cancer diagnosis at forty. Another sister with ovarian cancer at forty-seven. Plus a bunch of aunts and cousins with breast cancer over the years. It had to be genetic. How could it not be? "Didn't anybody test your sister?" Hutton asked. Kathleen nodded. "Eileen was tested," she explained. "We're negative." Hutton shook her head in disbelief. Something was wrong. They should have tested Kathleen too, back when Eileen was first diagnosed. Kathleen's cancer should have been avoidable.

When she could walk again, Kathleen returned to Roswell for a post-surgical appointment. Dr. duPont discussed the pathology report and outlined Kathleen's chemo regimen. None of it sounded good. As they were wrapping up, duPont said, "You know, Kathleen, your mother mentioned that your sister had breast cancer."

Kathleen knew where duPont was heading and put up her hand palm outward, like a school crossing guard. "Oh, no, Dr. duPont," she said, cutting off the discussion. "Our family had the test and we're negative."

DuPont nodded. Then she asked, "Do you think we could see your sister's paperwork?"

Kathleen shrugged. "Sure," she said, "but that's got nothing to do with it."

As Kathleen's husband drove her home from the hospital, she began to think. Both Hutton and duPont seemed to believe there was more to the genetic story than she knew.

Kathleen called her father from the car. "Dad, do you remember Eileen had that testing?"

"M-hmm," he murmured.

"I think they think it was wrong," she said. "But it can't be, right? Because there was only one company in the country that did it and they were that really special company in Utah."

Her father didn't know. He would call Eileen.

Kathleen's intraperitoneal chemo started two weeks later. It was like a torment invented by some medieval sadist—they pumped chemicals, so many chemicals—directly into her belly, through a plastic port that pierced every layer of her skin. Then they rolled her over, first on one side, then the other—to make sure the chemicals sloshed around inside her. The full course of treatment was six doses. She made it through four before giving up and switching to an intravenous alternative.

At some point during this treatment, Kathleen's parents called. As usual, they were on the speakerphone. After the pleasantries, her father said, "Your sister went back to her genetic counselor." He paused, as though thinking about how to phrase a piece of exceptionally bad news. "She does have the *BRCA1*." In the background, Kathleen could hear her mother sobbing, saying, over and over, "It's no one's fault. It's no one's fault."

It took Kathleen a moment to process what she was hearing. If Eileen *did* have the *BRCA1* mutation, then Kathleen probably had it, too. And if Kathleen had the mutation, then she could have had one of those prophylactic operations that people got when they knew they were carriers of the mutation. In fact, Eileen was now scheduled to get her own ovaries removed—to avoid the hell that Kathleen had just gone through, that she was *still* going through.

But if Kathleen had known that Eileen carried the mutation two years ago, Kathleen could have had her ovaries removed too, before the cancer spread throughout her gut. She could have done that *if* she had known that she carried the mutation. But Kathleen never got tested—no one ever suggested that *she* be tested—because Eileen was negative. The sisters just assumed that their cancers weren't genetic—they must have been caused by too many dental x-rays or inhaling cigarette smoke at a bar or some carcinogens leaking into the local water system. But genetic? What had gotten screwed up?

Kathleen could still hear her mother sobbing that it was no one's fault. But Kathleen knew. This was somebody's fault.

Del Exon 3

WHEN THE DUST settled, after a lot more sobbing and a lot more questions, Kathleen learned from her parents that back in 2007 Eileen had been given the most common *BRCA* test, which was called BRACAnalysis.

That test was negative. But there was *another* test, an extra test, that could detect other kinds of mutations. It was called BART—BRACAnalysis Rearrangement Test—and it cost around $700. Eileen did not originally get the BART test. And when she finally did, they learned that Eileen had a mutation called "del exon 3." It was a large-scale rearrangement of the nucleotides of the gene, rather than a single mistaken letter, that could only be detected using BART. And when Kathleen finally took the BART test, del exon 3 appeared for her as well. And, as they learned, it had caused both of their cancers.[1]

So why—WHY?—Kathleen wondered, hadn't Eileen initially been given the BART test? It took a while for Kathleen to dig out the answer to this question, and it made her more than angry.

At first, Kathleen assumed that Eileen hadn't wanted to spend the extra $700 for BART. But that was ridiculous. Eileen, who ran a variety of successful pet-related businesses in Atlanta, had plenty of money. She said she would have taken whatever test was offered if she had known about it.

The truth was that Eileen's genetic counselor in Atlanta had never even mentioned the BART test to her. Why? Because Eileen didn't meet the criteria for BART. To qualify for BART at that time, a woman diagnosed with breast cancer before the age of fifty needed to have either two first- or second-degree relatives who were also diagnosed with breast cancer before the age of fifty, or one who was diagnosed with ovarian cancer at any age. In 2007, when Eileen was first tested, Kathleen's ovarian cancer had not yet been discovered, and their Great-Aunt Marie, who had died from early-onset breast cancer, was too distant a relation. Marie's daughter Gale didn't develop cancer until her sixties, and the diagnoses of their other relatives were too unclear. So the criteria were not met. As

---

1 The *BRCA1* gene is large and messy, with an abundance of repeated segments, which made it particularly difficult to sequence. Ordinary sequencing of the gene for diagnostic purposes will detect the deletion or addition of particular bases in the normal gene sequence (e.g., the common del 185 AG mutation affecting Ashkenazi populations). But some mutations, such as those carried by Eileen and Kathleen, consist of a large-scale shift of bases within the gene. These mutations are difficult to detect using ordinary sequencing, and require additional techniques to detect. By way of analogy, a simple word processing spell-checker can easily identify misspelled words within a paragraph, but has no way to determine whether whole paragraphs have been deleted or switched around on a page. It has been estimated that approximately 10 percent of *BRCA1* mutations fall into this category.

a result, Eileen did not get a BART test until Kathleen was diagnosed with ovarian cancer—that's when Eileen finally met the criteria for BART testing. And once BART helped to diagnose Eileen's *BRCA* mutation, Kathleen also qualified for the test.

So the real question in Kathleen's mind was why BART wasn't available, even to people who would gladly pay for it. She looked at her own genetic testing report, which confirmed the presence of the BART mutation, and saw a help line at the bottom: PLEASE CONTACT MYRIAD PROFESSIONAL SUPPORT AT 1-800-469-7423 TO DISCUSS ANY QUESTIONS REGARDING THIS RESULT. She did.

Everyone Kathleen spoke with at Myriad was friendly and seemed to be trying to help. She learned that a test for large-scale *BRCA* rearrangements was first developed in 2002, and BART had been available commercially since 2006. Someone sent her a fact sheet that listed the testing criteria. Eileen's genetic counselor had been right—Eileen hadn't qualified for BART back in 2007. But none of this answered the real question: why not? Why would a company develop a test and then refuse to offer it to people who were willing to pay for it? Why wouldn't they even tell people about the test? It was crazy. When did a company withhold the availability of some product that you could pay extra for? Would you like some fries with that burger? A fabric protector for your new couch? An upgraded stereo for your SUV? Why not a BART add-on to go with your BRACAnalysis? Just $700 to spare your relatives from the living hell of Stage 3B ovarian cancer? Who wouldn't take that deal if they could afford it? And, afford it or not, shouldn't insurance be covering this add-on for everyone taking the *BRCA* test?

Myriad's customer help line never answered these questions. But that's not surprising, as the answers are wrapped up in several complex considerations well beyond the script of any telephone help desk. What Kathleen and her sister had stumbled onto were a set of antiquated policies that had been developed during the early days of genetic diagnostic testing. The bioethicists, genetic counselors, and physicians who first realized that *BRCA* testing could reveal potentially life-threatening conditions worried about what women might do with this alarming information. Might they suffer from depression or anxiety? Would they

rush to obtain medically unnecessary procedures—needlessly removing
their breasts and ovaries? Might they even contemplate suicide? Citing
these concerns, one influential set of guidelines signed by fifty-three
bioethicists in 1996 expressly discouraged women from getting BRCA
testing. Under this traditional, paternalistic view of medicine that somehow survived into the genomic era, it was felt that professionals knew
best and patients—women in particular—should be shielded from potentially upsetting information.

As a result, BRCA testing was not available to everyone. To get a
BRCA test, even without an insurance claim, one needed to meet strict
eligibility criteria, and to qualify for BART, the criteria were even stricter.
Though Myriad initially resisted these limitations (after all, it wanted to
sell as many tests as possible), it eventually acceded to the demands of
the medical community and agreed to test only individuals deemed to
be at high risk for hereditary cancer.

But how was risk assessed before a test was given? Testing criteria
were based largely on family history: how many of an individual's close
relatives had died of breast or ovarian cancer at a young age? It is no
coincidence that the BRCA genes were first discovered in Utah. Thanks
to meticulous record-keeping by the Mormon church, combined with
the state's uncharacteristically big families, Myriad could access data
from a large population of individuals carrying hereditary cancer-causing
variants. Not surprisingly, after the BRCA genes were discovered, testing
criteria were developed using assumptions based on these same families.
But the reality was that many American families, especially those who
had immigrated in the past couple of generations, were much smaller
and did not keep extensive family records. Nevertheless, restrictive testing
criteria were developed as though everyone had hundreds of known
relatives and a five-generation family history, whereas few actually did.
In the case of Eileen, this meant that she could not get the BART test
even after she was diagnosed with breast cancer, and Kathleen could not
get any test until she herself was diagnosed with ovarian cancer.

The final major factor affecting BART was coverage. BART involves
additional assays beyond the standard BRACAnalysis. Apparently, insurers didn't want to pay the extra $700 to test patients for BART

rearrangements unless there was a really, really good reason to do so, and they negotiated those strict testing criteria into their contracts with Myriad. The upshot being that now Myriad wouldn't offer the BART test unless a patient met strict testing criteria. There was no self-pay option, because offering one would imply that the insurers were refusing to pay for something that patients themselves thought was necessary. The combination of these factors—historical, systemic and financial—frustrated and maddened Kathleen, and made her determined to do something about it.

CHAPTER 12

Patents and Plaintiffs

BY FALL OF 2008, the ACLU had settled on Myriad Genetics as its target. The company owned or had exclusive rights to dozens of patents covering the *BRCA* genes. It had effectively cornered the market on *BRCA* testing. Its pricing was high, and there were many women who were unable to be tested either because they couldn't afford the testing or their insurance did not cover it. And a number of clinical researchers said that, but for the patents, they would be willing to offer testing at more affordable prices.

But now they needed to get down to details. They needed to identify specific Myriad patent claims that were vulnerable to attack, but which would also serve as exemplars of gene patenting throughout the industry. As Ravicher knew well, judges would not allow dozens of patents to be challenged in a single case. For case management purposes, only a handful of patent claims could be squeezed into a lawsuit. So they had to choose their targets carefully.

Hansen reserved a workroom at the ACLU to go over the patents and pick the exact claims that they would challenge. Despite the fact that they had been talking about gene patents for the last four years, until now Hansen had never actually read a patent cover to cover. In theory he knew what a patent was, but he envisioned the document itself as something akin to a car title or the deed to a house—a fancy legal document embossed with a seal that laid out the claims awarded by the Patent Office in a page or two.

In advance of the meeting, Hansen had asked an intern to find and make five photocopies of each *BRCA1* and *BRCA2* patent issued to Myriad or the University of Utah.[1] He was not prepared for what he saw the next morning, a can of Diet Mountain Dew in his hand. The three-ring binders piled on the workroom table were dauntingly high. For a moment, Hansen's heart sank. What had he gotten them into?

Simoncelli, Ravicher, Park, and Mason soon filed into the workroom, each carrying a cup of coffee or bottle of water. They, too, looked despairingly at the binders piled on the table as they took their seats. Though each of them had reviewed some of Myriad's patents over the past few months, seeing them piled up this way was daunting.

"Where to begin?" somebody asked.

Hansen stretched over to pull the first binder off the top of the pile closest to him. "At the top," he said.

They each turned to the first patent and began to read. Mason translated the science, while Ravicher parsed the claim language, occasionally scoffing or shaking his head at the breadth of what the lawyers had claimed for the company. At frequent points, the group paused to discuss. Hansen knew next to nothing about the science when they began. Simoncelli had a background in science, but she was at sea in the turgid patent language. Park was a bit further along the learning curve. As a Harvard undergraduate she had taken some pre-med biology courses, and had assisted with biotech patent cases in law school, both as a summer intern and as a clerk for a federal judge. But even for Park the forced march through so many patents was grueling. And, as Simoncelli observed, trying to learn science from a patent document was far from ideal.

They spent two days in the windowless conference room. Unlike the single-page document that Hansen had envisioned, a patent—which often runs to more than a hundred dual-column pages—consists of three main parts: a lengthy background section, a detailed description of the invention, and, finally, a set of legal "claims" that stake out what the inventor owns. The first part of a patent reads like a scientific paper written by a lawyer, and the last part reads like a legal document written by a scientist. In both cases, you get the worst of both worlds. The language of

---

1 Even though several of the challenged patents were owned and co-owned by the University of Utah, NIH, and other entities, all were licensed exclusively to Myriad. For the sake of simplicity I will refer to them all as "Myriad" patents.

patents is dense, laden with jargon and full of meandering, impenetrable sentences snaking across the page. There were words (*eukaryotic, exon, oligonucleotide, heterozygous*) that Hansen had never heard before—lots of them. But it was a language that he had to master in order to fight those who were versed in it: the patent lawyers and the Patent Office. Until now, they were the only ones who understood what was buried in these devilishly complex documents. And they were the ones with a vested interest in maintaining the status quo. That's why the system had survived for the past two hundred years without a serious challenge. A few rogue actors like Ravicher had been picking off weak members of the herd using *inter partes* reexaminations and other procedures. But Hansen's ambitions were much larger. His case wouldn't just target Myriad's patents. It would be grounded in Section 101 of the Patent Act—subject matter eligibility. Human genes were products of nature—they could not be patented. The case would strike at the heart of not just Myriad's patents, but all gene patents everywhere. Hansen's goal was nothing short of bringing down that entire system.

So line by line, claim by claim, they soldiered on. That first day they brought in lunch, then dinner. Then they went home to get some sleep and came back a few hours later to start it all over again. It was dark when they left and dark when they returned.

Going into that meeting, Hansen didn't know the difference between a chromosome and a gene. He had taken one introductory biology course at Carleton and passed only because the professor didn't want to submit the paperwork necessary to fail a senior. And even if Hansen had aced the class, it hadn't covered a fraction of the material spread out on the conference room table. But now that he had a purpose, Hansen embraced the complex information. By the end of the interdisciplinary crash course in human genetics and patent law, he understood, roughly, what these terms meant. It was an awakening. Hansen would later say that that marathon eye-straining, brain-numbing session was his favorite part of the entire case. The others were completely wiped out, but Hansen had loved every minute of it.

He was continually surprised, and sometimes enraged, by what they discovered lurking in the voluminous claims of the Myriad patents. One claim was particularly egregious—it gave Myriad the rights to a strand of DNA anywhere in the human body that shared a sequence of just fifteen base pairs (called a 15-mer) with the *BRCA* genes. That was outrageous—the human genome had 3.2 billion base pairs and the *BRCA* genes between them had about 200,000. The odds that a 15-mer that appeared somewhere in a *BRCA* gene also appeared somewhere else in the genome were incredibly high. Did Myriad now own snippets of DNA all across the human genome? Even Ravicher, who had seen his share of broad patents, had to admit that Myriad's lawyer had been both pretty bold, and pretty good, to get that claim through the Patent Office.

Eventually they picked fifteen claims from seven patents that covered the *BRCA1* and *BRCA2* genes themselves, as well as any portion of those genes (such as the 15-mer claim). The patents also claimed methods of using the *BRCA* sequences to predict early-onset breast and ovarian cancer and methods of developing cancer therapies using the genes. It was a lot of claims, but Hansen thought they should attack a variety of claim types. If they were going to challenge gene patents, they didn't want to leave anything standing when they were done.

Gathering the Experts

THROUGHOUT THEIR PREPARATION of the case, the ACLU team consulted with legal and scientific experts across the country. In some cases, they tested out their ideas and theories. In others, they wanted to know the kinds of issues and concerns that researchers, patients, and healthcare workers were worried about. But as they progressed toward filing a lawsuit, they needed to identify a set of experts who could provide written declarations to the court laying out important factual background to support the case.

Some of these experts were the geneticists who had advised them from the beginning—Chris Mason, the postdoc from Yale, and Bob Nussbaum, a physician geneticist at University of California San

Francisco. The litigation team consulted both Mason and Nussbaum on a regular basis during the preparation of their case.

In addition, Hansen and Simoncelli recruited a host of prominent researchers, academics, and physicians to their cause. In doing so, they visited labs and campuses across the country. Sometimes they were greeted with enthusiasm, sometimes with skepticism. In one case, the busy academic they were scheduled to meet missed their meeting entirely, leaving them in the hallway outside her empty office for an hour before they gave up and left.

Surprisingly, the most prominent scientists were among the most eager to help. Sir John Sulston, who won the Nobel Prize for Physiology and Medicine in 2002, had led the UK arm of the Human Genome Project. Years before that, his lab had been involved in the race to find the BRCA2 gene, and he writes about this experience, including the patent controversy between his lab in Cambridge and Myriad, in his book *The Common Thread*. Simoncelli, thinking "why not?" reached out to Sulston by email. To her surprise, he responded almost immediately and volunteered to submit a declaration in the case.

Joseph Stiglitz, winner of the 2001 Nobel Prize in Economics and a former chief economist of the World Bank, was teaching at Columbia University. Anthony Romero introduced him to Simoncelli by email. When she called him, Stiglitz needed little convincing to join the cause. His declaration made a strong case that gene patents were detrimental to innovation and the economy.

Other academics that the ACLU team recruited were less prominent, but played an equally important role. One of these was Shobita Parthasarathy, then an assistant professor of public policy at the University of Michigan. Parthasarathy had been studying breast cancer diagnostics since her days as a graduate student. In 2007 she published her findings in a book that detailed Myriad's rise to prominence in the BRCA testing market due, in large part, to its patents. Simoncelli had met Parthasarathy years earlier, while Parthasarathy was still conducting research for her book. When Simoncelli called to see whether she would be willing to submit a declaration in the case, Parthasarathy was unsure. She didn't have tenure at Michigan. What would it mean to get involved in a lawsuit? Would it look like she was a biased researcher? Would it harm her career? Eventually, Parthasarathy decided to join the effort, and her declaration proved to be an invaluable historical account of the *BRCA* testing landscape.

Standing to Sue

THE FINAL STEP in preparing their case was deciding who to recruit as plaintiffs—the injured parties who would actually be named in the lawsuit. Historically, the ACLU often represented itself as a plaintiff in constitutional cases, on the theory that every American was harmed by the government's violation of civil liberties. This is the approach that Hansen took when he successfully challenged the online porn provisions of the Communications Decency Act in *Reno v. ACLU.*

But times had changed, and the Supreme Court had made it more difficult to bring a case over the last few decades. Moreover, as Ravicher explained, patent cases were different. They weren't supposed to be, but they were. The court that heard their case would demand that the plaintiffs show they were actually *injured* by Myriad's patents, that they had what is called "standing" to bring suit. This question of standing had to be addressed in every case, but it was especially tricky in patent cases. In *Metabolite*, the dispute had involved two companies—the plaintiff Metabolite, who held the patent, and the defendant LabCorp, who was accused of infringement. But the ACLU's case would not be for patent *infringement*. Nobody was being accused of infringing Myriad's patents— all of the other labs running *BRCA* tests had left the market by 2000.

So who was being injured by the *BRCA* patents? As Hansen and his team soon realized, a lot of people were. Everyone agreed that having a large and diverse group of plaintiffs would be helpful. It would show a broad range of support for their cause, it would provide fallbacks if one or more of the plaintiffs got disqualified on grounds of standing, and it would help tell the story beyond the narrow legal issues. Simoncelli and

Park divided the primary responsibility for recruiting plaintiffs to the case. One of the ACLU's fortes was building coalitions, and that is what they set out to do.

The Problem at Penn

SIMONCELLI WAS FAMILIAR with the history of Myriad's clashes with the diagnostic testing community. To her, the first and most obvious group of plaintiffs would be the genetic testing labs that Myriad had sued or threatened to sue under its *BRCA* patents.

By the time that Myriad had launched its first *BRCA* testing program in 1996, a handful of academic labs and IVF centers were already offering *BRCA* testing to the public. The first of these was the University of Pennsylvania. Dr. Haig Kazazian, chair of the university's Department of Genetics and a pediatrician by training, has the ready smile and twinkling eye of someone who has comforted many sick children over his long career. But Kazazian, the son of Armenian immigrants, was not just a physician. He was also famous in the genetics community for pioneering work on "jumping genes"—genetic elements that move from one location to another. At Johns Hopkins, the silver-haired Kazazian had run one of the country's first genetic testing labs for nearly two decades. In 1994, Penn scored a significant coup by recruiting Kazazian away from Hopkins to run its new center for genetic testing.

To oversee the center at Penn, Kazazian hired Arupa Ganguly, a researcher educated in Calcutta who had recently developed a new technique for detecting mutations in large genes. In 1995, under their joint direction, the Penn Genetic Diagnostic Laboratory (GDL) began to offer testing for a number of known disease genes, both to research subjects and clinical patients. About a year later, with the lab up and running, Ganguly began to develop a diagnostic test for the newly discovered cancer genes *BRCA1* and *BRCA2*. By late 1996, the Penn GDL began to administer a *BRCA* test, and was soon diagnosing around five hundred women per year.

In early 1998, Kazazian and a colleague had dinner with Mark Skolnick, Myriad's co-founder, after a scientific conference in New York. The night was cold and windy; the restaurant was Italian. It was not expensive, but authentic, with a good selection of hearty northern Italian fare. Skolnick, who had spent years conducting genealogical research in Italy, approved. They discussed their families, the talks they had heard at the conference, and the usual gossip in the field. It was only after dessert, as the restaurant emptied and the waiter processed their separate checks that Skolnick, almost as an afterthought, took Kazazian aside. He informed the older scientist that Myriad now had patents covering *BRCA1*, and that Kazazian would soon get a letter from Myriad's lawyers. Skolnick, as though offering friendly advice about the wine selection, advised Kazazian to stop offering *BRCA* tests. Kazazian smiled broadly. He told Skolnick that, patents or not, he had no intention of stopping anything.

A couple of months later, Myriad sent a formal cease-and-desist letter to Kazazian. Kazazian ignored it and authorized Ganguly to keep going. In August, Myriad's outside litigation counsel, O'Melveny & Myers, sent a more pointed letter to Kazazian: "It has come to Myriad's attention that you are engaged in commercial testing activities that infringe Myriad's patents." As with the prior letter, Kazazian did not respond. He and Ganguly went about business as usual. More tests were completed.

Clearly, letters—even threatening ones—were getting nowhere with Kazazian. But within Myriad, the impasse with Penn soon became more than a question of a few hundred customers per year. The reputation of BRACAnalysis itself was on the line. Beginning in late 1998, scientists at Myriad began to hear disturbing rumors. Patients who had gotten *BRCA* testing at academic centers were being misdiagnosed, sometimes with horrific consequences. One Maryland woman, whose hospital sent her DNA to Penn for testing, was diagnosed with a *BRCA* mutation. Based on that finding, she had both of her breasts and her ovaries removed. Now, it seemed that the test results were wrong. She did not have a *BRCA* mutation after all. She was suing the university.

There were other cases with similarly disturbing facts. One, involving a misdiagnosis by Myriad's one-time competitor Oncormed, had been reported in the *Washington Post* and was now generating serious concern among physicians, patient groups, and the public. To Myriad, the shoddy *BRCA* testing being performed by other labs could poison the entire testing market—for patients, for doctors, and, most importantly, for the insurers who were just starting to come on board.

Myriad sued the University of Pennsylvania. It accused the university—Kazazian's and Ganguly's Genetic Diagnostics Lab in particular—of infringing four of Myriad's patents that claimed the *BRCA1* gene and its uses. Among other things, Myriad asked the court for an injunction that would prohibit the university from performing *BRCA1* testing other than pure academic research. That got the attention of Penn's legal department.

Myriad's lawsuit put Penn in an awkward position. On one hand, Kazazian insisted that he and Ganguly were doing nothing wrong. On the other hand, thanks to the scientific contributions of a researcher named Barbara Weber (who would eventually leave the university to become a pharmaceutical executive and venture capitalist), Penn itself was a co-owner of the *BRCA2* patents. In March 1996, Penn had exclusively licensed its share of those patents to Myriad. In exchange, Myriad agreed to pay Penn a royalty on each *BRCA2* test that it performed. As such, Penn stood to profit handsomely from Myriad's *BRCA* tests, but without any liability for mistakes or botched results.

For Penn, the path forward was clear. Kazazian and Ganguly had to stop testing the *BRCA* genes. It took about a year, and a few more pointed letters, but eventually Penn did shut down all *BRCA* testing. To Simoncelli, this episode made Kazazian and Ganguly ideal plaintiffs in their lawsuit. And, to her relief, both were more than happy to join the litigation against the company that had shut them down. The University of Pennsylvania wanted nothing to do with the lawsuit, but it had the grace to allow Kazazian and Ganguly to participate in their individual capacities.

Cease and Desist

ANOTHER RECIPIENT OF a cease-and-desist letter from Myriad was Harry Ostrer, the head of NYU's Molecular Genetics Laboratory. Ostrer had done his pediatrics residency at Johns Hopkins, where Kazazian was one of his mentors. A physician-scientist with a strong interest in social justice, Ostrer had worked at a refugee camp on the Thai-Cambodian border in the early 1980s. In addition to caring for Khmer and Vietnamese refugees, Ostrer collaborated with Kazazian on a genetic study of hemoglobin mutations affecting the local population. After returning to the United States and getting his PhD, Ostrer returned to his native New York. He was particularly interested in the genetic lineage of the Jewish people, and thus the Ashkenazi *BRCA* mutations.

Ostrer modeled NYU's genetic testing program after Kazazian's programs at Hopkins and Penn, and offered both genetic counseling and testing to patients. But NYU's program was new—it didn't have the equipment to conduct DNA sequencing on its own. In the late 1990s, Ostrer was sending DNA samples for processing to Kazazian. The lab at Penn would analyze the *BRCA* genes for NYU and send the results back to Ostrer. So NYU wasn't actually infringing Myriad's patents . . . yet. Nevertheless, Myriad demanded that NYU and Ostrer cease and desist all further *BRCA* testing. NYU reluctantly complied.

Yet Ostrer remained interested in *BRCA* testing, particularly with the prevalence of mutations in the Ashkenazi community. Many of his patients at Bellvue, a public hospital in New York, could not afford the Myriad test and thus went untested for the *BRCA* mutations. Ostrer visited Myriad in 2004 to propose a way that they could reduce the price of their test, but its lab director politely told him "that's not our business model." So in 2007, when Hansen and Simoncelli first approached Ostrer about a potential lawsuit against Myriad, he was intrigued. But he had reservations—after all, he didn't want to get his hospital sued. Ostrer eventually came around. A few months later he called Simoncelli and

NYU physician-geneticist Harry Ostrer was threatened by Myriad Genetics and became one of the plaintiffs in the case.

asked if it was too late to join the suit. They were planning to file the next day. It wasn't, she said, and signed him up.

Wendy Chung at Columbia was also eager to join as a plaintiff and two prominent geneticists from Emory University, another major testing lab, also came forward. All of these researchers said that, absent Myriad's patents, they would be willing to offer *BRCA* testing for their patients. Like Kazazian and Ganguly, they joined the lawsuit as individuals. None of their universities was interested in suing Myriad or, frankly, in opposing gene patents at all.

Finding *BRCA*

ONE PROMINENT RESEARCHER who didn't join the suit as a plaintiff was the person who was, perhaps, more closely linked with the *BRCA* genes than anyone else in the world: Mary-Claire King. Like Hansen, King had grown up in the suburbs of Chicago and attended Carleton College, graduating two years ahead of him, though neither claims to remember the other. Beginning in the 1970s, King, then a young genetics researcher at Berkeley, began a seventeen-year search for the genetic link to breast cancer. In 1990, she found that link on human chromosome-17 long after most other researchers had given up. King's announcement sparked a scientific frenzy in which a dozen research teams around the world raced to pinpoint the exact location and DNA sequence of the elusive breast cancer gene, which King had named *BRCA1*.

King had spent much of her career tracking *BRCA1* using the tools of genetic epidemiology, carefully tracing disease occurrences through the generations of hundreds of families. But to sequence the gene itself, the firepower of a top molecular biochemistry lab was needed. King joined forces with Francis Collins at the University of Michigan. Collins, a six-foot-three Virginian with a drooping moustache and a folksy demeanor, was a few years younger than King but had already earned a formidable reputation as a gene hunter. In 1989 his lab had isolated the elusive *CFTR* gene linked with cystic fibrosis—a significant discovery meriting a cover story in *Science*. Together, King and Collins rapidly closed in on the location of *BRCA1*.

But King's announcement had also attracted the attention of Mark Skolnick, the researcher at the University of Utah, another gene hunting powerhouse. Skolnick joined forces with Pete Meldrum, a local venture capitalist, to form a company—Myriad Genetics—with the specific goal of finding and commercializing the breast cancer gene. The new company struck a deal with the university, which gave them exclusive rights to any discoveries coming out of Skolnick's academic lab. Meldrum, an astute businessman, also reached an agreement with pharmaceutical giant Eli Lilly, which provided Myriad with millions of dollars in funding and access to the latest DNA sequencing equipment. These advantages, coupled with the extensive genealogical and health records compiled by the Mormon church in Utah, soon put Myriad in the lead. In 1994, the company stunned the scientific establishment by announcing that it had discovered and sequenced *BRCA1*. A little over a year later, Myriad scored another victory with *BRCA2*. The company immediately filed patent applications covering both genes and through a savvy series of lawsuits and corporate deals was able to assemble all of the patent rights covering both *BRCA* genes and effectively corner the U.S. market for *BRCA* testing.

*Mary-Claire King,
while a researcher at
Berkeley, narrowed the
location of the* BRCA1
*gene to the long arm
of chromosome-17.*

King was not pleased, but took the news graciously. At the 1994 meeting of the American Society of Human Genetics, where Myriad scientists announced their *BRCA1* discovery, King received a prolonged standing ovation as her fellow researchers recognized the essential contribution that she had made to the field. A few years later, King reports that she also received a cease-and-desist letter from Myriad at the University of Washington, to which she moved in 1995.

Bob Cook-Deegan, the Duke researcher who occasionally advised the ACLU team, knew King through human rights circles—he was part of Amnesty International and King had used gene sequencing technology to help reunite Argentine families separated during the country's brutal dictatorship of the 1970s. In 2008, Cook-Deegan and King were both attending a scientific meeting at Cold Spring Harbor Laboratory on Long Island, a prestigious institute for genetics research. Cook-Deegan, knowing that he would see King at the event, asked her if she would be willing to meet with the ACLU team while she was in the area. King was more than happy to oblige.

Hansen and Simoncelli drove from the city to the wooded Cold Spring Harbor campus. Simoncelli was eager to meet King, who was viewed by many as the "real" discoverer of the *BRCA* genes. When they told King about the lawsuit, she was delighted. King knew about gene patents from her time at UC Berkeley. Berkeley itself was an aggressive patent filer and had applied for patents covering her early discoveries relating to

BRCA1 years before Myriad determined the full sequence of the gene. But King was no longer at Berkeley, and she encouraged Hansen and Simoncelli to pursue the suit. Human DNA shouldn't be owned by anyone, she said. Even so, the rivalries that had arisen during the race to sequence *BRCA1* still loomed large, and King did not want to open old wounds by joining a lawsuit against Mark Skolnick's company. She offered to advise the ACLU team but declined to join the suit in any official capacity.

The Associations

THE NEXT GROUP of plaintiffs that the ACLU team approached consisted of the professional associations that represented testing labs, pathologists, and others who had an interest in performing *BRCA* testing. Simoncelli's first contact was with Mary Steele Williams, the executive director of the Association for Molecular Pathology (AMP). The majority of AMP's members were clinical geneticists at academic medical centers—people like Kazazian and Ostrer. Like most professional associations, AMP had a full-time staff, led by Williams, as well as a rotating group of scientists and researchers who served as its elected officers. One of AMP's officers was Debra Leonard, who had served as the organization's president in 2001. Leonard, a trained pathologist, oversaw the genetic diagnostics lab at Penn's Abramson Cancer Center, which was unaffiliated with the GDL run by Kazazian and Ganguly. Though Leonard's lab did not offer *BRCA* testing, she had certainly received her share of cease-and-desist letters. These came from Athena Diagnostics (for patents covering genes associated with Charcot-Marie-Tooth disease, Alzheimer's disease, and spinocerebellar ataxia), SmithKlineBeecham (for a patent covering a gene for the blood disorder hemochromatosis), and Miami Children's Hospital (for its infamous patent covering the gene for Canavan disease).

When Simoncelli contacted Williams in 2008, AMP was in the midst of re-examining its position on patents covering the gene. Leonard, who was still fuming over the barrage of patents asserted against her lab at Penn, served on AMP's influential Professional Relations Committee. She helped to develop a 2008 policy statement in which the association explicitly stated that the practice of patenting genetic sequences should be discontinued, either through legislation or the courts.

Nevertheless, AMP's board members were surprised when Simoncelli approached the organization about joining a lawsuit. AMP had occasionally signed on to *amicus* briefs that advanced issues relevant to its membership. It had done so in *Metabolite* and was now considering doing so in a suit involving blood analysis patents being asserted against the Mayo Clinic. But in these briefs, AMP was one of a half-dozen *amici*. It was a very different thing to become a plaintiff in a lawsuit. AMP had never sued anyone before and the board was nervous about how such a suit might affect the organization, how its membership might react, and how much effort it would require of Williams, AMP's only senior staff member.

In late 2008, Simoncelli's friend Jennifer Lieb, who advised AMP on legislative issues, put her in touch with AMP's then president, Jan Nowak, who directed the Molecular Diagnostics Laboratory at the NorthShore University HealthSystem in Evanston, Illinois. Hansen and Simoncelli flew to Chicago to meet with Nowak, and they hit it off. Together with Leonard, he helped to persuade the Professional Relations Committee to recommend that AMP join the lawsuit.

Once AMP signed onto the lawsuit, other associations followed. The College of American Pathologists, where Leonard also served on numerous committees, as well as the American College of Medical Genetics and the American Society for Clinical Pathology, all agreed to become plaintiffs in the case.

Counselors

GENETIC COUNSELORS WERE also impacted by Myriad's patents. These professionals are integral to the world of genetic diagnostics. It is their job to explain the ins and outs of genetic testing to patients, help them understand their testing options, and interpret their often ambiguous test results. Because of Myriad's patents, some genetic counselors began to feel like they couldn't give a full range of advice to their patients, that they couldn't obtain second opinions when Myriad's test results were unclear, and, thanks to Myriad's many advertising campaigns, that they were being inundated by requests for *BRCA* testing by people who didn't need it.

The genetic counselor who had most vocally objected to Myriad's practices was Ellen Matloff, the head of Yale Cancer Center's genetic counseling program. Like Penn, NYU, and other universities, Yale had received a terse cease-and-desist letter from Myriad in the late 1990s. It caused Yale to stop *BRCA* testing for its patients, a decision that infuriated Matloff. She was particularly incensed by Myriad's national advertising campaigns, which, she said, encouraged women to request Myriad's *BRCA* tests by preying on their fear and uncertainty.

Matloff's last straw was a copy of *Stagebill* that she received for the Washington, DC, production of the play *Wit*. *Wit* is a heart-rending drama about a woman's struggle with Stage 4 ("there is no Stage 5") ovarian cancer. The play began its East Coast run in New Haven and won the Pulitzer Prize, a Tony Award, and a long list of critical plaudits. And right there in the program, amidst promos for pre-theater dinner deals and the Kennedy Center's next season, was an ad for BRACAnalysis featuring a head-and-shoulders shot of a woman, left hand over her right breast, bearing the tagline KNOWLEDGE IS POWER. AND HOPE. Matloff could barely contain her rage. To her, the ad exploited an emotional tragedy to sell an expensive test to many women who probably didn't need it.

When Simoncelli told Matloff that the ACLU was considering a lawsuit against Myriad, she was floored. Like Lori Andrews in Chicago, Matloff had been begging attorneys for years to bring a case challenging Myriad's patents, but to no avail. In an effort to help her patients get second opinions after obtaining *BRCA* testing, Matloff had written to Myriad in 2005, seeking permission to perform supplemental testing for patients who had already received a Myriad test. But Myriad had refused. Now, the ACLU appeared out of nowhere to take on this cause. Matloff says that the day Simoncelli called her was one of the best days of her life.

Ellen Matloff, the head of Yale Cancer Center's genetic counseling program, was eager to join the lawsuit.

Hansen, Ravicher, and Simoncelli drove to New Haven to meet with Matloff. They agreed that she would make an ideal plaintiff. But when Matloff told officials at Yale that she was interested in joining the lawsuit, they gave her a cold reception. Yale held more than a few gene patents of its own, and the university was still smarting from a 2001 incident in which the high price of the AIDS drug stavudine—covered by a Yale patent—led to campus-wide protests and an international scandal. One university official told Matloff that it would be "unfortunate" if her involvement in such a lawsuit precluded her from continuing her employment.

Matloff couldn't believe it. Was Yale, one of the most leftist universities in the country, threatening to fire her for taking a stand? The veiled threat gave Matloff pause. She was thinking of starting a family—she needed her job. She discussed her concerns with Chris Mason, then a postdoc at Yale. Over lunch at the medical school cafeteria, he asked her what could really happen. Did she think that Yale would fire her just for putting her name on the lawsuit?

"Well, maybe," she said.

"I don't think you'll lose your job," Mason said, "but even if you do, you'll be on the right side of history."

Matloff also discussed the matter with Yale's general counsel, who was sympathetic to Matloff's position, but also protective of the university's interests. She put Matloff in touch with a local Connecticut attorney who agreed to represent Matloff on a *pro bono* basis. He negotiated an arrangement with the university that allowed Matloff to become a plaintiff in her personal capacity, so long as she did not mention Yale in the lawsuit. Matloff became one of the first plaintiffs to sign up for the lawsuit.

A Compassionate Profession

AT NYU, HARRY Ostrer suggested that the ACLU team speak with the genetic counselor who saw his patients, Elsa Reich. They met Reich at her small office, and the story she told them opened a window onto a world they had barely known about. As a freshman at Radcliffe College in the 1950s Reich had planned to go to medical school. But those plans were deferred when she married a Harvard student. They followed his career to Chicago where Reich, who had been only one year short of graduation, transferred to the University of Chicago. She changed her major to biology, graduated with honors, and was accepted to medical school. She was scheduled to enroll in the fall of 1956, but when matriculation day came around, the school refused to admit her. She was pregnant.

Reich was incensed, but eventually she accepted the school's decision. She had no choice. It was the 1950s. Reich put her ambitions of medical school aside and instead fulfilled the role of suburban wife. The family soon moved from Chicago to Montclair, New Jersey, where her husband commuted to Manhattan. One afternoon, while her three children were out of the house, Reich came across an interesting ad in the *New York Times.* It described a new master's degree program at Sarah Lawrence College in a field called genetic counseling.

The program, which was run out of the college's continuing education arm, was aimed at "married women in their thirties with two to four children living at home." Launched in 1969, it was the nation's first formal degree program in genetic counseling. A half dozen other colleges and universities soon followed suit. These programs laid the groundwork for the emergence of a new profession, separate and distinct from both that of the physician and the research geneticist.

Genetic counselors, who at the time were almost all middle-class wives and mothers, initially encountered resistance from those who thought that only PhDs and MDs were qualified to speak about genetic issues. In many clinics, genetic counselors were viewed as little more than glorified secretaries. But the demand for their skills was growing, thanks to the increasing use of prenatal testing and the rise of family planning. Genetic counselors—compassionate women with children of their own—were ideally situated to explain the risks and options available to parents facing difficult procreative decisions.

To Elsa Reich, this sounded like the ideal profession. It would allow her to combine her interests in medicine and biology with a genuine desire to help others. Most important, the part-time schedule offered by Sarah Lawrence would allow her to continue to care for her family while she studied. In the fall of 1971, Reich, along with eleven other women, matriculated in the program.

When she graduated three years later, Reich was hired by NYU Medical School as its first genetic counselor. She had been there ever since, and saw the evolution of the profession from one that dealt primarily with prenatal issues to one that encompassed a wide range of diseases and genetic conditions in infants, children, and adults. And, like Ellen Matloff, Reich felt the impact of Myriad's patents on her patients and in her everyday work. She signed up to be a plaintiff.

In addition to individual genetic counselors, Simoncelli thought it would be useful to seek support from the professional associations that represented the genetic counseling field, particularly the National Society of Genetic Counselors (NSGC). But Matloff warned her not to expect much from the NSGC. The profession had changed since the 1990s, when Matloff first began raising doubts about BRCA testing to a sympathetic audience of fellow counselors. In 2007, the largest employer of genetic counselors in the country, by some counts, was Myriad Genetics itself. Even genetic counselors, Matloff had found, tried to avoid biting the corporate hand that fed them. Matloff experienced the effects of this shift in allegiance when she started to criticize gene patents on the NSGC's public listserv. To her surprise, a moderator told her that her postings had been reported as offensive, and that she would be banned from the list if she did not stop. As Matloff had predicted, NSGC was not interested in participating in the lawsuit.

Activists and Advocates

BUT THERE WERE other organizations that were interested in joining the lawsuit as plaintiffs. Lori Andrews and Simoncelli had connections to the Boston Women's Health Book Collective, a non-profit that published the widely read women's health manual *Our Bodies, Ourselves*. As a publisher, the Collective embraced the ACLU's defense of free speech. Their publications told women about healthcare options, which might include non-Myriad *BRCA* testing. The Collective didn't want to follow in the footsteps of LabCorp, which was penalized for telling doctors how to infringe a patent in the *Metabolite* case. As a women's health organization, the Collective was also concerned about the lack of *BRCA* variant information for African American, Latina, and Asian American women. They were in.

Since the case would focus on breast and ovarian cancer, it would also be important to engage cancer advocacy organizations. The most vocal of these was Breast Cancer Action (BCA), a group with longstanding ties to the ACLU. BCA, twenty-thousand members strong, had grown from a small San Francisco-based patient support group into an activist organization whose self-proclaimed goal was to "transform breast cancer from a private medical crisis to a public health emergency." BCA first made national headlines in 1994 for staging an attention-grabbing protest at Genentech's South San Francisco headquarters. The protesters carried signs, honked car horns, and lay down in the road to draw attention to the lack of access that dying patients had to the company's then-experimental drug Herceptin. Now, BCA was willing to act as one of their plaintiffs.

Some, however, viewed BCA as a fringe group whose tactics were more closely aligned with radical AIDS activists than the major corporate-style charities that supported breast cancer research. Thus, the National Breast Cancer Coalition and, the biggest of them all, the Susan G. Komen Breast Cancer Foundation, creator of the ubiquitous pink ribbon campaign, wouldn't join the ACLU's suit.

These organizations harbored a number of doubts about the suit. First, *BRCA* was believed to be a factor in only 5 to 10 percent of breast cancer cases. But the "big umbrella" organizations represented all breast cancer patients. So putting resources behind such a narrow effort could be seen as favoring one small subset of their constituencies. They preferred to support causes that would benefit *all* breast cancer patients.

Second, several of the large organizations, led by the National Breast Cancer Coalition, had taken positions in the 1990s *against* predictive genetic testing, citing concerns that low-risk individuals would seek unnecessary testing and the traditional fear that women receiving positive test results might not be able to handle them emotionally. As of 2009 NBCC's website still displayed a policy statement recommending that *BRCA* testing be performed only in research studies, and not for clinical patients. By eliminating gene patents, it seemed that the ACLU wanted to make clinical *BRCA* testing more broadly available, a position that NBCC just wasn't ready to support.

Finally, some were of the view that, far from being a villain, Myriad offered a valuable service to their communities. True, the test was not affordable by everyone, but chemotherapy and radiation therapy were also expensive, and cancer advocates had worked to get those therapies covered by almost every insurance carrier in the country. Diagnostic testing would eventually be covered too. And the patents? While NBCC had vocally opposed Myriad's patents, other advocates suggested that companies needed patents to fund their research. What cancer drug was ever developed without patents? In the eyes of some within the major breast cancer foundations, putting Myriad and its ilk out of business could be a step in the wrong direction. So they politely told Simoncelli and Park "no, thank you."

Previvors

ONE GROUP THAT was more open to their overtures was a small organization called FORCE—Facing Our Risk of Cancer Empowered. FORCE was an advocacy and support group for "previvors"—women who had, or were likely to have, a *BRCA* mutation. Since the discovery of the *BRCA* genes, women who tested positive needed a place to go for information and support, but, because they didn't yet have cancer, didn't fit into existing cancer support groups. Patients in those groups wondered why these seemingly healthy women were there—without a cancer diagnosis, how could they relate to patients who were suffering through the disease and its brutal treatment?

FORCE offered a safe space for *BRCA* mutation carriers who knew that their day might come, but hadn't yet arrived. It offered information about *BRCA* testing. It helped women evaluate their options if they tested positive. Should they have prophylactic surgery? What were the options for breast reconstruction? Should they rush to have children before getting their ovaries removed? And what about preventative chemo like tamoxifen, an estrogen suppressor that had been shown to reduce the risk of breast cancer? What were the side effects? What plans covered it? Most importantly, FORCE provided a community where these women could interact and discuss their fears, experiences, and hopes without being judged.

Given the obvious connection between the ACLU's lawsuit and FORCE's mission, Park reached out to FORCE's founder, Sue Friedman. At first, Friedman was intrigued. Anything that could make *BRCA* testing more accessible seemed like a good thing. But she soon became nervous. FORCE was tiny; it didn't have a political focus like Breast Cancer Action. FORCE depended on maintaining good relations with the giants—NBCC and Komen. Could joining a lawsuit create some kind of backlash? In the end, Friedman offered Park a compromise: FORCE would get word of the suit out via its listserv, but would not formally join the lawsuit as a plaintiff. For Park that was enough. The connections that FORCE made for them proved to be invaluable.

The Patients

THE FINAL, and arguably the most important, plaintiffs in the lawsuit were individual patients. Researchers and diagnostic labs, while undoubtedly affected by Myriad's patents, were not the most sympathetic plaintiffs. Nor were trade associations or advocacy groups. What the ACLU needed were ordinary women who had been injured by Myriad's patents, either because they couldn't afford testing, they didn't have insurance, or something about the testing went wrong.

Hansen had known this from the outset, and his intuition was confirmed by Anthony Romero. Romero agreed that breast cancer was very real to many people. His own mother, in fact, was battling breast cancer at the time. Romero insisted that a civil rights case was about people, not abstract legal doctrine. They needed real people, real faces, real stories to animate the case, convey the narrative, and gain broad support for their cause.

Recruiting a group of affected women to the case began as Park's assignment, but it soon became her passion. The challenge that Park faced was finding these women. She was not an oncologist or a genetic counselor, and she had no access to hospital records or any other source of information about BRCA testing. If such records existed, they were probably at Myriad, and she obviously wasn't going to ask them. So Park's first approach was through FORCE. The organization, while small, had an active online presence. Many of its members participated in chat rooms and message boards, sharing information and tips about treatment, doctors, surgery, and insurance.

To reach these women and find out which of them had experienced issues with BRCA testing, Park developed a brief survey that FORCE distributed through its network. The survey did not mention a lawsuit or patents; it just asked a few questions about the recipient's experience with BRCA testing. Through the survey, Park was able to identify women who had been unable to get BRCA testing, who had been unable to afford it, or who had experienced other problems. Soon, leads began to filter into Park's mailbox. Many of them came from genetic counselors who knew of women who should have been tested, but were uninsured or underinsured. Park approached each potential plaintiff individually, listening to her stories and gauging her willingness to participate in a lawsuit against Myriad.

Most of these women knew nothing about patents. Some were

supportive, but afraid to get involved—could there be retaliation against them? Against their families? Could it affect their insurance coverage? Others, though, were outraged when they learned about the *BRCA* patents. These were the women that Park wanted.

One of the most outraged, and most determined to join the lawsuit, was Lisbeth Ceriani. Given that she already had cancer, she was not a member of FORCE. But in her own research on the test and the genes, she had read about Myriad's patents. In one of Ceriani's frequent calls to her genetic counselor, usually to harangue her about the lack of insurance coverage and Mass General's inability to do anything about it, Ceriani mentioned the patent on the *BRCA* genes. That triggered the genetic counselor's memory, who had seen one of Park's postings to an online group. "If you're interested in the patents, you should call this person at the ACLU," she told Ceriani.

The ACLU? Ceriani knew them, of course, but didn't understand what they had to do with genetic testing. So she called the number that the genetic counselor had given her and soon reached Park. They arranged to meet at Penn Station the next week, when Ceriani was going to be in New York for a conference.

Park met Ceriani at a small café near the Amtrak terminal, talking over the tinny announcements of delayed arrivals and departures. She explained the lawsuit. It all made sense to Ceriani. She just wondered why it had taken someone so long to challenge these patents. Ceriani signed the papers and became a plaintiff. To some extent, Ceriani's participation was self-interested. If the ACLU knocked out Myriad's patents, then maybe she could finally get the *BRCA* test. But even more than that, to Ceriani the concept of a patent on a human gene made no sense—it seemed inherently wrong. And she wanted to do her part to fix it.

Park eventually recruited six women to join the lawsuit as plaintiffs. Their stories had similar themes, but each was different in its own way. Ceriani was unable to get tested because Myriad had not entered into a contract with her Medicaid insurer, MassHealth. Patrice Fortune faced a similar situation with Medi-Cal, the California Medicaid provider. Runi Limary, Kathleen Raker, and Vicky Thomason all tested negative for known *BRCA* mutations. But due to their family histories and other circumstances, they had each been advised to seek BART testing for large-scale *BRCA* rearrangements. Their insurance did not cover it.

Genae Girard, a veterinary supplier from Austin, Texas, was diagnosed with breast cancer and tested positive for a mutation on *BRCA2*. Before undergoing surgical removal of her ovaries, Girard wanted a second opinion. But because of Myriad's patents, no other lab could confirm her results. Until she joined the lawsuit, all that Girard had been able to do to help the cause was volunteering her horse as a part-time therapy animal for cancer patients.

Myriad's *BRCA* patents had affected all of these women, effectively concealing from them the risks that they and their families might face, when the information they needed was encoded in their own DNA. With the addition of these six women, the ACLU's roster of plaintiffs was an even twenty. By the spring of 2009, they were ready to sue.

CHAPTER 13

Pulling the Trigger

THERE ARE NINETY-FOUR federal judicial districts in the United States. Because patents are created by a federal statute, all patent lawsuits are federal, and a plaintiff can generally choose to sue in any district where the defendant conducts business (e.g., selling products, advertising, or offering services). Choosing the district in which to bring a lawsuit may seem like a mundane procedural decision, but the choice can have surprisingly far-reaching implications.

In deciding where to file their suit, Hansen wanted to avoid districts with excessively patent-friendly reputations. Texas, Virginia, and Delaware were all known as attractive venues for patent litigation—they were out. By the same token, he wanted a judge that was used to handling complex cases with lots of witnesses and documents. That ruled out most districts outside of major metropolitan areas. And, finally, he didn't want to spend a fortune on hotels and air travel for the legal team and all of their witnesses (the District of Hawaii, while nice, could get expensive). In the end, Hansen went with his first choice: the Southern District of New York, just three subway stops from the office.

But the Southern District had more going for it than convenience. Nicknamed the "Mother Court," it is the oldest district court in the nation—established even before the U.S. Supreme Court. Its judges had heard some of the most complex and high profile cases of all time: from financial meltdowns to tax fraud, from the sinking of the *Titanic* to the antitrust case against Apple. Over the years, its courtrooms had seen an endless parade of the famous and the infamous: Ethel and Julius Rosenberg, Martha Stewart, John Gotti, Bernie Madoff, and even

Osama bin Laden (in absentia). It was also the court in which Judge Learned Hand issued his fateful 1911 opinion in *Parke-Davis*, one of the "product of nature" precedents the ACLU team would have to overcome. For better or worse, the Southern District of New York was where they would sue.

Raising Awareness

FROM A LEGAL STANDPOINT, the case was ready. The plaintiffs were assembled, the arguments were marshaled, the complaint was drafted and ready for filing. Everyone at the ACLU from Anthony Romero and Steve Shapiro to the National Board had signed off.[1] But one key step remained before Hansen could pull the trigger: public awareness.

PR was a key element of the ACLU's litigation strategy. As in all "impact litigation" it was important to have public support. Judges and juries were people, and people were influenced by the zeitgeist, by what others were saying around them. Public support had historically played a key role in cases involving civil rights, women's rights, and due process—it would be important here, too.

The ACLU's design department created a logo for the case—a black silhouette of a woman overlaid with a DNA molecule, standing beside an ominous warning in block characters: DO NOT PATENT MY GENES. The logo would be reproduced on web pages, flyers, solicitation letters, video messages, posters, and placards, and its simple message would become the rallying cry of the case and its thousands of supporters.

The ACLU's PR effort was helped by a small but growing body of popular articles that criticized gene patents—Myriad's in particular. One of the first appeared in 2002, when the *Boston Globe Magazine* ran a 5,300 word exposé on the *BRCA* patents. The story included interviews

---

1 By 2009, the ACLU's Patent Committee had still not completed its deliberations over the internal policy basis for a patent suit. However, given the First Amendment and women's health aspects of the case, Hansen, Shapiro, and Romero felt it was acceptable to proceed with the lawsuit. The Patent Committee eventually proposed an official ACLU Patent Policy, which was approved by the National Board in June 2010, about a year after the case was filed.

The ACLU's litigation logo for AMP v. Myriad.

with Debra Leonard, the geneticist from Penn, and Lori Andrews, among the most vocal opponents of gene patenting. In 2006, Michael Crichton's novel *Next* appended an anti-patenting screed to its sci-fi depiction of genetic engineering run amok. More press stories appeared during the debate on the Becerra bill in Congress, including Crichton's *New York Times* op-ed criticizing gene patents.

The Story Finder

IN 2004, the ACLU hired a former journalist and network news producer named Joel Engardio to work with its LGBT Rights Project. Engardio's job was to identify individuals with compelling personal stories to act as plaintiffs in ACLU cases that challenged state same-sex marriage bans. His unofficial title was "Story Finder." Part of this project involved producing short YouTube videos to tell the plaintiffs' stories in a compelling and convincing manner. The project was so successful that the ACLU soon created a video communications department under Engardio's leadership. He periodically checked in with each of the ACLU's internal units to see how his new group could help promote their agendas. When he

ACLU attorney Sandra Park and plaintiff Lisbeth Ceriani
discussing the case in Ceriani's kitchen.

met with the Women's Rights Project, Park told him about the case they were planning against Myriad. Engardio knew just what they should do.

In early 2009, Park and Engardio, video equipment in hand, embarked on a two-week cross-country tour to film the cancer patients who had agreed to be plaintiffs. They flew to Lexington, Kentucky, and drove deep into the back country to meet with Vicky Thomason, the ovarian cancer patient who had been unable to afford the BART test. When Park and Engardio finally arrived, they found Thomason tending a mare who had just given birth. Engardio set up his camera and lights in the barn and Park interviewed Thomason there.

Next they flew to Austin, Texas, where they met with Runi Limary and Genae Gerard, two friends who had encouraged each other to join the suit. When Park recruited Limary to the case, she had been a sixth grade math and science teacher. Now she worked for a breast cancer advocacy organization, helping patients navigate the complex world of insurance, hospitals, and recovery. When they were done with the interviews, Engardio and Park set out to find the best TexMex food in Austin. They ended their trip in Boston, where they interviewed Lisbeth Ceriani with her young daughter. Overall the trip was a success—they had their stories.

The masterful five-minute video that Engardio produced opens with an image of the Statue of Liberty over a stately chord progression from "The Star-Spangled Banner." Barbara Brenner, head of Breast Cancer Action, emphatically asks, "How can it be that a company controls genes? How is that possible?" Next, Park and Simoncelli make short appearances. The screen fades to black, and the words LIBERATE THE BREAST CANCER GENE materialize. A neutral male voiceover explains the significance of the *BRCA* genes while Ceriani makes lunch for her daughter. Hansen, conservative-rep tie loosened like a television DA, sits in a book-filled office and explains that patents can't cover products of nature. There are ominous panning shots of the Patent Office and Myriad's headquarters while a discordant Hitchcockian *sostenuto* plays in the background. Wendy Chung from Columbia, in wire-rimmed glasses and a white lab coat, peers at translucent gels and pipettes liquid into a test tube. Girard and Limary make brief appearances from Austin. The video hits all of the issues that are raised in the case: products of nature, impediments to research, the right to a second opinion, a woman's right to health, all in a package ready for mass consumption over YouTube.

In the Family

DESPITE ALL OF the ACLU's careful media and outreach planning, one of the most effective pieces of public storytelling was completely unexpected: a documentary film called *In the Family* that was independently produced by a thirty-one-year-old filmmaker named Joanna Rudnick. A few years earlier, Rudnick had tested positive for a *BRCA* mutation. The film chronicled her emotional journey through the implications of this finding, including the brutal details of prophylactic breast and ovarian removal surgery. She interviewed cancer patients and survivors, members of FORCE, other previvors, as well as a range of oncologists, breast surgeons, and cancer advocates.

But Rudnick's biggest coup was getting access to Myriad itself. In the film, she narrates over a panning shot of Myriad's high-tech headquarters spread across the rolling hills above Salt Lake City: "When it comes to who's getting tested and who isn't, the conversation leads to Myriad Genetics." The scene cuts to Myriad employees unloading packages at a company shipping dock, lab techs pipetting blood samples into tubes, and researchers peering at oversized computer monitors.

About an hour into the film, Rudnick interviews Mark Skolnick, Myriad's co-founder and scientific director.

It is unclear what possessed Skolnick to do the on-camera interview with Rudnick. Perhaps she told him that Myriad's test had saved her life, which it probably had, or perhaps he was charmed by the eloquent young filmmaker. But whatever the reason, Skolnick's appearance cannot have helped the company's public image.

In the film, Rudnick first asks him about the race to sequence the *BRCA* genes. Instead of offering the typical pablum about helping humanity and curing disease, Skolnick is surprisingly frank. "We did it to win the race," he says as Rudnick smiles uneasily. "We did it to beat the other team. Period." He shrugs, then continues, "And we won." Rudnick's expression is telling. Here is a company that is the steward of life-and-death information for thousands of women, herself included, and its sole motivation was some combination of competitive spirit and scientific hubris. Unstated, of course, is the money that Myriad stood to gain by winning the race to discover the most important disease gene in history, what the *New York Times* called "a genetic trophy so ferociously coveted and loudly heralded that it had taken on a near mythic aura."

The shot then cuts to a long wall hung with dozens of mounted plaques, like platinum records adorning the walls of a recording studio. Each of these plaques commemorates the issuance of a U.S. patent. Rudnick reads the number and title of one of the first plaques on the wall and begins to ask Skolnick whether this is the controversial patent—the first one to claim the *BRCA1* sequence. But he cuts her off. "There is no controversial patent," he says sharply. "It's all very easy to understand if you take the time." Rudnick laughs nervously.

Mark Skolnick admiring the wall of patents at Myriad Genetics headquarters.

Back in the lab, she pushes him further on the patents. He dismisses her questions with a scowl, commenting that people aren't upset about the patents that cover their iPods or telephones or computers or cars. "If it weren't for patents," he says, "modern society wouldn't exist." Rudnick, increasingly skeptical, asks him about the women who can't be tested because Myriad's prices are so high. She reminds him that when the *BRCA* test first came out, Skolnick said that its price would eventually drop to a few hundred dollars, yet it is now over $3,000 and the price has actually gone up over time. Why?

Here, Skolnick genuinely appears surprised. He is speechless for several uncomfortable seconds as the camera holds him in its pitiless frame. After a couple of false starts, Skolnick admits, "That's a good question." Then, in an unscripted statement that probably induced severe angina in his business partners, Skolnick slowly continues, "I think there's a point at which we have to start looking at decreasing the cost of the test." There it was. The one and only recorded admission, by a founder of the company, that the price of Myriad's testing was unjustifiably high, that it was probably preventing women from being tested, and that Myriad's patents enabled it to charge excessive rates for its life-saving technology.

Rudnick produced *In the Family* before the ACLU had actively reached out to the *BRCA* community, so the film did not include any mention of the lawsuit or the ACLU's efforts. Nevertheless, the release of the film coincided beautifully with the ACLU's publicity efforts in advance of the case. The documentary was shown at film festivals as early as February 2008, but its big moment was a PBS airing in October of that year, just as the lawsuit crystallized. Anyone who watched the PBS broadcast knew about Myriad and its patents.

Pay Dirt

IN ADDITION TO generating support for the lawsuit, the ACLU's public relations efforts served a second equally important function: fundraising. As a non-profit, the ACLU depended on contributions large and small for its very existence. It did not charge clients obscene hourly rates like private law firms; it did not sell products or services; it did not offer genetic testing. It brought lawsuits to defend civil liberties, but doing so wasn't free. So fundraising was a critical part of everything the ACLU did.

Romero was a master of the art of raising money. But 2008 and 2009 were not kind years for charitable organizations. The global financial crisis was in full swing and many dependable, long-term donors were facing crises of their own. Some were victims of the subprime mortgage debacle and the collapse of Wall Street titans like Lehman Brothers and Bear Stearns. Others had been swindled by Ponzi scheme operator Bernie Madoff. Plenty of old-line trusts and charitable foundations had seen their asset values plummet along with the capital markets, making them wary of new investments. And many small donors, the bread and butter of the ACLU's fundraising efforts, were feeling pinched as jobs disappeared, mortgages went under water, and savings dwindled.

The great recession hit the ACLU hard—particularly the loss of one wealthy donor who had dependably contributed more than $20 million every year. As a result, in late 2008 Romero was forced to implement a painful set of staff layoffs and salary cuts. In the midst of this gloomy situation, Hansen's gene patenting case offered a ray of hope. It gave the ACLU access to an entirely new set of donors beyond their traditional civil rights/civil liberties supporters. The case opened up new possibilities with foundations, wealthy donors, and ordinary individuals who cared about breast cancer—a vast and untapped reservoir of financial support.

Sign Here

IN THE WEEKS and days before filing the lawsuit, the ACLU's PR team alerted all of the major print and television news outlets that the case was imminent. They lined up the morning news shows, prepped their anchors, gave them talking points and B-reel footage. Interviews were taped in advance. Documents were strategically leaked. Soon, all was ready. The attack would commence at dawn.

Early on Tuesday morning, May 12, 2009, the ACLU and PubPat teams crowded into Hansen's office. Sunlight reflected off the building across the street. Like the president signing a particularly momentous piece of legislation, Hansen paused for a moment, his pen hovering above the signature line tabbed with a red Sign Here sticker. He lowered his hand and scribbled his name on the thirty-one-page complaint that they had labored for so long to produce.

After signing a half dozen more copies, he handed three neatly bound sets to Simoncelli. She grinned broadly. It had all started with her, Hansen said, smiling. He congratulated the team on a job well done. There was scattered applause. He told them that they had done a monumental amount of work to get to where they were today. Now the real work would begin.

Simoncelli could hardly believe it. She had started down this road five long years ago. Now she, Park and Ravicher would take a short subway ride to file the complaint with the court clerk. At last, they were on their way.

PART II

LITIGATION

EDWARD R. MURROW: Who owns the patent on [the polio] vaccine?

JONAS SALK: Well, the people, I would say. There is no patent. Could you patent the sun?

—SEE IT NOW (CBS, APR. 12, 1955)

CHAPTER 14

The Big Guns

IN MAY, snow still clings to the peaks of the Wasatch Range on the western fringe of the Rocky Mountains, taking on a soft blue-gray cast in the hour before the sun rises over the peaks. A series of interconnected brick buildings with dark windows spreads up the gently rising foothills toward the northern rim of the Salt Lake Valley. From their upper floors, one gets unparalleled views of the sunrise over the glistening snowcaps and the sunset over Salt Lake City, the spires of the ornate Mormon Temple now dwarfed by neighboring office towers. This is the headquarters of Myriad Genetics.

Rick Marsh, Myriad's general counsel, enjoyed arriving at the office in the early hours of the morning. It was a quiet time to get work done, to plan the coming day and think before the building filled with people. Early on the morning of Monday, May 11, Marsh was sitting at the small, round work table in his office, the pages of a contract spread out in front of him. He almost didn't notice when the white "outside line" light began to flash on his office telephone. Marsh, a square-jawed, broad-shouldered man with cropped graying hair, had joined Myriad as its top lawyer in 2002. A former tax attorney and accountant, Marsh now oversaw the company's small legal department. Assuming that the call was from one of his staff, he pressed the speaker button.

"Hello," he said, his eyes still focused on the contract spread across his table.

"Mr. Marsh?" said a voice on the other end of the line. The speaker identified himself as a reporter from the *New York Times*.

Marsh raised an eyebrow. Every general counsel in America dreads an unscheduled call from the *Times*. It seldom means good news.

"Mr. Marsh?" the reporter asked again.

"Yes?"

"Hey," the reporter began, adopting a friendly tone. "I just learned that Myriad Genetics is going to be sued. Would you like to comment for the record?"

Marsh frowned. He shut his office door, picking up the phone in his other hand. He told the reporter that he knew nothing about a lawsuit.

"I have a draft of the complaint in my hand," the reporter said. "It's scheduled to be filed tomorrow in New York."

Marsh was taken aback. He struggled to think who might be suing them in New York. An insurance carrier who thought it was being over-charged? A university with a patent on one of their new tests? A disgruntled employee? Marsh didn't know. "I'm not aware of any such lawsuit," he told the reporter. "I haven't been served. I have no knowledge about this."

"You're not aware of a lawsuit involving Myriad's patents?" he pressed.

"I'm sorry, no," Marsh said.

The reporter paused to write something down. "OK," he said.

Now Marsh's curiosity was piqued. He asked the reporter for a copy of the complaint. The reporter said he would email it.

Marsh was baffled. The *Times* reporter had implied that Myriad was about to be hit with a patent suit. But Marsh had received no demand letters, no offers to license, no threats. This wasn't how patent disputes worked. Going straight to litigation, without passing GO, was a bizarre and expensive tactic. He mentally ran through a list of Myriad's competitors. Which of them was behind this? LabCorp? Ambry Genetics? A newcomer with some fancy new technology to tout?

After hanging up with the reporter, Marsh told Myriad's CEO, Pete Meldrum, about the call, but without knowing more, he didn't want to start a fire drill. Meldrum agreed. Wait and see what it's all about.

Then Marsh received the email from the *Times* reporter. The draft complaint was attached. Marsh looked at the list of plaintiffs—twenty of them. He didn't understand it. The Association for Molecular Pathology?

Breast Cancer Action? A list of individuals whom he had never heard of. Then he scrolled to the bottom of the document. The lawyers who would be filing the complaint were listed as being from the ACLU and a group called the Public Patent Foundation. What was going on?

More reporters called as the day wore on. Marsh answered them all in the same manner: "At this point, we've just been made aware of the lawsuit, and we've had no opportunity to review it. No comment."

Around 4:00 p.m., Marsh got a call from a producer at the *Today* show. She told him that they would be airing a story about the ACLU's lawsuit against Myriad the next morning. Did Marsh want to comment? He couldn't understand it. *Today?* He still hadn't read the entire complaint, which seemed to be challenging several of Myriad's *BRCA* patents from the 1990s. He told the producer that they had barely received the complaint, which hadn't even been filed yet. But whatever it was about, Myriad firmly believed that all of its patents were valid and enforceable.

Marsh went home that evening with a bad feeling about the case.

The next morning, he got up early to watch the East Coast broadcast of *Today*. It was unbelievable. This wasn't a thirty-second news flash about a lawsuit that had just been filed. It was a fully reported story, fifteen or twenty minutes long. There were scripted lines, interviews, edited footage from multiple sources. Several of the plaintiffs and the attorneys from the ACLU appeared on the set in New York. Clearly, this piece had been in the works for weeks, if not longer. And Marsh had only gotten a call for comment yesterday?

The formal complaint was served on Myriad later that day. Marsh made photocopies, then called a meeting of the company's senior staff and counsel.

Jay Zhang, head of Myriad's patent department, briefed them on the status of the seven patents that had been challenged. First, he assured them that the patents had been validly granted in full compliance with the Patent Office's policies. Myriad wasn't an outlier here. There were thousands of gene patents, and the Patent Office had been approving them for decades. This was well-settled law.

Nevertheless, this lawsuit was like nothing that Zhang had ever seen before. There was plenty of patent litigation in the biotech industry, and even some litigation relating to gene patents. But all of the other lawsuits claimed that one party was infringing another party's patent, or that a patent was invalid because it covered something that was already known in the literature, or that it was an obvious improvement over some prior technology, or that it should not be enforceable for some other reason.

But this case, Zhang said, holding up his copy of the complaint, was different. Here, the ACLU and its twenty plaintiffs seemed to be attacking the entire system, including the Patent Office, and not just Myriad.

What's more, Zhang added, the challenged patents were old—they had all been filed in the mid-1990s, meaning that they would expire soon, between 2015 and 2018. Whatever happened, it wouldn't have a major effect on the company. In fact, they had been preparing for the expiration of the *BRCA* patents for years by entering into more than a dozen new product markets outside of breast cancer, all of which were heavily patented.

Zhang also noted that this lawsuit was different from the challenges that Myriad had faced over its patents abroad. Gene patenting was controversial not just in the United States, but in many countries. Their patents had been opposed in countries including Canada, France, England, Germany, Australia, and Japan, yet even in the most difficult of these jurisdictions, nobody had challenged the very notion of patenting human DNA.

As Zhang went on, Mark Skolnick, sitting at the end of the conference table, paged through the complaint. It said several times that Myriad's patents enabled the company to prevent others from conducting research on the *BRCA* genes. That, Skolnick observed with quiet intensity, was ridiculous. Myriad had never enforced its patents against researchers. In fact, thousands of scientific papers about *BRCA* had been published in the last decade. He looked at the list of plaintiffs and saw the name of his old friend Haig Kazazian. Penn, he suspected, wanted to run a *commercial BRCA* testing business. They weren't doing scientific research—they were out to make money, just like Myriad. This was a case of sour grapes by a frustrated competitor.

To some, the lawsuit looked like it was mostly about pricing and access to the test. But Myriad had worked hard—extremely hard—to get BRACAnalysis covered by almost every insurance carrier, health plan, and HMO in the country. It was clearly in the company's self-interest to do so, and today nearly every test that they ran was covered. And for those indigent patients that didn't have insurance, Myriad had a financial assistance program. It gave tests away for free. Thousands of them. Could there really be that many people who somehow fell between the cracks? Is that what this lawsuit was about?

Pete Meldrum flipped through the complaint as well. Two pricing issues had been raised: Medicaid and BART. Meldrum scoffed. Medicaid wasn't his fault. The list price for BRACAnalysis was around $3,100. Most of the insurance carriers were paying that or something close to it. But Medicaid carriers like MassHealth wanted to pay only $1,600. They were playing hardball, using their large base of insureds to pressure Myriad to slash its price in half. But if he did that, his business would be crushed. The Medicaid price would be publicly announced, and every insurer out there—Kaiser, Blue Cross, Aetna, you name it, would be screaming at him to lower *their* price to $1,600. How could he justify that to the shareholders? Myriad's stock would plummet. Meldrum would not be the CEO who allowed his company's profits to tank. He wouldn't be on the receiving end of a request by the board that he "step down to pursue other interests," or a lawsuit by disgruntled shareholders.

No, the smart thing to do was to wait out MassHealth. Eventually if its insureds, or NIH or the medical community, demanded that Medicaid patients get *BRCA* testing, MassHealth would come back to the bargaining table and offer something closer to Myriad's list price. In the meantime, a few Medicaid patients might suffer. But that was the fault of MassHealth, not Myriad.

Meldrum's hands were also tied with respect to BART. It was an expensive add-on. He would have loved to offer it as a standard part of BRACAnalysis, with an appropriate price increase. But the carriers wouldn't pay more than the price they had negotiated. So BART had to be offered separately. And to cover it, the carriers had insisted on strict eligibility guidelines. Too strict in Meldrum's opinion, but he could work on expanding the eligibility criteria once the value of the test had been proven. What did the ACLU want here? Kill the patents so that they would give BART away for free? Allow academic labs like Penn to offer *BRCA* testing and botch it like they had in the past? It was absurd.

Some felt that the suit was a publicity stunt. The heavy news coverage supported that view. The ACLU was trying to make a point, using them as a scapegoat for the ills of the healthcare system, for things that were not their fault. Ben Jackson, a young patent attorney who had first worked for Myriad as a law student and was now on the full-time legal staff, felt that Myriad was doing nothing wrong. Their patents complied with every rule handed down by the Patent Office; they were no different from any other biotech company—and better than most because their tests actually saved people's lives.

The most common reaction among Myriad's rank and file employees was surprise, then hurt and indignation. How did their tests harm the ACLU or anyone else? "I've been contributing to the ACLU for years," one employee said. "Never again!" Few of Myriad's employees understood the suit or why it had been brought. But one thing was crystal clear to Marsh and the executive team: they were facing a carefully planned assault that had probably been years in the making. They needed to take it very seriously.

Beauty Contest

THE NEXT DAY, Marsh began to receive calls from lawyers across the country. Was Myriad being represented in this new lawsuit? Was it prepared for a legal fight with the ACLU? Could law firm X, Y, or Z make a presentation to management about its deep expertise in cases just like this? For the most part, Marsh's answer to these unsolicited inquiries was "No, thank you." Nevertheless, it was clear that Myriad would need serious legal firepower to defend itself in the suit.

As a company with revenues that would exceed $300 million that year, Myriad had been blessed with comparatively few legal problems. Other than an occasional employment claim, the only major litigation that the company had faced since Marsh arrived was a 2003 royalty dispute with the University of Utah. Early litigation with Oncormed and NIH over ownership of the *BRCA* patents was well over a decade old, as was the suit against the University of Pennsylvania. As a result, Myriad didn't have trusted outside counsel to which it regularly turned for litigation advice. Even Myriad's patent work was now largely handled in-house by Zhang and his team.

Marsh discussed the matter with Pete Meldrum. The CEO's principal concern was cost. He hated few things as much as he hated legal bills. Over the years, Myriad had used a number of law firms for litigation matters. They were fine, Meldrum said, but expensive. He hoped they could find a more economical option this time. Meldrum did, however, remember a lawyer that had impressed him during the Oncormed litigation. Meldrum suggested that Marsh contact that lawyer.

Marsh pulled out the old Oncormed files and found that the Gaithersburg company had been represented by a New York law firm called Pennie & Edmonds, one of the best-known patent boutiques in the country. The attorney who had represented Oncormed was named Brian Poissant, a veteran intellectual property (IP) litigator who had racked up significant victories for clients like pharmaceutical giant Bristol Myers Squibb. But the legal landscape had been changing, and many stand-alone patent firms were either being absorbed by big general practice law firms or going out of business entirely. In 2003, the venerable Pennie & Edmonds had closed its doors and a hundred of its attorneys moved *en masse* to Jones Day, one of the largest law firms in the world. As part of that move, Poissant became head of Jones Day's IP practice. Jones Day would not be cheap. But Marsh had a feeling that, notwithstanding Meldrum's desire to pinch pennies, this was not a case to be trifled with.

Marsh called Poissant later that day and introduced himself. Poissant vaguely recalled the Oncormed litigation, but Oncormed was long gone and certainly no longer a client. Marsh mentioned the ACLU suit, and Poissant became interested. He had read about the suit in the *Times*, as had everyone else in the office. Poissant invited Marsh to New York to discuss a strategy for the case.

Marsh flew to New York later that week. To be diligent, he contacted another firm as well. He would meet with them both and then decide who to hire. It was what the legal industry called a beauty contest. But by the end of the week, Jones Day had clearly won the tiara.

The firm occupied the top several floors of Four World Financial Center, just across the street from Ground Zero. Construction of the new Freedom Tower overlooking the National September 11 Memorial was under way, and the entire area was surrounded by chain-link fences and blue plastic tarps. Mourners still left flowers in sad little clumps along the fencing. It was a sobering sight.

Inside the building, the decor was sleek and modernistic. Poissant himself was imposing, a big, bull-necked chemical engineer with a booming voice and a dominating presence. He had been litigating IP cases since the 1970s and was considered one of the giants in the field. He also had an excellent record in the Southern District of New York, where their case would be heard. But the courtroom litigator was only half the story.

Jones Day's secret weapon was Laura Coruzzi, a cell biology PhD who had been recruited by Pennie & Edmonds out of a prestigious postdoc at Mount Sinai School of Medicine. In those days, top patent firms often hired promising young scientists, putting them to work during the day and paying for them to attend law school at night. Coruzzi graduated in 1985 from Fordham's evening law program and became a partner at the firm just four years later. After Pennie & Edmonds's merger with Jones Day, she was put in charge of the global juggernaut's biotech patent practice. Over the years, Coruzzi could claim a long string of successes. Among her many clients had been the University of Utah.

That ancient history, however, had little to do with the current lawsuit. What won Marsh over was the incredible thoroughness of the presentation that Poissant and Coruzzi made. In just a few days, based on little more than the ACLU's short complaint and copies of the patents downloaded from the Patent Office's website, they had compiled a complete case strategy and plan, mapping a proposed path from the district court in New York all the way to the U.S. Supreme Court.

"The Supreme Court?" Marsh asked at the end of the presentation.

Poissant smiled broadly. "You can never be too prepared," he said. Marsh agreed. He hired Jones Day the next morning.

The View from Alexandria

SOON AFTER THE ACLU filed the complaint in New York, Park and Ravicher called Ray Chen, chief solicitor of the Patent Office. Ravicher knew Chen, and the call was a courtesy to let him know the lawsuit was coming. Chen thanked them for the heads-up and said he would look out for the suit. About the same time that Rick Marsh received the ACLU's complaint, it landed on Chen's desk.

The Patent Office, which spent the first two centuries of its existence in the large neoclassical edifice that now houses the National Portrait Gallery, has since 2005 occupied a new 2.4 million square foot complex in Alexandria, Virginia. Its gleaming glass-and-steel atrium soars, cathedral-like, ten stories above visitors' heads, intentionally designed to awe and inspire with the might of American innovation. From his spacious office, Chen oversaw all litigation involving the Patent Office. And like any other federal agency employing more than ten thousand people, there was a lot of it.

What Chen read in the ACLU's complaint, though, was new to him. Many of the people suing the Patent Office were kooks who were upset that their patent applications for new perpetual motion machines and time travel devices had been rejected. Or, of course, disgruntled employees and contractors. Or any of the other routine disputes and controversies that affected every large government agency. But this was definitely something new.

Chen, an NYU law grad with an electrical engineering degree, had spent the last decade climbing the legal career ladder at the Patent Office. It was a great job. Just a few years out of law school, he was arguing at least two cases per year at the Federal Circuit, an experience that few of his classmates could match. The previous October, he had appeared before the full twelve-judge court in the *Bilski* case, which was probably headed to the Supreme Court. Chen had steadfastly defended the Patent Office's decision to deny a patent on the energy brokering equation, and the court had agreed. In the wake of Chen's exemplary performance, he was promoted to solicitor of the Patent Office, the number two lawyer at the agency.

But even with a decade's experience, Chen couldn't figure out the Myriad suit. If (and that was a big "if") the ACLU had a legitimate grievance with Myriad, they should fight it out in court. Why was the Patent Office involved? Chen thought it significant that Dan Ravicher from PubPat was listed on the complaint. Ravicher was a patent attorney, and had a unique "number" signifying that he had passed the demanding Patent Bar Exam, qualifying him to practice before the Patent Office. He, of all people, should know that once a patent was issued, the Patent Office was out of the picture.

Given all this, Chen didn't think it would be difficult to get the Patent Office dismissed from the suit. He assigned the case to one of his junior staff lawyers, who would work with a local Assistant U.S. Attorney in New York to make the required filings (this certainly wasn't a case that warranted sending one of his own attorneys to New York).

With the morning's routine business taken care of, Chen turned back to the big question on everyone's mind: would the Supreme Court agree to hear *Bilski*?

We Regret to Inform You

ONE EVENING IN late May, the phone rang at Mark Skolnick's Salt Lake City home, a massive Italianate villa overlooking the valley below. He had an unlisted number, and so far reporters had not been able to find it. But the call wasn't from a reporter; it was Skolnick's mother.[1] She had received a letter in the mail, she told him. It was from the ACLU.

Skolnick knew of his mother's long personal involvement with the ACLU. A lifelong social activist, she had organized protests against school segregation in Topeka, Kansas, leading to the ACLU's landmark case, *Brown v. Board of Education*. She founded the local ACLU chapter in

---

1 Emily Marks Skolnick (1915–2017).

San Mateo, California. She can be seen in old photographs beside civil rights heroines like Rosa Parks. She led protests against the U.S. war in Iraq (her most famous appearance had been lying in the street dressed like Saddam Hussein).

"What did the letter say?" he asked.

It was very polite, she said. An officer from the ACLU of Northern California thanked her for her years of service to the organization. He then regretfully informed her that the ACLU was about to sue her son's company. The writer assured her that the lawsuit was no reflection on the ACLU's continued esteem for her and her many contributions over the years.

Skolnick told his mother that it was true. Much to his own surprise, the ACLU had sued Myriad.

She paused for a moment. "So whose side am I supposed to be on?" she asked.

Skolnick paused. Which side of justice was he on? He didn't answer. Instead, he told his mother that she could decide for herself.

Enter the Blogosphere

THOUGH INITIAL NEWS coverage of the lawsuit had been primed by the ACLU, it did not take long for Myriad and its supporters to mobilize a media counterattack. Not surprisingly, the two main constituencies that rallied to the company's aid were the biotechnology industry and the patent bar. Together, their representatives launched a spirited, often histrionic defense not only of Myriad, but of gene patenting itself.

One of the more frequent, and measured, voices in the public debate was Hans Sauer, IP counsel to the DC-based Biotechnology Industry Association (BIO), the main lobbying and trade group for the biotech industry. A German-born neurobiologist, the soft-spoken Sauer earned his law degree at Georgetown and oversaw the patent groups at two biotech firms before joining BIO in 2006. Just a few weeks after the ACLU's complaint was filed, Sauer appeared on public radio's popular *Kojo Nnamdi Show* along with Joshua Sarnoff, the bearded environmentalist-turned-patent-scholar from American University, and Shobita Parthasarathy, the sociologist from the University of Michigan, both of whom had consulted with the ACLU during the preparation of its case. Sauer retained his cool and logical tone throughout the show, notwithstanding a clear sense of being outnumbered.

Another frequent commentator on the case was Kevin Noonan, a Chicago patent attorney and blogger. He, too, had his moment on NPR—*Science Friday* with Ira Flatow—paired opposite Dan Ravicher. But these public radio debates were generally polite, if sometimes tense. It took the unique inhibition-loosening, gloves-off atmosphere of the internet to generate the most spirited debates concerning the case. Noonan, for example, lays out the arguments in favor of gene patenting in a blog post titled "Falsehoods, Distortions, and Outright Lies in the Gene Patenting Debate." The post, which begins with a reasonable summary of the science and case law, ends by referring to the ACLU's arguments as "propaganda" arising out of "fear, ignorance, or dishonesty" and likening them to quasi-religious "vitalism" arising from "the same contrariness that motivates some to deny evolution or other doctrines that make them uncomfortable."[2]

UNC law professor and blogger John Conley asked whether the ACLU's suit was a legitimate patent challenge or a mere "publicity stunt" and warned that "the whole biotechnology industry would be turned upside-down if the courts were to agree with the ACLU." This warning was heard by more than a few biotech investors, one of whom took it upon himself to call Anthony Romero directly. The investor proceeded to harangue Romero, insisting that this was the worst possible suit for the ACLU to bring, and that it was going to seriously damage medical research. During his years at the helm of the ACLU, Romero had heard much worse. He politely explained to the investor that the ACLU had considered the issues carefully and that it was proceeding in a responsible manner. The caller was clearly not satisfied.

---

2 Noonan's comparison of the ACLU to anti-evolution creationists is perhaps unintentionally ironic, given that in 1925 it was the ACLU (through famed attorney Clarence Darrow) that defended John Scopes for teaching the theory of evolution to his high school biology class in defiance of a Tennessee law.

But perhaps the most fuel was added to the fire by outraged members of the patent bar, exemplified by patent blogger Gene Quinn. Quinn is the Rush Limbaugh of patent law—hilariously and infuriatingly outspoken, with a large and loyal following around the world. Two days after the suit was filed, Quinn wrote in his popular blog *IP Watchdog*, "This lawsuit is nothing more than grandstanding, it presents frivolous arguments and outright lies." He even called for the ACLU lawyers who brought the case to be sanctioned by the court for unprofessional conduct. Soon, calls for judicial sanctions and even disbarment of Hansen, Ravicher, and the other ACLU/PubPat attorneys became regular rallying cries for Myriad's defenders, though Jones Day and the company itself never went so far.[3]

It might be seen as odd that the most vehement and vocal critics of the ACLU's suit were patent attorneys who had no personal stake in either Myriad Genetics or the diagnostics industry. But obtaining patents in the United States is big business. About 700,000 U.S. patent applications are filed every year. This work is done by about 3,500 law firms employing roughly 50,000 patent attorneys and agents. Add to that the 10,000 examiners at the Patent Office, and you get a market worth about $8.5 billion per year. Any market that size will fight to protect itself.

More importantly, though, many members of the patent bar seemed to view the ACLU suit as a personal affront, an assault on their personal dignity and livelihood. To make matters worse, they also must have felt that the ACLU and its allies—novices who dared to judge longstanding practices without truly understanding them—were intruding on their turf. The insularity of the patent bar contributes to this "us versus them" mentality. To become a patent attorney, in addition to a law degree, one must have a degree in science or engineering. Non-science majors don't even qualify to take the patent bar exam. Thus, there is a strong sense of pride and élan among patent attorneys—they have a unique qualification that "regular" attorneys don't possess.

But in some cases that pride also masks feelings of insecurity and

---

3 Under Rule 11 of the Federal Rules of Civil Procedure, an attorney may be sanctioned by the court for filing any claim that is brought "for any improper purpose, such as to harass or to cause unnecessary delay or needless increase in the cost of litigation" or that is frivolous or unsupported by any evidence.

resentment toward mainstream lawyers, especially those who graduated from "elite" schools and work in top law firms. Because many patent attorneys came to law as a second career, after they already had jobs and families and mortgages, they had to attend law school at night or juggle work with part-time programs at lower-ranked schools. Their professional pride comes from their technical skill and knowledge, not their legal pedigrees.

As such, patent attorneys can be fiercely defensive and deeply resent intrusion by inexpert outsiders. Particularly if those outsiders think they can improve the patent system or how it is practiced. And most especially if they want to tear down the patent system, or a major part of it. For this, the patent bar never forgave the ACLU or its allies.

CHAPTER 15

SDNY

EVERY TUESDAY MORNING at the federal courthouse in Lower Manhattan, the court clerk noisily wheels a metal mail cart down the high-ceilinged corridors, stopping briefly at each set of judicial chambers. This is not the court's ordinary mail run, but the highly anticipated delivery to each judge of the complaints and other matters that have been assigned that week. These case assignments can sometimes be of greater interest to the judge's law clerks than to the judge, so the mail cart is often intercepted in the hallway by eager young attorneys who want an early glimpse of what they will be working on in the coming weeks and months.

Each judge in the Southern District of New York generally has two law clerks. Unlike court clerks, who are career bureaucrats employed by the court, a judge's law clerks function more like his or her personal research assistants. They review the briefs submitted to the judge, summarize the law, make recommendations, and even produce draft opinions, depending on the judge's preferences. These positions, which generally last only one or two years, are coveted by recent law school graduates and, as a result, are extremely competitive. A judicial clerkship is considered a stepping stone to some of the most prestigious legal jobs in the country—positions at major law firms, the honors program at the Department of Justice, and the judiciary itself. Many judges receive hundreds of applications for only one or two openings per year.

On Tuesday, May 19, 2009, a week after the complaint was filed in *AMP v. Myriad*, Herman Yue waited outside the entrance to Judge Robert Sweet's chambers. He could hear the mail cart rattling in the distance. A graduate of NYU law school, Yue had started his one-year stint as

Judge Sweet's law clerk just a few weeks ago. The judge was out of town—spending time at his vacation home outside of Sun Valley, Idaho—so Yue was wearing shorts and flip-flops. His co-clerk, Shannon Rebholz, was on the internet in their small, shared office. With nearly fifty judges on the court and cases assigned at random, it was always fun to speculate about which matters would end up on their desks that week. At the Southern District, anything was possible.

Just two weeks before Yue arrived, financier Bernie Madoff had pled guilty to orchestrating the largest investment swindle in the history of the world. Then, at the end of April, news leaked that Supreme Court Justice David Souter would soon retire. The leading contender to fill his seat was Sonia Sotomayor, a Second Circuit judge who had served for years in the Southern District. One of her closest friends and mentors on the court had been Robert Sweet, and she was still seen in the corridors of the courthouse. When Sotomayor was formally nominated to the Supreme Court by President Obama, there was an upswelling of support, most noticeably from the courthouse staff, many of whom wore giant buttons bearing the words I SUPPORT SONIA.

This week, however, the buzz at the court concerned a modern-day pirate—Abduwali Abdukhadir Muse, a Somali national who was in custody somewhere in New York. Apparently, Muse and his shipmates had attempted to seize a U.S. container vessel off the Somali coast in April. Navy Seals, in recapturing the ship, took Muse into custody after killing the rest of his gang. Muse would be tried for piracy on the high seas—one of the first such prosecutions in over a century—and all of the clerks were secretly hoping that the case would be assigned to their chambers.

When the mail cart reached Yue, he casually asked if there were any new cases for Judge Sweet. The mail runner consulted the hanging file folders on the cart. Behind the tab for SWEET, ROBERT W. was a thin stack of papers. He pulled them out and handed them to Yue. "Enjoy," he said as he continued down the hall.

Yue flipped through the complaints, glancing at their cover pages. Halfway through the stack he paused. No way, he thought, re-reading the long list of parties. But it was there, right in front of him. "No way!" he shouted, his voice echoing through the long marble hall. "No! Way!"

Rebholz popped out of the office. She, too, was wearing shorts. "Did we get it?" she asked, barely suppressing the excitement in her voice.

"What?" Yue asked. He realized that she was talking about the Somali pirate case. "Oh, no. Not that," he said, handing her the complaint. "We got Myriad Genetics—gene patenting!"

Rebholz looked at the page crowded with names. Association for Molecular Pathology? The United States Patent and Trademark Office? The University of Utah Research Foundation? "Oh," she said, audibly sighing as she went back to the office. Yue could definitely have this one.

Though Rebholz was disappointed (Muse's piracy case was assigned to Judge Loretta Preska), Yue was fascinated by the case against Myriad. That Sunday, he had read a story about it in the *New York Times* and was already thinking about its implications. As it turned out, Yue knew more than a little about patents.

As the son of an atmospheric scientist and a librarian, Yue had always been a science geek. He loaded up on biology and chemistry courses in college, then earned a PhD in immunology from Berkeley. Part of his graduate research resulted in a patent application that the university filed in 2004. It covered a method for altering a cell's immune responses by activating a particular protein receptor that he had first studied in mice. Yue's wife, Gretchen, was also a named inventor on the patent, their marriage the result of a lab romance. But Yue eventually realized that the lab life wasn't for him, and he applied to law school. He got into NYU at the same time that Gretchen got a postdoc in immunology at the university's medical school. They moved to New York together, and immediately after law school Yue took a job at Patterson Belknap, the prestigious New York law firm where Dan Ravicher had once worked.

So Yue was very familiar not only with genetics, but with patents. That said, the patent on the immunology discovery that he and Gretchen made was not a gene patent. By the time they were in graduate school, the human genome had already been sequenced and the heyday of gene hunting was long past. It wasn't actually until law school that Yue came across gene patents for the first time. And there, in Professor Rochelle Dreyfuss's patent law class, he couldn't believe what he was hearing. He visited Dreyfuss during her office hours to ask whether it was true that entire genes could be patented. She explained, in her typically implacable manner, that they could. But Yue remained skeptical. It was insane, he thought, and it bothered him. In his NYU thesis he followed the lead of scholars like Rebecca Eisenberg at Michigan, who had proposed limiting patents on basic scientific research tools. But after he graduated, amidst the hubbub of his fledgling legal practice and then his clerkship with Judge Sweet, Yue largely put off the heady theories that had captivated him during law school. Someday, once his student loans were paid off, he might have the luxury to think deep thoughts.

Until now, that is. Yue couldn't believe his luck. New cases at the court were assigned to judges at random, based on their workload and availability. It was called assignment "off the wheel," named in honor of a spinning oak drum that was once filled with the different judges' names and used to assign them cases. Now the wheel was a computer algorithm, but just as random. Thanks to sheer dumb luck, Yue would get to work on a case involving both patents and biotech. And it wasn't just some patent infringement case between pharma giants. This case challenged the very notion of gene patents in America. For Yue, this was way, way cooler than pirates.

I'm Still a Plaintiff

BY JANUARY 2010, with chemo and radiation behind her, Lisbeth Ceriani was ready to have her ovaries removed. She still hadn't been tested for the *BRCA* mutations but, even so, her oncologist had a strong suspicion that her cancer was genetic. She urged Ceriani to have the operation, whether or not she was tested. The risk of ovarian cancer was just too high, and there was no good way to detect the disease early—the consequences were too great to take the risk.

Ceriani understood, but was hesitant. Removal of the ovaries induces what is called surgical menopause. Unlike natural menopause, which allows the body to acclimate over time, the surgical version arrives suddenly and agonizingly. Ceriani was not looking forward to this, especially if she was negative for the *BRCA* mutation. But she didn't know whether she was or not, and her oncologist was right—the risk of ovarian cancer was just too great. She should have her ovaries removed before she turned forty-five.

Ceriani scheduled the prophylactic surgery for the February school break, dreading what would follow. And then, in early January, she received a message on her home answering machine. It was from somebody at Myriad.

In the message, which Ceriani replayed a dozen times, a woman, trying to sound upbeat, told her that Myriad had just donated five hundred BRACAnalysis tests to the Cancer Resource Foundation in Massachusetts, a charity that would make testing available to Massachusetts patients who couldn't otherwise afford the test. The program was being administered at Mass General, and Ceriani should call them as soon as possible in order to be tested.

Ceriani was stunned. How was Myriad calling her home number? Weren't they adverse parties in a lawsuit? She was no lawyer, but even she had watched enough tv legal dramas to know this didn't seem kosher.

Ceriani dialed Sandra Park. Park agreed that the call was unorthodox and possibly in violation of the rules of legal ethics. But that wasn't the important thing. The important thing was that Ceriani could now be tested. Ceriani agreed. She told Park about the surgery she had scheduled in a few weeks, and Park urged her to get the test before the surgery. Park also told her to be sure to get the BART add-on. Ceriani had never heard of this. Park explained that BART was an additional test that Myriad could do. They charged extra for it, but there was no reason that, once they had her DNA sequenced, they couldn't run that analysis, too. Park also told her that she would have to meet strict criteria to be eligible for BART testing—like a clear family history of breast or ovarian cancer.

This gave Ceriani pause. Her extended family was in Italy. She wasn't in touch with them. Her paternal grandmother had died, she thought, from stomach cancer, but nobody knew for sure. In the end, Ceriani didn't care. She told the genetic counselor at Mass General that there was definitely a history of ovarian cancer on her father's side of the family. That worked. The genetic counselor didn't ask questions and sent her blood sample off to Myriad. The results came back a couple of weeks later. She was positive for one of the BART rearrangements.

Ceriani breathed a sigh of relief. Now at least she knew. The surgery and aftershocks that she was about to endure wouldn't be for nothing. And, even more importantly, her daughter had a good reason to get tested once she got a little older.

Shortly after she received her test results, Ceriani got another call from Myriad. Again, the woman on the phone sounded sympathetic, understanding. She was so glad that they were able to help Ceriani to get tested. Mm-hm, thought Ceriani. She knew where this was headed. So now that Ceriani had been fully tested, the woman continued, could they assume that she would no longer be a plaintiff in the lawsuit?

Ceriani laughed. "It took me three years to get the test, asshole," she said. "I'm still a plaintiff." She hung up.

CHAPTER 16

Chicken and Egg

New York's Daniel Patrick Moynihan United States Courthouse is located on Foley Square, a sterile neighborhood between Chinatown and the Brooklyn Bridge that is crowded with courts and government offices. Opened in 1996, the twenty-seven-story marble-and-granite edifice pays architectural homage to the art deco towers of Rockefeller Center. But unlike the Rock, home to NBC studios and the tony Rainbow Room, the million square-foot Moynihan building holds more than fifty judicial chambers and forty-four courtrooms, most of them bustling with the ebb and flow of litigants, attorneys, police officers, jurors, and judges.

Judge Robert W. Sweet's courtroom was on the eighteenth floor of the massive judicial complex. He liked the courtroom—it had good acoustics, plenty of space, ample lighting, and decent air-conditioning. It was a far cry from the antiquated chambers in the old Foley Square Courthouse where he began his judicial career more than thirty years earlier. Now, at the age of eighty-eight, Sweet appreciated the modern amenities.

February 2, 2010, was a cold, clear day in the city. Most of the snow had melted during a warm snap the week before, leaving behind a gray-black grit that clung to the streets and sidewalks. Judge Sweet had begun the morning, as he often did, at the Sky Rink—a large indoor ice-skating arena with a panoramic view over the Hudson. As skaters ranging from preternaturally gifted preschoolers to seniors like Sweet himself glided to the music of *Harry Potter* and Andrea Bocelli, they caught glimpses

*Judge Robert Sweet relied
on his law clerk Herman
Yue when grappling with
the scientific aspects of
the case.*

of the opalescent ice coating the river below. Sweet had been working
on a new dance routine with his trainer, a former Soviet Olympian. Some
found it curious that a man approaching ninety would indulge in such
a pastime, but Sweet was stubborn. He wore a helmet and knee pads,
and had never been seriously injured on the ice. Most importantly, he
found it an excellent way to keep in shape while clearing his head before
a long day of hearings and motions. Today would undoubtedly be such
a day.

The judge had only one case on his calendar, and it was a doozy—
Association for Molecular Pathology v. Myriad Genetics. The stack of
briefs in his office was at least four feet high. He had spent untold hours
poring over the papers and discussing the science, not to mention the
legal arguments, with his talented law clerk Herman Yue. This was only
the second time in thirty years that Sweet had hired a clerk with a PhD.
And by an amazing coincidence, it had been at exactly the right moment.

In person, Judge Sweet more resembled an English rare book dealer
than a New York political operative. Always dressed impeccably, sporting
a silk pocket square and trimmed white moustache, one could easily
envision him reclining in a leather-backed chair at his club, pipe in one

hand, Scotch in the other. Yet Sweet's avuncular nature concealed a mind both sharp and probing. He had a reputation as a fair judge—sometimes stern, but open-minded and pragmatic. He did not tolerate grandstanding or theatrics in his courtroom, but was generally interested in hearing what the parties had to say.

Though he had no formal scientific training, Judge Sweet was not a stranger to intellectual property cases and had heard a handful of patent disputes during his years on the bench.

But Sweet was probably best known for his opinion in *Pelman v. McDonald's*, in which two obese teenagers sued the golden arches for contributing to their poor health. After considering an abundance of nutritional and health-related evidence, Sweet ruled for McDonald's, famously reasoning that "It is well-known that fast food . . . contain[s] high levels of cholesterol, fat, salt, and sugar . . . If a person knows or should know that eating copious orders of supersized McDonalds' products is unhealthy and may result in weight gain (and its concomitant problems) . . . it is not the place of the law to protect them from their own excesses."

Like the *McDonald's* case, *AMP v. Myriad* seemed to be generating considerable public notoriety. Numerous parties had filed *amicus* briefs in support of one side or the other. Coming in on the side of the plaintiffs were a number of women's health and disease advocacy groups, as well as medical and scientific associations. One brief submitted by the American Medical Association and the American Society of Human Genetics had been prepared by two of the law professors who had advised ACLU, Lori Andrews and Josh Sarnoff. Some briefs were filed by big-name non-profits like Greenpeace and the March of Dimes, while others were joined by obscure groups like the Indigenous Peoples Council on Biocolonialism. By and large these *amici* simply repeated or emphasized the arguments made by ACLU and PubPat in their own court filings.

Several parties also weighed in on behalf of Myriad, including the industry association BIO, a number of biotech companies, the Boston Patent Law Association, and Kevin Noonan, the patent law blogger. One surprise entrant on Myriad's side was the Genetic Alliance. The Genetic

Alliance was a patient group or, rather, a group of patient groups. Its founders were parents of children with a rare genetic disorder—pseudoxanthoma elasticum, or PXE. Their principal goal was encouraging anyone with the technical ability—universities, biotech companies, the government—to look for cures for the diseases that the now-expanded group cared about. If patents were needed to spur that innovation, so be it. For them, the only thing that mattered were cures. Judge Sweet would take all of these *amicus* viewpoints into consideration, but of course his primary focus was on the actual parties to the litigation. He was there, after all, to resolve their dispute, not to solve the world's problems.

When Hansen heard that they had drawn Judge Sweet, he was encouraged. Hansen hadn't argued before him, but he knew that the judge was sympathetic to First Amendment issues. In a notorious decision in the early 1990s, Sweet had invoked the First Amendment to strike down a New York law that made it a misdemeanor to "loiter, remain, or wander about in a public place for the purpose of begging." In striking down what was commonly known as the panhandling ban, Sweet invoked the First Amendment's guarantee of freedom of expression: "Walking through New York's Times Square, one is bombarded with messages. Giant billboards and flashing neon lights dazzle; peddlers hawk; preachers beseech; the news warily wraps around the old Times building; the Salvation Army band plays on. One generally encounters a beggar, too. Of all these solicitors, though, the only one subject to a blanket restriction is the beggar." Hansen could relate to this judge. Park, who had worked as a law clerk for a different judge in the Southern District, also had a favorable impression of Judge Sweet. The clerks at the court had a saying, "Sweet was sweet."

Nevertheless, some lawyers at the ACLU worried about the judge's age. At eighty-eight, they questioned whether Sweet would understand the cutting-edge science at the heart of the case. But Hansen wasn't worried. He hadn't understood the science either, but that didn't stop him.

The case was moving along as quickly as could be expected. The

ACLU had filed the complaint, laying out the plaintiffs' grievances, last May—nine months ago. Since then, more than 250 documents had been filed. There were the usual declarations and stipulations, scheduling orders and notices of appearance, as well as some more substantive filings.

In November, Judge Sweet had heard a motion to dismiss the case filed by Myriad. Brian Poissant argued that none of the twenty plaintiffs had "standing" to challenge the patents. That is, because they hadn't actually been injured by the patents, nor did they face any real threat from Myriad, the court shouldn't waste its time hearing their case. It would have been a convenient way to make the case go away without ever discussing the patents, and a large number of cases are thrown out every year based on preliminary motions like these.

Dan Ravicher, at least thirty years younger than Poissant, argued against the motion. He contended that each of the twenty plaintiffs had been directly affected by the patents in some way—either by not being permitted to carry on clinical or research activities, by lacking alternative testing labs to which to refer patients or, in the case of the individuals, by being unable to obtain testing for potentially life-threatening conditions. The judge agreed with Ravicher. The plaintiffs' injuries, though not yet proven, were alleged with enough specificity and plausibility that the plaintiffs should be permitted to make their case.

Which brought them to today's hearing. This morning, the ACLU was seeking "summary judgment." That is, they asked the court to rule in their favor without the need for a jury or witness testimony, which were required only to establish controverted facts. Since, they argued, the questions before the court—whether patents should be allowed on human genes—were purely questions of law, there was no need for a full trial. Sweet was open to this approach, but was interested in hearing the parties arguments. On one wall of his spacious chambers, hanging among decades of old photographs, memorabilia, and awards, was a framed plaque bearing a quotation from Judge J. Edward Lumbard, one of Sweet's early mentors: "Never assume a god damn thing." These were words that the judge lived by.

Never Assume

AT PRECISELY 10:00 A.M., Judge Sweet entered the courtroom through a small door connected to his robing room. The bailiff announced his entry with a deep "All rise!"

Courtroom 18C had three large, north-facing windows with heavy, green draperies tied back with thick cords. The view from the courtroom was spectacular in the clear morning light—carrying the eye from the ziggurat-like criminal court complex across the street to the green-domed beaux arts basilica that once housed the New York Police Department (and was now a high-end residential co-op) to the graceful Empire State Building rising above the Midtown cityscape.

The first thing the judge noticed was that his courtroom was filled to capacity—unusual for a patent case. Even after he motioned for everyone to take their seats, people remained standing at the back of the courtroom. Surveying the many unfamiliar faces, Sweet assumed that several of them were individual plaintiffs—patients as well as the scientists and counselors. He guessed that others—the ones in suits—were attorneys for the many *amici* who had submitted briefs, and perhaps some of the *amici* themselves. He also picked out several members of the press scattered throughout the crowd. This was definitely a case that people cared about.

Judge Sweet took his seat behind the bench. Because this was a hearing on summary judgment, no jury had been empaneled, leaving the wainscoted enclosure with two rows of six seats—the jury box—empty. Looking at the crowded courtroom, the judge invited any licensed attorneys in attendance to sit in the jury box, to make space for others to be seated in the gallery. There was a general shuffling and movement of people until the jury box was occupied by twelve lawyers. But even with this accommodation, people remained standing at the back of the room.

The plaintiffs were entitled to argue first. Hansen sat with Park and Ravicher at the counsel table. Simoncelli joined them. Although she was not an attorney, Hansen had requested that she be permitted to sit with counsel, and the court had agreed. The rest of the ACLU and PubPat legal team sat in the gallery directly behind the counsel table. Harry

Ostrer and Elsa Reich from NYU also watched anxiously from the gallery. Bob Nussbaum, one of their genetics experts from the University of California at San Francisco, had taken the overnight "red eye" flight from California to attend the hearing that morning. Chris Mason was there as well. After years of work on the case, the geneticist was eager to see how its first prime-time run went. And it was none too soon— Mason's wife was expecting their first child in a few months. She was already annoyed by all the time that he had spent on the lawsuit—an activity that paid nothing and didn't even reward him with academic credit. Once the baby came, he could expect to have far less time for pursuits like this.

Hansen rose and strode to the podium directly in front of the bench. He stood tall in a freshly pressed gray suit and dark tie. He lay a sheaf of papers on the podium, then began, speaking clearly, authoritatively, convincingly in a flat Midwestern accent that was refreshing in the New York courtroom.

He started with DNA: the role that it played in the body, the information that it carried from one generation to the next. He then spoke of DNA that was extracted from human cells. Yes, it was isolated and plucked out of its native environment, but that did not transform it into some new substance *invented* by the one who isolated it. Hansen paused for a moment, then extended his right index finger over the podium. "As I am standing here right now," he said, "if I pricked my finger and let the drop fall on the podium, the blood on the podium would be isolated from my body. It would still be blood. It would not be an invention of any kind."

The judge nodded thoughtfully, then gestured toward his necktie, the knot of which was visible above the collar of his black robe. It was printed with a subtle DNA pattern—a memento from a charity event underwritten by Sweet's wife. Judge Sweet asked Hansen to explain why it didn't matter that Myriad had chemically altered the DNA by purifying and isolating it, the crux of their argument.

Hansen smiled, appreciating the judge's game spirit. While the extracted DNA might be chemically different from the DNA in one's cells, he said, it wasn't different in any meaningful way. The information that the DNA carried was not altered by the trivial changes made when the DNA molecule was removed from the cell, and the information was what was important.

Hansen continued, turning to the free speech issues and the problems with Myriad's methods claims, which amounted, he said, to no more than mental comparisons made in a physician's mind. The judge nodded and interposed questions occasionally, but remained quietly attentive during Hansen's presentation.

Hansen then moved on to address the larger economic issue underlying the case—whether patents are necessary to incentivize researchers to make new discoveries. The answer, he argued, was no: "A patent is not a reward for effort—there is no question that Myriad has engaged in effort in order to uncover the significance of the *BRCA1* and *BRCA2* genes. It's also not a reward for uncovering nature. To use the language of the United States Supreme Court, 'You cannot patent an ancient secret of nature now disclosed.' That is what Myriad has done, and that is all Myriad has done, and that is why these claims were not properly granted and your Honor should find the claims invalid."

The courtroom sat in silence. Hansen's delivery had been impassioned, relentless, eloquent. Even some of the patent attorneys who had come to observe were impressed. With that, Hansen collected his papers and returned to the counsel table. Sandra Park then rose and approached the podium. She had largely been responsible for drafting the summary judgment motion and papers that were submitted to the court. This morning, her job, far less glamorous than Hansen's, was to describe in detail the many prior cases on which their arguments relied. She did this efficiently and authoritatively. The judge had no questions for her.

When Park finished, Judge Sweet turned to the defendants. Several observers from Myriad were in the gallery—Rick Marsh, who was managing the litigation, and Ben Jackson, who now headed Myriad's in-house patent group. No one from the University of Utah was present.

The Jones Day team sat at the counsel table: Poissant in the center, flanked by Laura Coruzzi and an associate. Coruzzi had been taking notes throughout the ACLU's argument, occasionally pointing something

Myriad's general counsel Rick Marsh (right) at a Yale School of Medicine panel on AMP v. Myriad *in February 2012 (pictured with Dr. Allen Bale, Yale Medical School).*

out to Poissant. He responded by nodding and scribbling his own notes. When the judge signaled him, Poissant rose and approached the podium. Though he had not previously argued before Judge Sweet, he was a veteran of the Southern District. He felt comfortable here and exuded a respectful but confident presence.

After briefly praising the Supreme Court's decision in *Chakrabarty*, Poissant challenged Hansen's claim that gene patents had hindered cancer research or access to genetic testing. The evidence, he said, pointed in precisely the opposite direction. And, contrary to Hansen's arguments regarding the effect of gene patents on medical care, it was the *abolition* of gene patents that would cause catastrophic effects for the industry and for patients. "They are talking about the invalidity of thousands of gene patents. This could unravel the foundation of the entire biotech industry. Numerous therapeutic drugs and diagnostic tests that are now in development will never see the light of day." Poissant denigrated Hansen's First Amendment argument as mere "frivolous atmospherics"—an argument so absurd that he had never heard it before.

He then made a valiant effort to counter the argument that removing a gene from a human cell didn't really change the gene in any meaningful way. As an analogy, he used an egg. "Think of a large egg with a very,

very, very long thread embedded in it. That is what we are talking about. The egg is the cell, the thread is the genomic DNA all bundled up and it's thousands of miles long. You have to go in there and find it and unravel it and then you have to look for the part of it you are really interested in. That is what isolated means. Simply by opening up the egg and finding this long gene—the battle is just beginning. You have to now go along the DNA and find what part of it is really the interesting part, what part is the critical part you are looking for, and that is what they did."

Judge Sweet listened attentively. When Poissant mentioned Learned Hand's 1911 *Parke-Davis* case, holding that purified adrenaline was patentable, the judge joked, "You have affected my adrenaline." Poissant pointed at the spot on the podium where Hansen's hypothetical drop of blood had landed, laughing that there was probably some adrenaline in there, as well.

But Poissant quickly became serious again. For well over an hour he took the court through the challenged patent claims, explaining the significance of the inventions they embodied. He came back to *Chakrabarty*, which allowed the patenting of GE's oil-eating bacterium, and the cases that Park had cited, showing how each of them supported his arguments, and not theirs. He ended by predicting the grievous harm that would befall the biotech industry if gene patents were suddenly abolished. Patents, he claimed, had promoted cancer research, clinical development, quality assurance, patient access, and even affordability. Without them, medical science would not have advanced nearly so far.

When Poissant finished, he returned to his seat. The court next heard from Ross Morrison, the Assistant U.S. Attorney representing the Patent Office. When Morrison concluded his brief remarks, it was Hansen's turn for rebuttal. He redirected the court's attention to Poissant's egg metaphor.

There is an art to using metaphors in legal argument. Especially when the facts and the law are complex, simple metaphors can help a judge, and a jury, to understand what's at stake. Hansen, Park, Simoncelli, and Ravicher had spent hours arguing about how to describe the genetics and patent law arguments that would be all-important to their case. In their brief, they compared gene sequencing to proofreading a book and using a microscope to read tiny letters; they also argued that DNA sequences were not machines like carburetors. When arguing that isolating a gene from the larger cell did not change its essential nature, they compared the gene to a vein of gold found in the ground and a drop of blood from a person's body. Some metaphors, however, they rejected. At one point, Hansen thought that he had found the perfect metaphor for DNA: the gold thread spun by the fairy tale character Rumpelstiltskin. Simoncelli and Park howled with laughter and told him, "No, absolutely not!" Rumpelstiltskin was a terrible analogy, and they did not include him in their brief. Now, standing before the judge, Hansen realized that Poissant's egg metaphor was equally awful.

"Cracking open the egg and separating out the yolk," Hansen began, "does not make the yolk an invention. I would have thought it would have been clear to all of us that the yolk was the invention of the chicken, not of the cook." There was laughter in the courtroom. After a brief pause, Hansen continued, "They want to say there is this little string hidden somewhere deep in the yolk and they had to dig really deep. If the string is in the yolk the chicken put it there, the cook didn't put it there, and no matter how hard it was for the cook to find it within the yolk, it's still an invention of nature. It's not the invention of the cook."

In the time that it had taken Poissant and Morrison to finish their arguments, Hansen had developed a response of such folksy common sense and homespun wit that it stuck with those who had heard it long after the hearing had ended, becoming one of the most famous catch phrases in the case: *The yolk was the invention of the chicken.*

"These Poor Women"

THE BIOTECH LOBBYIST Hans Sauer had come up to New York to attend the hearing. He was surprised by the crowd in the courtroom—it was unusual for a summary judgment hearing. He was also surprised that he didn't know most of the faces around him—another unusual occurrence, as Sauer usually knew 90 percent of the people attending any given event involving both biotechnology and intellectual property. As he scanned the crowd, he did see some reporters that he recognized from the major news outlets. There were a lot of women in the gallery—some wearing head scarves or hats—the clear insignia of cancer patients. Sauer had a bad feeling about this.

Sauer also spotted an unfamiliar face sitting in the "dock" with the court officer. This was an area reserved for court personnel, but this young woman was wearing some kind of name badge on a lanyard that looked like it was from the NYU medical school. He couldn't make out her name, but wondered who she was. A scientist of some kind? Had the court hired its own scientific expert? Sauer was concerned about this. The woman, it turns out, was Gretchen, Herman Yue's wife, there to watch the case that he had been talking about for so many weeks. Because the courtroom was full, the judge allowed her to squeeze into the dock with the court officer.

Sauer's fears were confirmed once the hearing got under way. He had submitted an *amicus* brief to the court on behalf of BIO. For the most part, it repeated, in slightly different words, the arguments that Jones Day had made on behalf of Myriad—isolated human genes were patentable and they had been for decades. It wasn't until the final page of the brief that Sauer made his most important point—this case wasn't just about Myriad or its patents. Nor was it just about *gene* patents, which, to be honest, were not that important to the majority of Sauer's constituents in the biotech industry. This case was about patenting natural products of *every* kind.

The BIO brief described a compound derived from the bark of the *Gingko biloba* tree, which was useful in preventing blood clots. He described an antibiotic derived from *microbispora,* which was effective against several drug-resistant bacterial strains. Would all of these valuable inventions suddenly become unpatentable if the court invalidated Myriad's *BRCA* patents? If so, it would spell disaster for the industry and would throw healthcare back by decades. Sauer worried that Myriad's lawyers were so focused on DNA and *BRCA* that they were not making these arguments, which were far more important to the industry and his constituents.

Outside the courthouse at the Southern District of New York.
Back row, left to right: Dan Ravicher, Robert Nussbaum,
Tania Simoncelli, Chris Hansen, Sabrina Hassan, Aden Fine, Chris Mason.
Front row, left to right: Sandra Park, Lenora Lapidus, Elsa Reich.

Sauer's concern grew as he listened. Poissant had a big courtroom presence, but he seemed to have difficulty with the science. Maybe this was OK, since the judge didn't know it either, nor did the ACLU's lawyer, Hansen. But the scientists in the room could tell when a lawyer was faking it. Sauer looked over at Laura Corruzzi, the Jones Day PhD sitting at the counsel table. He knew that *she* knew this wasn't going well. But she never even blinked, even as Poissant offered nonsensical explanations like, "You have to open up the cell and go in and find it and when you find the *BRCA* DNA . . . you amplify it to make a lot of it so you can look at it." Huh? This sounded like tiny scientists driving their mini-spaceship through the human body like *Fantastic Voyage*. It in no way resembled what actually happened in the lab.

But the worst moment came when Poissant tackled the human angle. First, he patronizingly referred to the individual plaintiffs as "these poor women," and then had the gall to argue that they should actually be grateful to Myriad. "These poor women," he said, oblivious to the rolling eyes throughout the courtroom, "they wouldn't even know they had a *BRCA* gene problem if it hadn't been for the inventions and discovery of

Myriad." Talk about tone deaf. In Sauer's mind, it couldn't have
gone worse.

As he left the courthouse, buttoning his overcoat against the frigid
wind, Sauer noticed a gathering on the courthouse steps below him. At
first he thought that somebody had fallen down, but then he saw video
cameras and boom mikes, in the center of which stood Hansen and Park.
It was a press conference on the courthouse steps—like in the movies,
like John Gotti was about to be led out of the courthouse in handcuffs.
Sauer couldn't believe it. This wasn't a patent case. Patent cases were
dull, technical, long, and complicated but, in the end, logical. This was
something else entirely. This was theater.

What Hearing?

AFTER THE HEARING, Marsh called Pete Meldrum back in Salt Lake
City. The hearing went fine, he said. Jones Day did a good job. They
couldn't read the judge—he could go either way.

Meldrum acknowledged it all without saying much. The case wasn't
a priority. That morning, Myriad had reported its financial results for
the last quarter of 2009. The news was good. Quarterly revenues were
up to $93 million, and the company had over $450 million in liquid
assets. The genetic diagnostics business was running a jaw-dropping 88
percent profit margin. Later that afternoon Meldrum would have his
quarterly conference call with the Wall Street analysts. He was confident
that they would be more than impressed. At that point, the lawsuit wasn't
even a blip on the radar screen.

But amid the positive financial news, there was one announcement
that caused Meldrum a twinge of nostalgia. That morning, he had
announced the resignation of Mark Skolnick, co-founder of the company.
After nearly two decades of service, Skolnick would be stepping down
from his positions as chief scientific officer and director and assuming
the honorary title of "Senior Scientist Emeritus and Scientific Visionary."

It was time, Skolnick had said. All the members of his old research team had moved on. Now, the only research being conducted at the company was to improve the testing process incrementally and to add additional genes and markers to the service. It wasn't groundbreaking science. In retirement, Skolnick and his wife planned to spend more time back in Italy, close to her family, and to become more involved in their sons' lives.

In Chambers

THIS WAS NOT a simple case, and Judge Sweet would not reach a decision overnight. Over the next several weeks he met regularly with Yue to discuss the issues, go over the briefs, and review the arguments made at the hearing. Sweet later called Yue his translator and his guide—an invaluable and providential aid in this important case.

One thing Sweet insisted on doing was discussing each of the many *amicus* briefs that had been received. Many judges just ignored these, but Sweet wanted to be sure that he understood all of the policy issues wrapped up in this technical patent question. The work was painstaking but fascinating. For Yue, it was a dream come true—the best possible way to end his one-year clerkship. He dug into the science, the precedents, the enormous written record. His clerkship would be over at the end of March, when he would go back to Patterson Belknap and the practice of law. They had less than two months.

60 Minutes

THE FIRST FEW DAYS after a big hearing, everyone tries to read the tea leaves and predict what the judge will do. Some, like Dan Ravicher, are naturally pessimistic. Others, like Hansen, are cautiously optimistic. In any event, they all agreed that, no matter how the judge ruled on their summary judgment motion, the case was far from over.

They had to draft press releases for both possible outcomes; they had to think about the schedule for the inevitable appeal to the Federal Circuit; they had to update their constituents and prepare for whatever public statements Myriad might make. And, of course, they all had other work that had been put on the back burner in the lead-up to the hearing. Now was the time to catch up, because once the decision was handed down, the pressure would be on to prepare for the appeal.

One major activity during this period involved the press. The case had attracted some attention when it was filed, but after the hearing in New York, it was major news. Park was fielding almost daily calls from reporters. Everyone, it seemed, was interested, from national and local newspapers to women's and health magazines to scientific and medical journals. Park also reached out to the individual plaintiffs, seeing whether any of them were willing to be contacted for the show. Though some of the plaintiffs did not wish to be in the limelight, Ceriani, Limary, and Girard were eager tell their stories to the public (Limary, for example, was featured in an article in the popular *SELF Magazine*). One of the most exciting press opportunities came from the iconic CBS news show *60 Minutes*.

Before the case was filed, Joel Engardio, the ACLU's story finder, sent the five-minute video that he and Park had made to a *60 Minutes* producer. Though the video served an important purpose in telling the patients' stories to the public, Engardio had also created it as an audition reel for *60 Minutes*. Morley Safer, who had interviewed Lori Andrews about gene patents back in 2001, took the bait. He now wanted to produce a segment on the case. This was big, even for the ACLU. In November 2008, the show's interview with president-elect Obama had attracted a record 25 million viewers. If they could get even a fraction of that audience, their cause would be advanced tremendously. Romero encouraged the ACLU team to cooperate fully.

Park coordinated with the producers, arranging for a film crew to tape the legal team as it entered the New York courthouse and to interview them at the ACLU offices.

The segment also included Lori Andrews, reprising her 2001 appearance on the show. Hansen, conservative tie askew, was filmed pretending to work at his desk while Safer's voice-over described the lawsuit. For balance, Safer interviewed Kevin Noonan, the Chicago patent lawyer and blogger who had unofficially been designated Myriad's go-to guy for pro-patent commentary. Noonan, looking ultra-corporate in a pinstriped suit, rimless glasses, and distinguished graying beard, sounded eminently reasonable until Safer trapped him into a comparison of testing DNA and selling cars. "Whether it's cars or health," Noonan said, taking the newsman's bait, "there has to be a profit for some investor to invest." It was logical, but hardly sympathetic, particularly when followed by Ceriani's and Girard's heartfelt appearances. The show was taped and edited by mid-March, but the ACLU told CBS that the court's decision would be imminent. They should wait until it came out. Reluctantly, Safer agreed.

The End of the Beginning

JUDGE SWEET ISSUED his decision on March 29, 2010, two days before the end of Herman Yue's clerkship. A whopping 156 pages long, it laid out, in clear and concise prose, a description of the prevalence of breast and ovarian cancer; the search for the BRCA genes; Myriad's patenting of the genes; the disputes over pricing, coverage, and access; and each of the parties' legal theories. But of course what interested readers the most were the court's conclusions regarding the patents.

Sweet first addressed Myriad's composition of matter claims. He acknowledged that Myriad was the first to isolate and purify the BRCA genes, and that an isolated and purified gene is in some ways different from a gene residing within a human cell. Yet he ultimately found these arguments unconvincing. First, he referenced a long list of prior cases holding that purified forms of different natural substances—tungsten, uranium, vanadium, even pine needles removed from their sheaths—are not eligible for patent protection. Even purified, these substances are still "products of nature," which have long been held to be unpatentable.

Next, he confronted the contrary precedent in *Parke-Davis v. Mulford*, the 1911 case in which the esteemed Learned Hand upheld a patent on purified adrenaline. Sweet refuted that earlier decision, noting that it was "dubious" when rendered a century ago, and notwithstanding that, "certainly no longer good law."

Sweet next turned to Myriad's argument that the isolated and purified *BRCA* genes are "markedly different" from the genes as they exist inside an individual's body. In doing so, he considered not the chemical properties of the DNA molecule, but its unique function as an information carrier. "In light of DNA's unique qualities as a physical embodiment of information," he wrote, "none of the structural and functional differences cited by Myriad . . . render the claimed DNA 'markedly different.'" As a result, they covered "unpatentable products of nature."

Sweet was equally skeptical of Myriad's method claims, which were directed at diagnosing an individual's increased risk of cancer by determining whether she carried one or more of the *BRCA1/2* mutations listed in the patent. These claims are invalid, he wrote, because they describe mere "abstract mental processes." Under the Federal Circuit's precedent in the *Bilski* and *Mayo* cases,[1] inventions must be tied to some kind of machine or transformation of physical substances in order to be patentable. But simply analyzing and comparing two DNA sequences and drawing conclusions from them is a pure mental process divorced from any machine or transformation. As such, Myriad's method claims were also unpatentable.

In ruling that Myriad's patents were invalid, Judge Sweet tried not to denigrate the company's scientific accomplishments. "The identification of the *BRCA1* and *BRCA2* gene sequences," he writes, "is unquestionably a valuable scientific achievement for which Myriad deserves recognition." However, he went on to explain, "that is not the same as concluding that it is something for which they are entitled to a patent." Convincing the Patent Office to grant patents on these scientific truths and products of nature, the judge concluded, was no more than an elaborate "lawyer's trick."

Given that the court invalidated all of Myriad's challenged patent claims, Judge Sweet did not need to address the constitutional issues brought against the Patent Office. Hansen's First Amendment arguments never made it into the opinion, and Judge Sweet granted the Patent

---

1 At this time, both *Bilski* and *Mayo* were working their way through the judicial system, each on its way to the Supreme Court.

Office's motion that it be dismissed from the case. While this might have seemed like a victory for the Patent Office, it caused tremendous problems for the government later, as we will see.

Pigs Fly!

THE DECISION WAS a total victory for the plaintiffs. All fifteen of the patent claims they had challenged were struck down. The score was 15-0! The years of preparation and hard work had paid off. But they knew that they could not rest on their laurels. Ravicher had convinced them of that. No matter how favorable Judge Sweet's decision was, he said, an appeal was certain and they would almost certainly lose at the Federal Circuit. So, after a quick meeting, they got back to work on their other cases and began to think about the appeal.

The reaction to Judge Sweet's decision by the patent blogosphere was swift and vehement. "Pigs Fly!" wrote one commentator, calling Judge Sweet's ruling "radical and astonishing in its sweep." Another blogger expressed his exasperation with the ruling in an article colorfully titled "Foaming at the Mouth—The Inane Ruling in the Gene Patent Case." But Gene Quinn topped them all in terms of sheer manic vitriol. After spending nearly a paragraph comparing Judge Sweet's opinion to the *Twilight Zone*, accusing the judge of "almost unfathomable intellectual dishonesty," and parodying the opinion à la Homer Simpson ("la la la la la . . . I'm not listening . . . la la la la la la . . . the patent is invalid"), Quinn makes an impassioned appeal to the public: "For crying out loud, WAKE UP people!!!! Where and when will this assault on the patent system stop?"

In addition to this impassioned fringe, there were more measured responses to Judge Sweet's decision, among them Myriad's. In a press release issued the day after the decision, Myriad's CEO calmly stated, "While we are disappointed . . . we are very confident that . . . the Federal Circuit will reverse this decision and uphold the patent claims being challenged . . . More importantly, we do not believe that the final outcome of this litigation will have a material impact on Myriad's operations . . .

Francis Collins, Director of the NIH, playing his hallmark guitar with a DNA double-helix inlaid in the fretboard.

Notwithstanding today's decision, we are extremely proud of what Myriad has been able to accomplish over the years in promoting women's health in the area of hereditary breast and ovarian cancer. Countless lives have been saved as a result of our efforts in concert with the healthcare community."

By and large, the public reaction to Judge Sweet's decision was positive. The *New York Times* declared it a victory in a "war over human nature." Joseph Stiglitz and John Sulston, the two Nobel laureates who had submitted declarations in support of the ACLU's case, applauded the decision in a *Wall Street Journal* op-ed. They called it a "major victory for science and innovation." The editorial board of *Nature Biotechnology* heralded the decision as a welcome blow to a "petrified" patent system.

Francis Collins, who had not been focusing on the case, heard about Judge Sweet's decision via multiple news feeds on the day it was released. Collins had stepped down as the director of NIH's Genome Institute in August of 2008, allegedly to work on a book and serve as an informal scientific advisor to candidate Obama. It was no surprise when, in July 2009, President Obama appointed Collins as director of NIH. With his sterling scientific reputation, his evangelical Christian credentials and his folksy persona—he was routinely photographed riding his Honda

Nighthawk motorcycle and playing an oversized acoustic guitar—Collins's Senate confirmation went off without a hitch.

Now, less than a year into his job as head of the world's largest scientific funding agency, Collins fondly recalled his lab at Michigan and frantically racing to sequence the *BRCA1* gene with Mary-Claire King. It had been twenty years since King had mapped the gene to the long arm of chromosome 17 and accepted his offer to collaborate. Since then, the entire human genome had been sequenced—under his leadership. But for most of the intervening years, Myriad's patents had given it a monopoly on *BRCA* testing. Collins couldn't hide his pleasure at the thought that those patents might now be eradicated.

The *60 Minutes* segment entitled "Patented," with its appearances by Hansen, Ceriani, Girard, and others, aired on April 4. At the end of the segment, host Scott Pelley reported that Judge Sweet had ruled in favor of the plaintiffs, and, over the audible *tick-tick-tick-tick* that signified the end of the episode, noted that Myriad was appealing the decision. Two months later, the appeal was filed.

Moving On

OF EVERYONE WHO read Judge Sweet's decision, perhaps none was more affected than Tania Simoncelli. She had planted the seeds for the case back in 2004. Now, six years later, they had not only sued someone in federal court, but they had won. Simoncelli never dreamt what would be involved in the process—the hours, days, and weeks spent poring over articles and old cases, formulating theories, dissecting patent claims, meeting the plaintiffs and hearing their moving stories. The case had consumed her and it had changed her. When Simoncelli first moved to New York, she had kept up with the cello, playing professionally with ensembles and orchestras and as a solo accompanist to the Merce Cunningham Dance Studio. One 2006 *Village Voice* review praised her as playing with "verve." During the lead-up to the district court filing, Simoncelli worked every night at the ACLU until seven, came home, ate a bowl of cereal, then practiced for another three hours. It was grueling and unsustainable. Simoncelli completed a final set of performances with choreographer Christopher Caines's troupe and, wishing to end on a high note, put her cello away for good.

Ironically, though, Simoncelli was not at the ACLU offices when Judge Sweet's decision was handed down. Instead, she was in Washington, DC. In February, two months before the decision was released, she left the ACLU to become special assistant to a commissioner at the Food and Drug Administration. Leaving the ACLU was a difficult decision for Simoncelli—so much of her career was wrapped up in the work she had done there. But the opportunity to serve in the new Obama administration, in a position where she could influence policy directly, was hard to pass up. At the ACLU, Simoncelli's role had always been ambiguous— she was a non-lawyer in an organization of lawyers. One senior ACLU attorney affectionately referred to her as a "square peg" among a bunch of round holes. She could never file a lawsuit herself. Her role was limited to persuading others to care about her causes. That had happened in a big way with gene patenting, but it was not clear that it would happen again. Nevertheless, Simoncelli would always be cheering on her ACLU colleagues from the sidelines. And, though she didn't know it at the time, the center of gravity of *AMP v. Myriad* would soon shift back to Washington, where things were already beginning to move behind closed doors.

CHAPTER 17

We're from the Government

THE NATIONAL ELECTIONS OF 2008 and the arrival of Barack Obama at the White House crashed over Washington like a tidal wave. Obama, himself a graduate of Columbia and Harvard, surrounded himself with intellectual overachievers, graduates of elite schools, holders of advanced degrees. And his direct subordinates did the same, so that throughout the Capitol there was a steady influx of bright and remarkably diverse new faces from the academic and technology hubs of the country. For the most part they were idealistic, they were hardworking, and they were extremely smart. Many compared the new administration to John F. Kennedy's Camelot. These were people who wanted to get things done.

President Obama's pick for solicitor general was Elena Kagan, the dean of Harvard Law School. The solicitor general, or SG, is the government's top litigating lawyer. Some of the nation's most prominent attorneys—Thurgood Marshall, Ken Starr, Robert Bork, to name just a few—have held this coveted office. The SG, with a staff of only four deputies and sixteen assistant solicitors, represents the United States government before the Supreme Court and otherwise oversees the federal government's appellate litigation.

During her fifteen-month tenure as solicitor general, Elena Kagan argued before the Supreme Court six times—more than most elite attorneys would over a lifetime. Despite, or perhaps because of, her sterling performance, Kagan's term was cut short when President Obama nominated her to replace retiring justice John Paul Stevens on the Supreme Court. Kagan resigned as SG on May 17, 2010, and her top deputy, Neal Katyal, was named acting SG.

House of Cards

LIKE KAGAN, KATYAL was an overachiever. Only forty years old, the Indian-American lawyer and academic was already one of the fastest rising stars in the administration. After getting his law degree from Yale, Katyal clerked for Supreme Court justice Stephen Breyer, then spent a year as a national security advisor to the Department of Justice. As a young Georgetown law professor he represented Al Gore in contesting the results of the 2000 presidential election. Katyal went on to defend Osama bin Laden's chauffeur, who was being detained without cause at Guantanamo. In 2006 Katyal achieved a landmark Supreme Court victory for the hapless driver and his fellow detainees, toppling the Bush administration's entire program of offshore detainment as a violation of due process.

Katyal's unexpected victory gave him instant celebrity. He made appearances on *The Colbert Report* and even did a cameo on the Netflix drama *House of Cards*. One popular blog dubbed Katyal the "Paris Hilton of the Legal Elite," and an old Yale mentor proclaimed him "the Thurgood Marshall of his era." All of this embarrassed Katyal, but also made him an easy choice when President Obama had to pick a top deputy SG to serve under Elena Kagan.[1] And now, with Kagan busily preparing for her Senate confirmation hearings, Katyal was placed in charge—acting solicitor general of the United States of America.

Katyal knew that his new role would be different, but he never thought the difference would manifest itself so quickly. On the day he was promoted, while still packing up his old office on the fifth floor of the sprawling Department of Justice building, Katyal got a phone call. It was Larry Summers, director of the National Economic Council. Summers was a larger-than-life figure—former president of Harvard University, treasury secretary under Clinton, chief economist of the World Bank. He had a reputation for being both brilliant and abrasive. On the line with Summers was Cameron (Cam) Kerry, John Kerry's younger brother,

---

[1] Of the four Deputy Solicitors General, three are career Department of Justice employees and one—the Principal Deputy SG—is a political appointee.

Neal Katyal, former acting solicitor general of the United States,
appearing on The Late Show with Stephen Colbert.

now serving as general counsel of the Department of Commerce. What these two wanted, Katyal couldn't begin to guess.

After a perfunctory introduction and congratulations, Summers cut right to the chase. There was a case, he said, that Elena Kagan had been monitoring. It was important. Now that Kagan was gone, it would be part of Katyal's portfolio.

"What case?" Katyal asked, mentally reviewing the host of financial meltdowns, government bailouts, and Bush-Cheney atrocities that were making their way through the courts. He wasn't even close.

Last month, Summers explained, a federal district judge in New York invalidated a biotech company's patents. The patents covered a genetic test for breast cancer. The case had been brought by the ACLU—they said that the patents prevented women from getting tested. Now the ACLU wanted to abolish gene patents altogether.

The Patent Office was up in arms. A multi-billion dollar industry was at stake, they said, and the SG had to decide whether the government was going to intervene. It was still early, and the company hadn't filed its appeal yet. But Kagan knew it was coming. She had already circulated a memo soliciting feedback from the heads of governmental agencies that might have an interest in the case. And, Summers added, it was very likely that not everyone would agree. In fact, some groups within the administration seemed to be sympathetic to the ACLU. Summers, for his part, was skeptical of the patents. What, exactly, had been invented? But clearly the Patent Office disagreed. So, he concluded, it would be a brawl. Welcome to the job.

Katyal listened as Summers spoke. The case must be important if Kagan had initiated an interagency process before an appeal was formally filed. But the issues it raised reverberated across the political spectrum. Healthcare was on everyone's mind. Less than a month ago, President Obama had signed the Affordable Care Act—Obamacare—a massive political compromise that was intended to extend health coverage to millions of uninsured Americans. And one of the president's rallying cries, both on the campaign trail and during his first year in office, was that his mother had died from ovarian cancer at the age of fifty-three, arguing until the end with her insurance company. And who could forget that in November, just days before the election, then-candidate Obama's grandmother had tragically been taken by breast cancer. So, yes, insurance coverage and cancer were squarely in the sights of the administration. Katyal would need to educate himself on these issues, and fast.

All the President's Lawyers

WHEN PEOPLE THINK of the executive branch of government, they usually envision the White House and the then-sitting president. It is true that the president sits atop the executive branch, co-equal with the legislative and judicial branches of government, and subject to a delicate machinery of checks and balances that is constitutionally designed to preserve the equilibrium of our democratic state. But in reality, the executive branch today would be virtually unrecognizable to the founding fathers. It is a vast bureaucratic apparatus sometimes referred to as the administrative state. It is comprised of hundreds of different agencies, bureaus, offices, and commissions, and they affect nearly every aspect of modern life—food inspection, housing, education, worker safety, student loans, power generation, national parks, air pollution, interstate highways, airport screenings, drug trafficking, international trade, public health, medical research, the military, consumer protection, and social security, to name just a few.

The United States Department of Justice (DOJ) is itself a massive, sprawling agency with over two hundred thousand personnel and offices in every state. The head of the DOJ is the attorney general of the United States (the AG), a powerful political appointee. Next in the hierarchy are the deputy AG, the associate AG, and the solicitor general. These are all political appointees subject to Senate confirmation, though usually less in the public eye than the AG. The DOJ has seven major divisions, each led by an assistant AG, as well as a number of subsidiary agencies like the FBI, DEA, and Bureau of Prisons. The Civil Division of the DOJ handles all non-criminal litigation involving the U.S. government and its hundreds of agencies, other than in a few specialized areas like tax, environmental, and antitrust law, each of which has its own division within the DOJ.

Most of this litigation is pursued by one of ninety-four local and regional U.S. Attorney's offices across the country. The attorneys in these offices appear regularly in court to defend the government in cases ranging from injuries at national parks to disputes over procurement contracts to allegations of harassment by federal employees. The Civil Division of the DOJ also includes a group of sixty or so attorneys who focus on appeals of cases involving the government. This elite appellate staff is based at DOJ headquarters, commonly known as "Main Justice," in Washington, DC. In addition to appeals of government cases, the appellate staff files *amicus* briefs in cases in which the government is not a direct party, but has an interest. The ultimate decision whether to appeal a case, or to file an *amicus* brief, rests with the solicitor general, but the case is first analyzed and a recommendation made by the Civil Division's appellate staff.

When Judge Sweet in New York ruled against Myriad, the Patent Office was dismissed from the suit. But it wasn't dismissed the way it had *wanted* to be dismissed—by winning the case. The attorneys from the Patent Office and the U.S. Attorney's Office in New York had argued that the court should uphold Myriad's patents. But, to their surprise, Judge Sweet ruled that Myriad's patents were *invalid*, and then dismissed the Patent Office from the case because he didn't need to address the ACLU's constitutional arguments. That left the Patent Office out of the picture; it could not defend the patents on appeal.

Unless, that is, the SG authorized the government to file an *amicus* brief. Then the government would become a friend of the court like the American Medical Association, Greenpeace, or any other member of the public. It wasn't the same as being a party to the suit, but courts generally give more weight to their "friends" in the government than to other interveners. The decision whether to file an *amicus* brief was the SG's alone. No matter how badly an agency wanted to speak its mind in court, it could do so only if authorized by the SG.

So, within days after Judge Sweet's ruling, officials from the Patent Office and its parent agency, the Department of Commerce, suggested to the solicitor general—then Elena Kagan—that the United States should consider filing an *amicus* brief defending the Patent Office's gene patents. That suggestion initiated an interagency process within the DOJ, and the person selected to coordinate it was a young attorney on the appellate staff named Mark Freeman.

Freeman was everything that a career DOJ attorney should be—smart, dedicated, and committed to playing everything by the book. He even resembled a quintessential government agent: clean cut, square jawed, resolute—an Eliot Ness for patents. Though Freeman came from a family of scientists, including a father who was a nuclear physicist, Freeman did not have a technical degree himself. As a student he had always been more interested in languages and culture. After earning a degree in East Asian Studies from Harvard, Freeman landed a job at Panasonic in Japan. He began as a glorified translator, but within three years Freeman was negotiating major patent licensing deals with companies like Apple, Microsoft, and Texas Instruments.

When he came back to the United States, Freeman returned to Harvard to get his law degree and then entered the prestigious Department of Justice honors program. Now, just six years later, he was a staff attorney in the appellate staff, responsible for defending the United States in cases across the country. And because of his work experience and general comfort with science—an attribute shared by few of his colleagues—the rest of the appellate staff was happy to let him sink his teeth into most of the patent cases that came along. As a result, Freeman, at a relatively junior level, became responsible for much of the government's brief in the closely watched case involving Bernard Bilski's energy trading patents, which the Supreme Court was likely to decide any day now.[2]

When Solicitor General Kagan initiated the interagency process for *AMP v. Myriad*, Freeman was tasked with two principal responsibilities. First, he would produce a draft of the Civil Division's recommendation to the solicitor general. Once drafted, this recommendation would be sent up the chain of command—first to a supervising attorney on the appellate staff, then to the deputy assistant AG in charge of the appellate staff, and then to the assistant AG in charge of the entire Civil Division. That individual would ultimately sign the recommendation and submit it to the solicitor general's office. While this system may sound painfully bureaucratic, it is designed to enable the government to arrive at a position that has been thoroughly vetted through multiple levels of accountability, and which eventually has the blessing of a high-ranking political appointee.

The second task assigned to Freeman was coordinating input from other governmental agencies that were interested in the case. This activity often involves fielding questions and organizing meetings, conference calls, and endless email threads to discuss the case. In other words, herding cats. Once other agencies have submitted their input and recommendations to the Civil Division, the line attorney bundles them together with the Civil Division's own recommendation and submits the entire package to the Office of the Solicitor General, where it undergoes a similar climb up the hierarchical ladder until reaching the desk of the solicitor general, who decides what the government's position will be.

---

2 Though the Civil Division typically prepares the first draft of briefs submitted by the United States government to the Supreme Court, the solicitor general's office has ultimate authority to argue before the Court. Thus, briefs may be revised by attorneys in the SG's office before filing, and oral argument is handled either by the SG or one of the four deputy SGs. The government's case in *Bilski* was argued by Malcolm Stewart, a deputy SG.

Herding Cats

JUDGE SWEET RELEASED his opinion on March 29. On April 8, after getting his instructions from Elena Kagan, Freeman sent an email to representatives of the agencies and offices that he thought would be interested in the gene patenting issue. These included, in addition to the Patent Office, the NIH, Department of Commerce, U.S. Trade Representative, International Trade Commission, and Federal Trade Commission. The question was whether the government should file an *amicus* brief defending the Patent Office's issuance of gene patents. They had until the second week of May, just before Mother's Day, to respond.

At least a dozen Patent Office officials got involved in preparing a recommendation to DOJ. These included Ray Chen, the Patent Office solicitor, and David Kappos, its director, together with numerous biotechnology and policy specialists. The Patent Office and Department of Commerce submitted a joint memo that recommended defending gene patents. The U.S. Attorney's office in New York concurred.

But as Summers had hinted to Neal Katyal, there was a contrary view as well. This view was outlined in a detailed memo from NIH, which described the probable effects of gene patents on "the future of medicine." These effects were dire, and NIH recommended making no filing. In other words, NIH wanted Judge Sweet's opinion to stand.

When they heard about the NIH memo, the team at the Patent Office, from Director Kappos to the most junior policy analyst, was stunned. They couldn't remember a time when another agency within the administration had formally opposed the Patent Office on what was clearly a matter of patent policy—a policy that had been in effect for nearly three decades.

Freeman knew that the NIH memo was coming, but hadn't tipped off anyone at the Patent Office. Attorneys on the appellate staff viewed themselves as something like priests—when officials from an agency shared their innermost thoughts, fears, and aspirations, the DOJ attorneys held their confidences as though protected by the seal of the confessional.

NIH had been communicating with Freeman since the beginning. Within days after he first reached out to the agencies, Freeman received a call from Kathy Hudson, chief of staff and right hand to NIH's director, Francis Collins. Hudson was familiar with both the Myriad patents and the ACLU lawsuit. She had first met Tania Simoncelli at the Center for Genetics and Society at Berkeley. At that time, Hudson was working on genetics policy at Johns Hopkins University. When Simoncelli moved to the ACLU and began to consider a lawsuit challenging gene patents, she continued to consult Hudson.

Now at NIH, Hudson relished the opportunity to get back into the game. She told Freeman that Collins wanted to arrange a meeting. Perhaps he and others at NIH could help the DOJ with the scientific nuances of the case. Freeman welcomed the help. Through these meetings, he learned that NIH was quietly assembling a coalition of agencies that thought human genes should *not* be patented.

NIH's engagement in the case was far from typical—it was not an agency that generally involved itself in government litigation, and certainly not patent cases. But *Myriad* was perceived not just as a patent case, but as a case implicating biomedical research, public health, and women's rights. As such, Collins, an opponent of gene patents since his days at Michigan, could not help but get involved.

As it turns out, NIH was not the only governmental body interested in upholding Judge Sweet's decision. The appellate staff attorneys began to receive calls from wholly unexpected quarters. At the center of the hundreds of different agencies, commissions, offices, and bureaus within the executive branch lies one office with undeniable influence and power—the White House.

The White House today represents more than the president, the First Lady, and the Oval Office. On this highest plateau of the executive branch—formally known as the Executive Office of the President—is nestled a little-known but large and powerful bureaucracy.

One of the lesser-known White House organizations is the Office of Science and Technology Policy—OSTP. Depending on the president's interest in science issues, the size and influence of OSTP has waxed and waned over the years. Under President Obama, a self-declared science nerd, OSTP grew to a staff of 135, its largest ever.[3] OTSP's director, John Holdren, was a renowned Harvard environmental scientist who also served as one of the president's top science advisors. With Holdren's support, OSTP participated in the growing interagency discussion coordinated by Freeman.

OSTP was not the only White House office that had an interest in the case. Before being appointed special advisor to the director of the Office of Management and Budget (OMB), Ezekiel "Zeke" Emanuel, older brother to White House chief of staff Rahm Emanuel, was a breast oncologist and the chief bioethicist at NIH's Clinical Center. Emanuel had a dim view of the U.S. healthcare system, calling it in 2007 "a dysfunctional mess." His job at OMB had largely been to develop and advance the president's healthcare reform agenda. Given his background in breast cancer and bioethics, Emanuel was well aware of Myriad and its patents. Many of his own patients had carried *BRCA* mutations, and he had advised them regarding both testing and measures they could take to reduce their risk of cancer. Emanuel soon became an active participant in the gene patenting debate within the administration.

In some respects, it is not surprising that science-focused offices and agencies were interested in *AMP v. Myriad*. But help came from a third, unexpected, quarter of the administration as well—economists. Under the best of circumstances, Washington is crawling with economists—they weigh in on questions ranging from the minimum wage to the prime lending rate, from fiscal policy to import tariffs. The Obama administration, which had the misfortune to enter the White House in the midst of the worst financial crisis since the Great Depression, had an even greater than normal need for economic advice, and economists practically flooded the capital.

Naturally, the Department of the Treasury, the Internal Revenue Service, the Department of Commerce and the Federal Reserve all have

---

3 I have been told by a reliable source that Jo Handelsman, the associate director for science at OSTP, brought President Obama a stuffed plush microbe each time she met with the president in the Oval Office, and that by the end of his second term, an entire drawer of the Resolute desk was overflowing with these fist-sized microorganisms.

huge staffs of professional economists. Dozens of other agencies that are not primarily charged with financial matters also have economists on the payroll. These include the Department of Justice and the Federal Trade Commission, which conduct heavy-duty economic analysis in antitrust and merger cases. Many economists also work within different White House entities, including the OMB, which develops the administration's massive budget, the Council of Economic Advisors, which has been referred to as "the president's personal stable of eggheaded economic analysts," and the National Economic Council, or NEC. The NEC is a little-known but powerful advisory body that includes the secretaries of state, labor, and commerce, plus other high-ranking administration officials. President Clinton created the NEC as an economic counterpart to the National Security Council—a sort of joint chiefs of the purse. Historically, the head of the NEC has been one of the president's closest economic advisors.

During the early years of the Obama administration, Larry Summers, the outspoken but brilliant Harvard economist, led the NEC. Summers, himself a cancer survivor, helped craft the massive governmental bailout that steered the nation clear of financial collapse. Treasury Secretary Timothy Geithner, no lightweight himself, said of Summers, "Few economists can claim as big an imprint on American history." If any economist had the president's ear, it was Summers. And within NEC, Summers and his staff made *AMP v. Myriad* a priority.

The Antitrust Division of the DOJ also took an interest in the case. Why? Antitrust law is particularly concerned with monopolies. But not all monopolies are illegal. Suppose that you discover the world's last salt deposit in your backyard. You will have a natural monopoly over the sale of salt, but you did nothing wrong to acquire this monopoly. Thus, you are not violating the antitrust laws. If, on the other hand, you already control 80 percent of the world's salt production and prohibit your customers from buying salt from anyone else, thus driving the remaining competition out of business, you would potentially have an antitrust problem. Patents, by their nature, give their owners exclusive rights to make, use, and sell inventions. These exclusive rights are like mini-monopolies, but they are legal ones. Nobody was accusing Myriad

Genetics of using its patents to break the antitrust laws, even though it had locked up the market for *BRCA* testing since the late 1990s.

Nevertheless, a growing number of economists, including Nobel laureate Joseph Stiglitz, had begun to ask whether the powerful exclusive rights afforded by patents were actually good for the economy. They were particularly concerned about innovation—did patents like Myriad's inordinately limit innovation and market entry by other firms, raise consumer prices, and otherwise stifle competition? Traditionally, the answers to these questions had been a resounding no, as the exclusivity conferred by patents was viewed as a tradeoff that was required to encourage firms to invest in new discoveries.

But that orthodoxy was increasingly being questioned by a new breed of economists who studied industrial organization, innovation, and law. Stiglitz, who had served as chairman of President Clinton's Council of Economic Advisors and then as chief economist of the World Bank, had been recruited to the ACLU's cause early in the case. His written declaration to the district court explained that "as a society we tolerate some monopoly power and some restrictions on the use of knowledge in the belief that they might spur innovation. But the social costs of these distortions and inefficiencies can outweigh the benefits." Stiglitz expressed particular concern for patents on human genes, including *BRCA1* and *BRCA2*, writing that these "provide an extreme example of the lack of balance that can occur in our patent system."

Other prominent economists, some of whom were then serving in the Obama administration, followed suit. One of these was Carl Shapiro, a professor of economics at Berkeley and the new deputy assistant AG for Economics in the DOJ's Antitrust Division. Shapiro, together with Christine Varney, the head of the Antitrust Division, shared a concern about the effect of patents on scientific research and innovation. They saw in *AMP v. Myriad* an opportunity to join a policy reform effort within the administration.

The Patent Office did not sit idly by while its sister agencies conspired against it. It formed its own coalition of agencies to support patents covering genetic materials. The Department of Agriculture, which was active in plant genetic research, was one of these. The Department of

Energy, whose interest in genetics went all the way back to the effects of atom bomb radiation on human cells, was also favorably inclined toward the Patent Office's position. The U.S. Trade Representative, which pushed the United States' aggressive intellectual property agenda around the world, also seemed to line up on the side of the Patent Office.

The growing number of governmental offices seeking to weigh in on *AMP v. Myriad*, not to mention the growing divide between the camps led, respectively, by the Patent Office and NIH, presented challenges to Freeman, who was simultaneously trying to manage the process and develop the DOJ's recommendation to the solicitor general. The result was meetings. Lots of meetings. And, of course, emails.

Between June and September of 2010, Freeman convened a series of meetings at Main Justice. The attendees usually included several representatives from the Patent Office, and one or more from each of NIH, OSTP, and NEC, as well as one or two higher-ranking attorneys from the appellate staff. But people from other agencies showed up as well. One attendee found some of these meetings so crowded, he joked that people were hanging from the chandeliers. And everyone seemed to have consulted Hans Sauer at BIO, who happily assisted however he could, and even attended some of the meetings.

On some days, it seemed that everyone in Washington wanted an update from Freeman, and he could have spent his entire day sending and responding to emails. The list of governmental entities that were involved, directly or via email, in the discussions of the case included:

Department of Justice—Office of the Solicitor General
Department of Justice—Civil Division
Department of Justice—Antitrust Division
Department of Justice—U.S. Attorney for the Southern District of
    New York
Department of Agriculture
Department of Commerce
Department of Defense
Department of Energy
Department of Health and Human Services (HHS)

Federal Trade Commission
U.S. Trade Representative
National Institutes of Health (NIH)
Centers for Disease Control (CDC)
Patent Office
U.S. Agency for International Development (USAID)
and
Executive Office of the President (White House), including:
  National Economic Council (NEC)
  Office of Management and Budget (OMB)
  White House Counsel
  Office of Science and Technology Policy (OSTP)
  Office of Energy and Climate Change Policy
  Council of Economic Advisors

Interestingly, the Food and Drug Administration (FDA), which is directly involved in issues relating to diagnostic tests and patents, does not appear on this list. It is possibly because Tania Simoncelli, who moved to the FDA after leaving the ACLU in 2010, scrupulously tried to avoid any appearance of a conflict of interest in her new job.

The final group that was involved in the formulation of the government's case was the ACLU. When evaluating complex cases, the Department of Justice sometimes asks the parties to clarify certain facts or arguments that were made in the court below. Accordingly, in June 2010, Freeman invited Hansen's team to Main Justice.

A Visit to Main Justice

MAIN JUSTICE IS formally known as the Robert F. Kennedy Department of Justice Building. From the exterior, the building, which occupies a full city block between Pennsylvania and Constitution Avenues, looks like any of the other massive federal structures that surround the National Mall. Its monumental proportions, colossal Ionic columns, and gray limestone façade evoke an imperial air, but on a scale that dwarfs even the grandest structures of ancient Rome. The burnished aluminum portals to Main Justice are twenty feet high, embossed with a pair of lions *sejant* representing watchfulness and strength. One cannot help but be humbled, and perhaps a little intimidated, when passing through them for the first time.

But once inside the imposing building, the severe gray façade gives way to an unexpected rush of color and light. Main Justice, completed in 1935, represents the culmination of a massive WPA decorative arts program. Its ten thousand art deco lighting fixtures were manufactured by the same company that outfitted the Waldorf Astoria Hotel in New York. Its five marble staircases spiral gracefully between seven spacious levels. Its high coffered ceilings are painted in bright shades of blue, green, and red interspersed with geometrical deco designs. But the most striking features of Main Justice are its murals. Rivaling in scale and color the masterworks of Rivera, Orozco, and Siqueiros, the murals of Main Justice depict a stunning array of scenes from American history and justice, some of which the *Washington Post* ranked "among the finest mural paintings of this country."

Hansen, Park, and Ravicher took the train to DC for the meeting at Main Justice. They met with seven people in a small conference room in the Civil Division: Mark Freeman; Scott McIntosh, the supervising attorney of the appellate staff; Beth Brinkman, head of the appellate staff; and three Patent Office attorneys who neither introduced themselves nor spoke during the meeting. The seventh government attendee was Jessica Palmer, a summer intern from Harvard Law School who held a PhD in molecular and cellular biology from Berkeley. Palmer was interested in science policy and had spent two years before law school on a fellowship at NIH. Freeman could not have asked for a more capable assistant on this complex case.

Much of the hour-long meeting, at which the ten attendees were crowded around a table designed for eight, consisted of Freeman and Palmer asking scientific questions to the ACLU team. Most of these were fielded by Park. When the meeting was over and Freeman thanked them for attending, none of the ACLU group really understood what had just happened. They assumed that the DOJ was planning to submit a brief supporting the Patent Office and the gene patents that it had issued. They were wrong.

CHAPTER 18

Splitting the Baby

As THE HOT, muggy DC summer wore on and the deadline for submitting a recommendation to the solicitor general approached, Freeman was still nowhere. He began to wonder whether there was any way to split the baby—to find a solution that would satisfy each side, at least partially.

The question whether an isolated and purified human gene is patentable hinges on whether it is more accurately described as a product of nature (not patentable) or a man-made substance (patentable). As Freeman pored over Myriad's patent claims with Palmer and half a dozen technical specialists from the Patent Office, NIH, and OSTP, they noticed a subtle difference. Some, like Claim 1 of Patent 5,747,282, covered isolated DNA that encoded a particular protein—the tumor suppressor produced by the *BRCA1* gene. Any properly functioning *BRCA1* gene would produce this protein and thus be covered by Claim 1.

Claim 2, on the other hand, covered isolated DNA that had a particular nucleotide sequence—a string of nearly eight thousand As, Ts, Gs, and Cs that spread across a full six pages of the patent document. This was the sequence of a "normal" *BRCA1* gene.[1] But there was a catch. These eight thousand bases were just the coding regions of the *BRCA1* gene. That is, this exact sequence of bases would encode the tumor-suppressing protein described in Claim 1. But the *BRCA1* gene itself, as it exists in the body, is much, much longer. It spans more than 110,000 bases. Of these, only 8,000 are needed to make or "code" the tumor suppressor

---

1 It isn't until Claim 8 that the three major cancer-causing variants of *BRCA1* make their appearance.

protein. These coding bases are collected in twenty-two groups called *exons*. The other 102,000 bases in the gene are called introns, and we don't really know what they do. If all we care about is encoding a protein, then we only have to worry about the exons. For that reason, Claim 2 of the '282 patent only covers the 8,000 coding bases in the exons of *BRCA1*.[2] But in a human cell, the exons are spread out across the full length of the gene. There could be 300 coding bases followed by 5,600 non-coding bases, then another 180 coding bases, and so on down the line. A DNA segment containing exons alone does not occur within the human body— the exons are always interspersed among the much larger introns.

To encode a protein, the exons in a gene need to be gathered together into a single contiguous strand, and this is done with the help of a single-stranded molecule called RNA, which is similar, but not identical, to DNA. So-called messenger RNA (mRNA) attaches to a strand of DNA and replicates it, then splices out the introns so that only the coding exons remain. But RNA is not DNA,[3] so Claim 2 of the patent, which covers "isolated DNA," would not cover an mRNA molecule even if had the same 8,000 bases listed in the patent.

What the claim *does* cover is a type of molecule called complementary DNA, or cDNA. Essentially, once a strand of mRNA has replicated the coding region of a gene, scientists can walk the process backward to construct a strand of DNA from that mRNA. The resulting cDNA is single-stranded and contains only the exons from the original gene. Thus, a cDNA molecule constructed from the *BRCA1* gene would contain only the 8,000 coding bases from the full 110,000-base gene. cDNAs are useful in the mass production of synthetic genes such as insulin and human growth hormone.

And here's the important part: cDNA only occurs in the lab. It does not exist in human cells. It has to be constructed using mRNA to reverse-engineer a single-strand of DNA that only includes the protein-coding regions of the original gene. It is not a product of nature. It is

2 The *BRCA2* gene is approximately 85,000 bases long, with 27 exons that include around 11,000 coding bases.

3 RNA is similar to DNA, except that it has only one strand instead of two and in it the base thymine (T) is replaced by uracil (U).

made by humans. Ergo—patentable. As it turns out, patents claiming laboratory-synthesized cDNA molecules had been around since the early 1980s, years before the first patents claiming entire genes.

This distinction between naturally occurring "genomic" DNA, or gDNA, and lab-created cDNA was just the compromise that Freeman needed. If the Federal Circuit agreed, it would result in some of Myriad's claims being upheld (cDNA) and some being invalidated (gDNA). Each side would get something that it wanted, though the outcome would doubtless please the ACLU far more than Myriad. After all, cDNA was not used in the diagnostic process; Myriad did not make cDNA constructs, nor did it have any use for them. Its cDNA patent claims were likely thrown in by the lawyers for good measure, on the off chance that Myriad might get the opportunity to license them to another company someday.

Of course, as with all good compromises, nobody in the government was entirely happy with the gDNA-cDNA distinction, least of all the Patent Office. In the view of Ray Chen and his staff, the "isolated" DNA that Myriad claimed just wasn't the same as the DNA within a person's cells. It was processed in the lab and became something very different from what existed in nature. Such a compromise would destroy decades of Patent Office precedent and endanger thousands of valuable patents. But Collins and the other scientists at NIH were also worried. To them, the coding sequence of cDNA also appeared in gDNA, so why distinguish between the two at all? If you considered the information-carrying properties of DNA, as Judge Sweet did, then there was no reason to treat cDNA any differently from gDNA. In other words, neither should be patentable.

Despite uneasiness at both the Patent Office and NIH, the White House contingent thought the cDNA compromise had legs. So if they could officially get the White House to back the proposal, it might fly, even in the face of the Patent Office's opposition. But getting a consensus view from dozens of offices within the White House, most of which had no background in science or patents, would not be easy.

As October approached, the NEC and OSTP staff planned a meeting. It would be for White House decision makers only—nobody from the

DOJ, the Patent Office, NIH, or other agencies would be invited. This summit was scheduled for the evening of Thursday, October 14. The deadline by which the solicitor general had to file an *amicus* brief in the case was October 29. A decision had to be made that night. Time had run out.

La Casa Rosada

JIM KOHLENBERGER, CHIEF of Staff at OSTP, walked to the meeting from his office in the red brick New Executive Office Building a couple of blocks from the White House. It was a cool evening and the sun had just fallen below the line of office high-rises to the west. There had been a heavy rain that afternoon, and Kohlenberger stepped gingerly to avoid the many puddles on the sidewalks and streets of the capital.

As he crossed the pedestrianized section of Pennsylvania Avenue and looked across the White House lawn, he was surprised to see a buzz of activity beyond the wrought iron fence that separated the White House from the demonstrators and tourists that usually crowded the area. Dozens of workers in hardhats were installing a metallic framework of some kind around the base of the building. Kohlenberger paused for a moment to get a better look. The sun had now set and a cloudy, gray dusk had fallen across the city. A minute later, one of the workers threw a switch in a steel equipment cage crisscrossed with electrical cables.

And then, just like that, the White House became pink. All around its perimeter, floodlights burst into life and bathed the famous building in a soft, carnation-hued glow. Then Kohlenberger saw the ribbons—pink ribbons—everywhere, on banners fluttering beneath street lights and office windows, even pinned to the uniforms of the guards stationed at the west entrance. He remembered that it was Breast Cancer Awareness Month, and that the White House would be commemorating it tonight, transforming itself into a luminescent symbol of solidarity with the millions of Americans who had been touched by the disease. He smiled and walked through the security checkpoint.

They met in the Roosevelt Room, located across a wide, carpeted hallway from the Oval Office itself. Teddy Roosevelt, astride a rearing stallion and decked out in full Rough Rider gear, stared down at them. Every seat was filled at the long walnut table. The attendees represented the full range of White House offices and organizations, all with policy-making authority. Some of these groups, like the Council of Economic Advisors, had been around since the forties, while others, like the Office of Energy and Climate Change Policy, were brand new. But whether the offices were old or new, the faces around the big table reflected the youth and diversity that had propelled Obama into office.

Once everyone was settled, Kohlenberger kicked off the discussion of whether the White House should back NIH and oppose gene patents. His comments loosely ranged across territory from Biology 101 to health policy, and then back again. For some, the link to breast cancer—the pink lights outside clearly playing their part—decided the question hands down. For others, the decisive factor was economics—wouldn't a patent on every gene keep the price of this technology beyond the reach of many? Zeke Emanuel and other strong voices supported Kohlenberger, while others worried about the effect of eliminating gene patents on medical innovation. After an hour of intense conversation, they were done. The White House would go along with the DOJ's compromise proposal. NEC— probably Summers himself—would let the solicitor general know.

Making the Call

AFTER THE UNEXPECTED call from Summers and Kerry on his first day as acting SG, Neal Katyal had kept a watchful eye on *AMP v. Myriad*. So far, the principal coordination effort was being ably handled by the Civil Division. So during the summer and early fall of 2010, Katyal focused on the many other cases that required his attention.

But by early October, the center of gravity in the case had shifted from the Civil Division to the Office of the Solicitor General. The assistant SG who had been assigned to the case, Ginger Anders, received a stack of memoranda making recommendations whether or not to file an *amicus* brief. In addition to the Civil Division, a number of other agencies had weighed in, including the DOJ's Antitrust Division, the Patent Office, NIH, USDA, and the U.S. Trade Representative—an unusual assortment if ever there was one. Anders would evaluate all of this material, then submit it along with her own recommendation to Malcolm Stewart, the assigned Deputy SG.

Stewart, a career government attorney who had been in the SG's office since 1993, was smart and very careful. He also had experience with patent cases. Stewart had argued on behalf of the government—the Patent Office—at the Supreme Court last year in *Bilski*. And in June, on the last day of the Court's 2009 term, the Supreme Court had sided with him.

But even with support from a top deputy, Katyal needed to understand the issues himself. It was his duty as acting solicitor general. And, fortunately, he had access to some of the best tutors in the country.

In the area of genetics, he received a crash course from Francis Collins and other senior scientists at NIH. Every Monday evening, Katyal would walk from Main Justice to the Metro Center Red Line station and take the train north to NIH's Bethesda campus. There, he got a personal tutorial in genetics from the man who had led the Human Genome Project. It was an amazing education, one that Katyal would never forget.

As for innovation economics, technology markets, and competition, Katyal's private tutor was the director of the National Economic Council, Larry Summers. At Harvard, Summers's graduate seminars were heavily over-subscribed. Now, Katyal had the preeminent economist—one of the youngest people ever to get tenure at Harvard—all to himself. Sitting in Summers's office, going over the impact of patents on product markets and healthcare, often with other NEC analysts and officials in attendance, the instructors outnumbered the student in this unique and high-stakes colloquium.

When Katyal received a call from Summers after the "pink lights" meeting at the White House, he knew that the time to make a decision was at hand. By Friday, October 15, Katyal's staff had wrapped up their final meetings with the Patent Office and submitted their recommendation to Katyal, along with the now-weighty pile of memoranda from across the administration. Katyal took the package home and spent the weekend studying it. On Monday, he met with his small team. He slept on it, and on Tuesday morning he carefully drafted an email to thirty-nine different officials across the administration. As he pressed SEND, he knew that he was making history.

Sent: Tue Oct 19 12:20:07 2010
Subject: Myriad

I am deeply appreciative of all the hard work each of you, and countless others, have put into this issue. The debate has been deeply interesting and done with superb skill on all sides. I wanted to let you all know that I am leaning toward a brief being filed next week, on behalf of neither party, that would argue that isolated human DNA is not patentable as a product of nature, but that cDNA and associated other inventions that create products not found in nature are patent eligible under [Section] 101. I have asked [the Civil Division] to prepare a draft brief along those lines, and my aim is to have it circulated to you late Friday or Saturday for your review. Again, I am grateful for the intensity and good faith each of you brought to our discussions on this tough case.

Neal

That was it. The decision had been made. Katyal would go with Freeman's Solomonic compromise: complementary DNA was patentable, genomic DNA was not. Everybody got something. Nobody would be entirely happy.

Katyal also made the difficult decision not to allow the Patent Office to voice a dissenting view. In other controversial cases involving a split among federal agencies, the solicitor general sometimes permitted an agency in the minority to submit its views in an appendix to the government's main brief. But that would not happen here. Katyal had listened to the evidence and the arguments. The government's position was set.

Now the ball was back in Freeman's court. It was up to him to draft the government's amicus brief. It was Tuesday. Katyal had promised a draft for the entire administration to review by Saturday. Taking into account the need for internal approvals, Freeman had no more than two days to draft the brief that would soon be filed with the Federal Circuit.

As Katyal expected, his note was not received warmly by the Patent Office. David Kappos, its director, seemed to be particularly incensed. Before joining the government, Kappos had led the intellectual property department at IBM, overseeing one of the largest patent portfolios on the planet. In the private sector, Katyal's behavior would have been unthinkable. The solicitor general was supposed to be the Patent Office's lawyer. Shouldn't the SG represent the position that the Patent Office had spent hundreds, if not thousands, of hours developing over the last six months? A position that also represented nearly three decades of Patent Office practice and policy, which had been reviewed multiple times by the courts and never challenged until now. And how could the SG not only ignore the Patent Office's position, but adopt an opposing view? A position advanced by the people who had sued his agency in the first place? The word traitorous must have crossed Kappos's mind more than once. And the compromise that Katyal had offered? Ridiculous. Of course cDNA was patentable, but so was isolated genomic DNA. They couldn't give that up.

Two days after receiving Katyal's note, Kappos delivered a keynote address at the Annual Meeting of the American Intellectual Property Law Association in DC. Speaking before a packed lunchtime audience, Kappos hinted that the Obama administration might intervene in *AMP v. Myriad* in a manner inconsistent with the Patent Office's longstanding policies on patentability. The audience was stunned. What was going on in Washington?

At 11:26 am on Friday morning, Katyal sent an email to Cam Kerry at the Department of Commerce, copying Kappos, Chen, and a few others. Twelve minutes later, he distributed Freeman's draft brief, produced in record time, to the entire administration working group. Katyal asked for responses by the following Tuesday.

The brief was a scientific tour de force, filled with examples of natural substances that, while difficult to isolate, could not possibly be patented. It explained that the element lithium naturally occurs only in combination with other elements (lithium chloride, lithium hydroxide, etc.) and was not isolated in its pure metallic form—a significant feat of ingenuity—until 1818. Would anyone argue that the third element in the Periodic Table, when isolated, should be patentable as a new composition of matter?[4] And what about the humble electron? Though electrons were theorized since the nineteenth century, one was not isolated until an ingenious ion trap was invented in the 1970s. But did Hans Dehmelt, who won the 1989 Nobel Prize for this accomplishment, invent "the elemental unit of negative charge in the universe"? The very notion was absurd. So it should be for DNA, the brief argued. If it was found in nature, then no matter how hard it was to isolate, it should not be patentable. It was a true product of nature.

The legal and policy teams at the Patent Office worked over the weekend to review the brief. By then, they knew that it was pointless to push back against the basic position that genomic DNA was not patentable. So they did what they could and submitted a few minor clarifications and corrections by the Tuesday deadline. But, perhaps more importantly, they developed a set of talking points. They knew that once the press, and the ever-watchful bloggers, got their hands on the government's brief, the Patent Office would be besieged by questions, if not outright anger.

In the talking points, the Patent Office remained defiant, making it clear that that it had "no plans to change its policy, its examination procedures, or its guidance to examiners," and definitively declaring, "We issued patents to isolated genomic DNA molecules as recently as last Tuesday, and we will continue to do so until binding Federal Circuit precedent is changed." Take that, Mr. Katyal.

---

4 It may be more than coincidence that Freeman's father, a professor of physics, conducted his doctoral research at Harvard on the spectral characteristics of lithium compounds. Two elements of the periodic table have been patented, americium (no. 95) and curium (no. 96), both in 1964 by Glenn T. Seaborg, the Nobel laureate after whom element no. 106 (seaborgium) is named. Unlike lithium, however, americium and curium are synthetic elements that do not exist naturally.

You're Bob Sweet?

ON THURSDAY EVENING, there was a cocktail reception at Rockefeller University in a large meeting room overlooking Roosevelt Island on New York's Upper East Side. It was a prelude to a symposium that would be held the next day—part of the university's centennial celebration. The keynote speaker was Eric Lander, one of the leading figures in the Human Genome Project. Lander now directed the Broad Institute, a unique joint venture between Harvard and MIT. Lander was also one of President Obama's top science advisors. Along with John Holdren, the head of OSTP, Lander co-chaired the President's Council of Advisors on Science and Technology—PCAST (reputed to be President Obama's "favorite" committee).

But Lander, a Brooklyn native, was best known in New York legal circles for work he had done in one of the first criminal trials to use DNA fingerprinting evidence. Though Lander was a mathematician by training and a geneticist by avocation, he always had a strong interest in the law.[5] His 1989 expert testimony in a Bronx courtroom led to the exoneration of a criminal suspect who had been prosecuted on the basis of a blood spot on his watchband. Lander showed that it wasn't the victim's blood. The experience led the lawyers that Lander had worked with, Peter Neufeld and Barry Scheck, to form a non-profit organization called the Innocence Project. The Project, based at Cardozo School of Law, has now exonerated more than three-hundred criminal defendants using DNA evidence, and Lander's place in the history of genetics law has been secured.

That evening's reception at Rockefeller included a number of university officials as well as local political figures and wealthy donors—required trappings at any university event. Lander, in his early fifties with a slight paunch, thick curly hair, and a bushy moustache, had an outgoing, friendly manner that made him both a popular TV guest and an effective Washington operator. In other words, he knew how to work a crowd. Lander circulated expertly, chatting with donors and university officials, a drink in one hand, a witty anecdote at the ready.

---

5 Lander reports that he met his wife in an undergraduate constitutional law class at Princeton.

At one point, Lander spotted an elderly, white-haired gentleman speaking with a group of senior researchers. Lander ambled over, assuming he was a major donor, maybe even a Rockefeller. As Lander inserted himself into the conversation, he glanced at the man's name tag: "Bob Sweet."

Lander, who had within the past couple of days been speaking with Francis Collins and the solicitor general about *AMP v. Myriad* was surprised. "You're Bob Sweet?" he asked the man.

The elderly gentleman, his silk pocket square obscured by the plastic name badge, looked up. "I am," he replied.

Lander smiled. "Are you the one that wrote—"

Sweet returned the smile, a twinkle in his eye. "Yeah, that's me." Sweet had always enjoyed public life and being recognized in a crowd. Starting his career as deputy to hizzoner, the mayor of New York, had helped in that regard. Sweet had shaken lots of hands and smoothed over innumerable difficulties in the Byzantine city government. But this evening he was at Rockefeller not on his own account, but as the guest of Adele, his wife of forty years. Adele Sweet, heiress to the *New York Post* fortune, ran the foundation endowed by her late mother, Dorothy Schiff, once reputed to be the most powerful woman in New York. Schiff's Foundation chose the advancement of science as one of its causes, and it generously supported Rockefeller and other scientific institutions across the city. Thus, over the years, the Sweets had become well-known fixtures at receptions and benefits for the elite scientific institutions of New York.

But even so, in the seven months since Judge Sweet had invalidated Myriad's gene patents, he had become something of a cause célèbre in scientific and medical circles. Now, people with PhDs in genetics and biochemistry wanted to shake his hand. And the judge, fondly recalling his days in City Hall, enjoyed being back in the limelight. Whether it was with union bosses or molecular biologists, he could put on the old "Sweet schmooze" that had once gained him notoriety throughout the five boroughs.

Lander nodded, sizing up the spry judge before him. "Well," Lander said, taking a sip from his glass. "You'll be interested in what happens tomorrow."

Sweet arched his thick white brows. He didn't know what Lander was talking about, but his City Hall instinct kicked in and he smiled anyway.

Battle Lines

TRUE TO HIS word, Neal Katyal filed the government's brief with the Federal Circuit on Friday. It was signed by Freeman and the other appellate staff attorneys up the chain of command. But in a sharp break with tradition, the entire legal staff of the Patent Office refused to sign the offending legal document.

The government's brief, and the fact that it contradicted longstanding Patent Office policy, did not go unnoticed in the blogosphere. Everyone understood the importance that such a brief could have on the proceedings to come. Court watcher Hal Wegner gleefully criticized Attorney General Eric Holder, who played no real role in the case, for "hijacking the patent system" and jibing that he must have flunked "Patents 101." The authors of the *Genomic Law Report* marveled: "Swine Soar Higher in Myriad Thanks to U.S. Government's Amicus Brief." And Kevin Noonan veered into political theory, musing that "It may be that many of the same people who contribute to the ACLU also support (and contribute) to the prevailing (for now) political party." The battle lines were drawn. In an unprecedented reversal of position by the U.S. government, the ACLU now had a powerful ally in Washington though they didn't yet know it.

CHAPTER 19

The Patent Court

VIRTUALLY ANY DECISION handed down by a trial court in the United States can be appealed. If the trial court is a federal district court, then the appeal will go to one of twelve regional Circuit Courts of Appeal, which are numbered one through eleven, with a twelfth covering the District of Columbia. There is, however, a major exception to this rule: patent cases. If a case involves patent law, then it is appealed to a special court that sits in Washington, DC—the Court of Appeals for the Federal Circuit. Despite the fact that the Federal Circuit hears an oddball assortment of other appeals relating to veterans claims, government contracts and trademarks, the court's most prominent docket has led to its well-deserved nickname "the Patent Court."

The Federal Circuit was created in 1982, after representatives of the technology sector expressed concern that the twelve regional circuits were deciding patent cases in inconsistent ways. As NYU professor Rochelle Dreyfuss, a long-time court observer, has noted, during the 1940s and 1950s, "a patent was twice as likely to be held valid and infringed in the Fifth Circuit than in the Seventh Circuit, and almost four times more likely to be enforced in the Seventh Circuit than in the Second Circuit." This degree of inconsistency, some argued, made it difficult to predict whether or not patents would effectively protect product markets. If they couldn't, investment in R&D might drop and economic growth would suffer. In addition, many—particularly members of the patent bar—felt that patent cases involved complex technologies and legal doctrines that might be beyond the ken of the generalist judges sitting on the regional courts of appeal. It would be far better, they thought, to have a single, expert appellate bench that could decide patent matters in a consistent and competent manner. So in 1982 President Ronald Reagan signed the Federal Courts Improvement Act, fusing together two existing appellate tribunals to form the Court of Appeals for the Federal Circuit.

Not surprisingly, a high percentage of judges appointed to the "Patent Court" were patent attorneys and others with scientific or technical training. And nobody appreciates patents more than patent attorneys. As a result, from its earliest days, the leanings of the Patent Court were noticeably pro-patent. In decision after decision, the court upheld patents against whomever dared to contest them. And it did so with pride, enjoying its reputation as "the de facto supreme court of patents."

It was for this reason that Dan Ravicher asserted, well before their case was even filed, that they would lose at the Federal Circuit. No matter what happened at the district court, the case would be appealed. And when it was, it would come before the Patent Court. And they would lose. Hansen and Park, less steeped in the lore and culture of patent law, did not share Ravicher's pessimism. Yet they, too, knew that appealing Judge Sweet's ruling would be difficult. They would have to work hard to ensure that his favorable decision did not become a historical footnote.

The Tail of the Elephant

RICK MARSH AND the Jones Day team had not been pleased, but also not entirely unprepared, for the decision by Judge Sweet. From Myriad's standpoint, the lawsuit was something to be taken seriously, but not the end of the world. They were relatively confident that the misguided reasoning of the district court would be corrected by the Federal Circuit, which, unlike Judge Sweet, respected the Patent Office and its longstanding policies.

So when Myriad released its Annual Report in September 2010, the ACLU lawsuit received little attention. Enough, of course, to comply with the securities laws, which required Myriad to disclose all material risks affecting the company to its shareholders, but no more than that. Thus, the lawsuit is first mentioned on page ten of the Annual Report, after two paragraphs describing Myriad's substantial portfolio of 175 issued U.S. patents, twenty-three of which covered BRACAnalysis. The lawsuit, Myriad noted, challenged only fifteen claims of seven patents, but those seven patents contained 164 *other* claims that were not being challenged. And, of course, the district court's decision was still at an early stage of appeal. In that light, the threat seemed small. A mosquito buzzing about the tail of an elephant.

And the elephant was substantial. In fiscal year 2010, Myriad's revenue exceeded $360 million, $150 million of which was profit. It had nearly half a billion in the bank. In addition to BRACAnalysis and its existing colon cancer and melanoma tests, Myriad released a genetic test for prostate cancer and had plans for pancreatic cancer next. Internationally, the company was expanding into Western Europe, Asia Pacific, and Latin America. Myriad was healthy.

As for the lawsuit, Jones Day had the appeal well in hand. After the district court decision, the legal team handling the case shifted from Brian Poissant's group in New York to an appellate team in Washington. Unlike the ACLU, in which the same small team would try a case from trial through all of its appeals and even to the Supreme Court, large law firms like Jones Day had special appellate groups that would swoop in once a case was bound for appeal. The theory was that trial work and appellate litigation were sufficiently different that they should each be handled by specialists. This approach supposedly gave clients the ultimate in expertise at every stage of litigation. It also enabled firms like Jones Day to recruit top new law graduates with the promise that they could spend their time working on complex appellate cases. The firm's marketing materials boasted a specialist appellate group consisting of seventy lawyers, forty of whom were former Supreme Court clerks.

Jones Day competed with a dozen elite firms in Washington to recruit as many recent Supreme Court clerks as possible. In addition, these firms vied with one another to hire high-level DOJ attorneys and Solicitors General on their way out of the government, often with the change of administrations. It was all part of the glorious revolving door of Washington—work for a few years as a prominent government lawyer, get to know the system, then ease into a comfy seven-figure income in the private sector. If a company had a case that was heading to the Supreme Court and it had the money, it needed one of these firms at its side. And few could match the firepower of Jones Day.

Jones Day's DC office occupies the massive old Acacia Life Insurance Building, a slate gray, art deco edifice just a block north of the Capitol. Its great bronze doors are flanked by two carved griffins—tributes to Acacia's nineteenth century origins as the Masonic Mutual Relief Association. But even the block-long Acacia building could not contain Jones Day's growth. Now, a giant glass-and-steel office building sprouts from the back of the Acacia, tethered to it via a futuristic network of catwalks and struts that evokes some kind of giant orbiting space station.

The Jones Day appellate partner assigned to Myriad was Greg Castanias. Though based in DC, Castanias was first and foremost a Hoosier—born, raised, and educated within driving distance of Indianapolis. In college Castanias double-majored in English and philosophy. In law school he never took a patent course. Yet after joining Jones Day in the early 1990s, Castanias fell into the world of high stakes patent litigation. He excelled at it. Now, twenty years later, still broad-shouldered and fit, his dark hair receding only slightly from his

Greg Castanias led Myriad's appellate litigation team at Jones Day.

pronounced brow, Castanias led the firm's Federal Circuit practice. He had argued before the Federal Circuit more than forty times. He (together with a team of Jones Day subordinates) wrote the annual survey of Federal Circuit cases for the *American University Law Review*. And just a few years earlier Castanias scored a decisive 9–0 victory at the Supreme Court in a complex international case on behalf of a Chinese company.

While Castanias took the reins from Poissant, Laura Coruzzi remained their technical expert. She had worked on Federal Circuit appeals before, and she would do the lion's share of the work around the product of nature arguments they would have to make when defending Myriad's patents.

Your Good Friend, Randy

JUDGE RANDALL RAY Rader—lantern-jawed with a penetrating gaze and a thick mop of hair bisected by a ruler-straight side part—was appointed to the Federal Circuit in 1990. A devout Mormon, Rader was a graduate of Brigham Young University in Provo, Utah, and had six children. He moved to DC to attend law school at George Washington University. The DC bug bit Rader, and after graduating from GW he stayed in town. First he worked for two members of the House, and then found his true calling as an aide to Utah's powerful senator Orrin Hatch. Rader spent seven years on Hatch's staff in various roles, including as counsel to the Judiciary Committee's Subcommittee on Patents, Trademarks and Copyrights. There, Rader started to make his mark on patent policy.

Despite his long tenure in Washington, the outspoken Rader cultivated an image as a maverick and an outsider. Among other things, he performed as the lead vocalist in a rock band satirically named *De Novo*.[1] Many in the IP field fondly remember legal conferences at which Rader, sporting dark sunglasses and a leather jacket, strode across the stage

---

1 In judicial parlance, "de novo" review is the review of an issue "from the beginning" without reference to prior decisions.

capably rendering classics by Neil Young, the Rolling Stones, and Van Morrison. He was personable and outgoing, signing even the most mundane business correspondence, "your good friend, Randy."

Dan Ravicher had a long history with Judge Rader. When Ravicher was a law student at the University of Virginia, Rader made the two-and-a-half-hour drive from Washington to Charlottesville every other weekend to teach a patent law class there. Ravicher got the highest grade in the class and went on to become Rader's student intern. But that was years ago.

On April 9, 2010, just a couple of weeks after Judge Sweet's opinion was released, Ravicher spoke on a panel at the annual Fordham IP Conference in New York. With him on the dais were Professor Rochelle Dreyfuss from NYU, Judge Pauline Newman from the Federal Circuit, a European patent judge, a member of the UK House of Lords, a law firm partner and an attorney from Microsoft. The panel was titled "Patentable Subject Matter," but as Dreyfuss noted, the Supreme Court's highly anticipated decision in *Bilski* had not yet been released, so the juiciest recent development they could discuss was Judge Sweet's ruling in *Myriad*.

The annual Fordham event does not follow the usual format for legal conferences. It includes a lot of people who have only a short time to speak, and there are no PowerPoint slides. The result is fast-paced and designed to elicit debate, if not outright shouting. Its organizer, Fordham's Professor Hugh Hansen (no relation to Chris), prides himself on being the IP version of a talk show provocateur à la Geraldo Rivera or Jerry Springer. Hansen did not moderate Ravicher's panel, but the free-for-all atmosphere of the conference was in evidence.

About a minute into his ten-minute talk, Ravicher began to make a point about the patentability of "purified" natural substances. He pointed to a plastic water bottle sitting on the table. "It says here right at the top . . . 'purified water.' So this is not water that exists in nature. . . . Man had to do something to take the water that is found in nature to make it purified. So the question we have to ask ourselves is: was that sufficient intervention between what God gave us . . . and what man created to merit a patent?"

At that point, someone in the audience called out in a loud voice, "How many people have died of water pollution over the course of human events?"

Ravicher looked up. It was his old professor, Judge Rader. "That's an interesting question," Ravicher answered, not sure where Rader was going.

Rader responded loudly, "Probably billions."

Ravicher decided to ignore the all-too-familiar heckler and continued with his presentation. By the end of the hour, Ravicher didn't feel so bad. Rader had also launched a random question at the panelist from the House of Lords, and there were plenty of other questions and comments from the active audience. But given that Ravicher was in the midst of litigating *AMP v. Myriad*, which was about to be appealed to Rader's court, the judge's open hostility to his position was not only annoying, but worrisome.

In Search of Calmer Waters

A MONTH LATER, Judge Rader attended BIO's annual conference in Chicago. The event was huge, with over fifteen thousand attendees and keynote speeches by Bill Clinton, George W. Bush, and Al Gore. David Kappos from the Patent Office spoke, as did high-ranking officials from the FDA and USDA. Judge Rader, who would soon be promoted to Chief Judge of the Federal Circuit, must have felt invigorated in the company of this stellar cast of IP stars.

Rader was scheduled to speak on two panels, one of which was titled "Patenting Genes: In Search of Calmer Waters." For this 8:00 a.m. session, he was joined by Lori Pressman, a consultant who had previously worked at MIT's licensing office, Bob Cook-Deegan, the Duke academic who had led a recent study of gene patents, John Whealan, associate dean of IP Studies at George Washington University and a former Patent Office solicitor, and Judge Annabelle Bennett from the Federal Court of Australia. The moderator was Jennifer Gordon, a partner at the Houston-based firm Baker Botts who had worked on BIO's *amicus* brief in *AMP v. Myriad*.

Gordon began the panel by engaging the audience of around two hundred and fifty people with a straw poll. Who in the room, she asked, agreed with Judge Sweet's decision in *AMP v. Myriad*? She waited a moment, but not a single hand went up. She nodded, then asked who disagreed with the decision. Almost every hand in the large conference ballroom was raised. That set the stage for the discussion, in which Judge Sweet's opinion came to resemble a doctrinal piñata that each speaker took turns whacking with mounting gusto and abandon.

Cook-Deegan, who appeared to be alone in sympathizing with the ACLU's case, offered a few words of praise for Judge Sweet, commending his command of the science and his logical reasoning. But Judge Rader quipped that writing a well-reasoned opinion wasn't hard to do. "Believe me, I do this for a living," he said. Whether the opinion was well-reasoned or not was neither here nor there.

When it came time for Rader to deliver his prepared remarks, he was even more critical of his fellow jurist. Judge Bennett, a small and energetic woman sitting a few places down from Rader, was taken aback by his comments. In public speeches, at least in Australia, judges usually exhibited some restraint in commenting on the work of their brethren, especially when a case is headed for the speaker's own court. Yet Rader unabashedly took Judge Sweet to task for failing to offer a clear legal standard for the patentability of natural substances. "This approach is subjective," he said, "and, to be frank, it's politics. It's what you believe in your soul, but it isn't the law."

Cook-Deegan pushed back, observing, "There is a criterion, and it's DNA. The court said that if it's DNA, it's unpatentable."

But Rader would not be swayed. "I can do that with anything you give me—biotechnology, software, engineering, anything. I can say it's just an information bearer, it's just a gear, and everything is from nature. That's not a test."

The panel ended with Cook-Deegan alone defending Judge Sweet's decision, though even he thought that Sweet had gone a little too far in his "DNA is special" thinking, and even he had raised his hand when Gordon asked whether Judge Sweet had "gotten it wrong" (a gesture that he later came to regret).

At the conference dinner that evening, the panelists sat together, exchanging war stories over glasses of wine and good food. Bennett, who would soon be dealing with Myriad's patents in Australia, let loose with a healthy dose of earthy antipodean humor. John Whealan, sporting an expensive suit and long, silver ponytail, sauntered in almost an hour late. Arm in arm with Whealan was an attractive, elegantly dressed woman with a gleaming Hollywood smile and luxuriant blonde hair. She was Sherry Knowles, chief patent counsel of pharmaceutical giant GlaxoSmithKline. Several people seated at the table knew Knowles, who was a vocal advocate for strong patent protection. In a recent *Science* article, Knowles bemoaned the fact that a simplistic invention like the "beerella" (an umbrella that snaps onto a beer bottle) could have stronger patent protection than a breakthrough cancer therapy.

Like Rader, Knowles would be experiencing a major career shift on the first of June. She was leaving GSK to start her own IP consulting firm. With these two watershed moments approaching, the dinner party became lively, if not raucous. One attendee described it as a "fun evening of suppressed tensions and papered-over doctrinal disagreements, but delightful conversation and story-telling." Little did the dinner companions know that the evening's tensions and the discussion at the morning's panel would gain even greater prominence in the coming months.

Shooting at the King

COUPLED WITH RADER's comments at Fordham, the BIO speech made it clear to Ravicher that Judge Rader—now Chief Judge Rader—did not think much of the district court opinion in *Myriad*. In fact, it seemed that he had already made up his mind that Judge Sweet was wrong. Which was a problem, given that the case was being appealed to the court over which Rader now presided.

Chris Hansen agreed. There were rules of judicial conduct that couldn't be ignored. Judges weren't supposed to decide cases until the parties had submitted their briefs and argued their positions. If a judge pre-decided a case, he was clearly biased. And federal law required that a judge must disqualify himself "in any proceeding in which his impartiality might reasonably be questioned." This seemed like such a case.

So Hansen and Ravicher filed a motion with the Federal Circuit seeking to have Chief Judge Rader recused from any involvement in their case. The move was a risky one, as it was not yet known which of the twelve circuit judges would be assigned to the three-person panel that would hear the case. Those assignments were random, and not revealed until the morning of the oral arguments. But by then it would be too late, so they had to oppose Judge Rader's involvement now, before the panels were assigned.

Ravicher and Hansen's motion argued that "Chief Judge Rader's statements in this case have created an appearance of partiality that calls into question his ability to engage in impartial legal analysis based on the record and the argument of the parties." They also accused him of violating the Code of Conduct for United States Judges, which states that "A judge should avoid public comment on the merits of a pending or impending action." These were fighting words. They called into question Rader's judicial impartiality, the very core of a judge's reputation. On the battlefield of litigation, this was a nuclear option.

Castanias was puzzled and surprised by the unorthodox motion. Such tactics were direct challenges to judicial discretion—bad practice for someone who regularly appears before the court. In his brief opposing the motion, he argued that Rader's comments at Fordham and BIO were general opinions about the state of the law, not specific opinions on the case. Nothing that Rader said at either conference implied that he had pre-judged the case, which was about DNA, not bottled water. The Federal Circuit Bar Association also weighed in against the recusal motion. They were concerned that such a motion could discourage judges from participating in public conferences like Fordham and BIO—a development that would hurt both the judiciary and the bar (and conference organizers like themselves).

Rader, privately seething, did not publicly comment on the recusal motion.

Yet the Federal Circuit never ruled on the motion. The court docket simply noted that the panel for the case had not yet been assigned, and that if Chief Judge Rader were on it, the motion would be then considered.

Castanias was not surprised. At Jones Day they had a saying— "If you are going to shoot at the king, you had best kill him." Ravicher had taken a shot at Judge Rader, and missed. Now he would suffer the consequences.

The entire recusal affair cast an air of hostility and personal animus over the *Myriad* proceedings. It also opened a door to tactics against specific judges that had not previously been deployed in the nerdy world of patent litigation. Admittedly, Judge Rader was among the more out-spoken and opinionated judges in the field. One waggish author, commenting on the incident, referred to Rader as the judge who "shoots from the lip." But whatever his faults, there were only nine judges in the country—the sitting justices of the U.S. Supreme Court—with more power over patent matters than Randall Ray Rader.

Machine or Transformation

ON THE DAY before Hansen and Ravicher filed their motion to recuse Judge Rader, a far more significant development occurred in Washington. More than seven months after the oral arguments, the Supreme Court released its opinion in *Bilski*. Few were surprised that the Court—like the Patent Office, the district court and the Federal Circuit before it—had denied Bilski and Warsaw a patent on their formula for pricing utility bills. All nine justices agreed that the formula was merely an abstract idea, not a patent-worthy invention.

More interesting, however, was the high court's rejection of the Federal Circuit's "machine or transformation" test. In an opinion written by Justice Kennedy, the great compromiser, the Supreme Court held that while an invention that constituted a machine or a transformation of matter was, indeed, eligible for patent protection, these two categories were not necessarily the *only* categories of patentable subject matter. For example, inventions directed to diagnostic medical techniques, computer programs, data compression, and the manipulation of digital signals might be patentable, even though they did not satisfy the Federal Circuit's "machine or transformation" test.

This rebuke was surprising. The Federal Circuit was supposed to be the expert court in charge of patent matters. It had developed the "machine or transformation" test over many years. The Supreme Court's rejection of the test, without any real alternative, puzzled and frustrated many, including, the judges of the Federal Circuit.

The teams at the ACLU and Jones Day both watched this development with interest. But even more interesting was what happened the next day. The Supreme Court, in a one-sentence ruling, decided *Mayo v. Prometheus*, a case that was also on appeal from the Federal Circuit. The previous September, a three-judge panel of the Federal Circuit considered the validity of a company's patent on a blood test for adjusting a patient's medication. The court's opinion was written by Judge Alan Lourie, a crusty seventy-four-year-old chemist who had served as Associate General Counsel of drug maker SmithKline Beecham before joining the court nineteen years ago.[2] Lourie wrote that the blood test satisfied the "machine or transformation" test because it resulted in changing the dose of a patient's medication—a "transformation" in the patient's body. A few months ago, Judge Sweet had relied on the Federal Circuit's ruling in *Mayo* to reject Myriad's method claims because they didn't cause such a transformation.

But now, the Supreme Court, which had just eradicated the "machine or transformation" test in *Bilski*, issued a one-sentence ruling that vacated Lourie's decision in *Mayo* and remanded the case back down to the Federal Circuit for reconsideration. This judicial move is known as a GVR—grant, vacate, remand. In other words, the Supreme Court said to the Federal Circuit, "You decided this case based on your old 'machine or transformation' test. We just eliminated that test. So go back and decide *Mayo* again, but this time do it right." The Federal Circuit would have to reconsider *Mayo*, but without applying its now-defunct "machine or transformation" test.

---

2 The other members of the panel were Chief Judge Paul Michel and Ron Clark from the District Court for the Eastern District of Texas, temporarily sitting on the Federal Circuit by designation.

Everyone who was watching knew that *Mayo* would form an important piece of the puzzle in *AMP v. Myriad*. The key issue in *Mayo* was whether using the result of a blood test to adjust a patient's drug dosage was effectively trying to patent a "law of nature" (i.e., the relationship between metabolite levels in the blood and the patient's ability to metabolize the drug). The patents in *Myriad* related both to laws of nature (diagnosing an elevated risk of cancer by observing mutations in the *BRCA* genes), as well as products of nature (the genes themselves).

As Hansen pored over the courts' opinions, he realized that three interrelated chess games were being played simultaneously on the table of patent eligibility: *Bilski*, *Mayo*, and *Myriad*.[3] And the moves in one game affected each of the others. *Bilski* had just been completed—checkmate. And that required the Federal Circuit to reconsider *Mayo*. What did it mean for *Myriad*?

The games were even more complex because it wasn't entirely clear *who* had actually won. In *Bilski*, the Patent Office got its way, but the Federal Circuit's key test for patentable subject matter had been overturned. Did that help the Patent Office or hurt it? And to make matters worse, the players seemed to shift sides from game to game—playing white in one match and black in another. In *Bilski*, the Patent Office won by refusing to grant patent for the energy traders' abstract idea. But in *Mayo* and *Myriad*, the Patent Office wanted its patents to stand.

Complicating things further, the kibitzers in the room were all shouting advice to the players, but the kibitzers in one game were the players in another. So just as Hansen had filed an *amicus* brief at the Federal Circuit in *Bilski*, so had Hans Sauer from BIO. Jennifer Gordon, the lawyer who had filed BIO's *amicus* brief at the district court in *Myriad* had also filed a brief for Novartis in *Bilski*. Myriad's in-house lawyer Jay Zhang had filed an *amicus* brief in *Mayo*, as did several of the medical association plaintiffs that Hansen was now representing in *Myriad*.

---

3 Though unknown to Hansen at the time, the fourth case in this tetralogy, *Alice v. CLS Bank*, was still being fought out in the District of DC. It would soon make its way up the appellate ladder as well.

And what about the United States government? Hansen had no idea what to make of the meeting they had attended in Washington. By far the most surprising *amicus* brief in *AMP v. Myriad* was the one filed by the Department of Justice. Hansen had assumed that the government would double down on the position that the Patent Office had taken in New York. But when Hansen read the brief, he didn't think he understood it. Had the U.S. government abandoned the Patent Office and taken their side? How was that even possible?

It was a complete reversal of position. Hansen, in all his years litigating cases against the government, had never seen such a turnaround; Park and Ravicher agreed. This was not only unprecedented, but unbelievably good news for their side.

Hansen knew how influential a DOJ brief could be in a controversial case. He recalled his work on *Brown v. Board of Education*, the continuation of the landmark 1950s school desegregation case. When that case was first heard by the Supreme Court, the most important brief—and the one that was probably responsible for bringing down school segregation—was submitted by the attorney general of the United States. So maybe the players in this game of chess weren't the parties at all, but the Patent Office, the Federal Circuit, the solicitor general and the Supreme Court. Which of these institutions would ultimately be the final arbiter of gene patents in the United States?

Perhaps the most puzzling aspect of the government's brief was the carve-out for cDNA. Despite arguing that the DNA found in human cells should not be patentable, the brief reasoned that cDNA did not occur in nature and thus *should* be patentable. Hansen convened the team to discuss. The question they faced was whether or not to accept the government's compromise and give up on cDNA. On one hand, agreeing with the Department of Justice would put them in a strong position. The courts respected the opinions of the government, and the fact that this was a significant turnaround was evidence that the government took the issue seriously. But, on the other hand, at the district court they had defeated *all* of Myriad's patents—both gDNA and cDNA. Mary-Claire King, the researcher who had begun the hunt for the *BRCA* genes, had advised them to challenge patents on cDNA as well as genomic DNA. Doing otherwise, she told them, would be like winning a battle while losing the war. Because, as Judge Sweet had written in his opinion, it was the *information* encoded by the DNA that mattered, and the coding regions of gDNA and cDNA were exactly the same. None of it should be patentable.

So Hansen decided to ignore the government's *amicus* brief. They filed their own brief a week after Thanksgiving.

A FEW DAYS before Christmas, the Federal Circuit released its "reconsidered" opinion in *Mayo*. Again, Judge Lourie wrote the opinion, joined on the panel this time by Chief Judge Rader and Judge Bryson. While the new opinion played lip service to the Supreme Court's decision in *Bilski*, little else changed. Judge Lourie again upheld the blood testing patent, noting that while the Supreme Court had ruled that the "machine or transformation" test was not the *only* test for determining whether an invention was eligible for patent protection, it was still *a* test that could be used. And, in this case, a patent claim covering the observation of a patient's blood metabolite levels to adjust her drug dosage caused a sufficient transformation in the body. Thus, the patent stood. Few were surprised.

CHAPTER 20

Magic Microscope

ON JANUARY 25, 2011, President Obama delivered his third State of the Union Address to the joint chambers of Congress. In an upbeat and inspirational speech, he called on America to "out-innovate, out-educate and out-build the rest of the world." He urged Congress to support new funding initiatives for basic research and science, to fulfill the promise of "our generation's Sputnik moment." He also called for reductions in healthcare costs, including payments made by Medicare and Medicaid. The administration's priorities were clear. Within the White House, spirits were high.

But Neal Katyal listened to the address with a tinge of regret. The day before, the president had announced his nomination for solicitor general. Though Katyal had been serving as acting SG since Elena Kagan resigned last May, a permanent SG needed to be confirmed by the Senate and installed in office.

Many, including Katyal himself, thought that he stood a good chance of being that person. He was, after all, a star. Not only that, history was with him. According to one court watcher, the last five times that a principal deputy SG was promoted to acting SG, that person eventually succeeded to the permanent SG position. Even prominent SGs like Charles Fried, Seth Waxman, and Ted Olson had attained their positions via this route.

But for Katyal, it was not to be. The day before his address, President Obama nominated Deputy White House Counsel Don Verrilli as the next solicitor general of the United States. Verrilli, of course, was eminently qualified for the job—head of the Supreme Court practice at law firm

Jenner & Block, he went on to high-level positions within the Department of Justice and White House. He had argued before the Supreme Court a dozen times and had won significant victories in cases involving telecommunications, the First Amendment, and copyright law. Verrilli had a reputation as an unflappable litigator. One observer said that he had a pulse like a metronome. Katyal couldn't begrudge Verrilli the nomination. It must have been disappointing, nonetheless. That being said, the Senate confirmation process was slow and painful. Katyal would remain acting SG until Verrilli was sworn in, probably months from now. And at the top of Katyal's docket was *AMP v. Myriad*. He knew that the case was important. The amount of press coverage, not to mention the drama within the administration, attested to that fact. And Katyal was keenly aware that his decision to buck the recommendation of the Patent Office had caused some waves. Had these waves scuttled his nomination? Maybe. Katyal didn't know. But it didn't matter. He had done the right thing.

Now, however, Katyal felt that he owed the administration, and the nation, an explanation for the position that he had taken. To that end, he would do something that no solicitor general, acting or permanent, had ever done before. He would personally argue the government's position at the Federal Circuit.

The realm of the solicitor general of the United States is predominantly the U.S. Supreme Court. Nicknamed the "tenth justice," the SG even has an office in the Supreme Court building. While the SG is nominally in charge of all litigation brought by the federal government across the country, most cases in the appellate courts are argued by senior lawyers from the agencies (like Ray Chen, Solicitor of the Patent Office), Assistant U.S. Attorneys in the region where the case is being tried, or appellate staff attorneys in the Department of Justice Civil Division (like Mark Freeman and his superiors). The solicitor general typically does not even have time to argue every case at the Supreme Court, and assigns Deputy SGs to argue many of these. It was for this reason that Malcolm Stewart, rather than Katyal, presented the government's case in *Bilski*.

Ultimately, the solicitor general argues only the most significant cases that the Supreme Court will hear. The Federal Circuit, a specialized appellate court dealing primarily with patents, was not on the radar screen of most SGs. In fact, it would be surprising if any SG before Katyal had ever set foot inside the court, or even knew where it was located. But Katyal felt that *AMP v. Myriad* warranted an appearance by the acting SG, and on February 10, 2011, he notified the clerk of the Federal Circuit that he personally wished to present the oral argument on behalf of the United States. The court was more than happy to accommodate him.

Panel B+

FOR THE UNINITIATED, the Court of Appeals for the Federal Circuit is not easy to find. It occupies an eight-story brick building that is entered from a pedestrian lane bordering Lafayette Square. Though the court is just a stone's throw from the White House, it is maddeningly inaccessible to taxis, limos, and Uber drivers. The court building is attached to a large yellow mansion with a wrought iron porch that is best known as the last residence of First Lady Dolley Madison. In this restored house, which is now used for court receptions and banquets, the former first lady hosted dinners, fêtes, and soirées for the crème of Washington's antebellum elite.

The entrance to the court building itself opens onto a nondescript brick courtyard enclosing a modest fountain—a setting where one might expect to find travel agencies and dental offices. But through another set of thick glass doors the telltale metal detector and security checkpoint unmistakably identify it as a federal facility of some importance.

Oral arguments in *AMP v. Myriad* were scheduled for 10:00 a.m. on April 4, 2011. The spring morning was cool and sunny, and the sidewalk was blanketed by a fragrant pink carpet of cherry blossom petals. The opposing legal teams—Hansen, Park, and Ravicher for the plaintiffs; Marsh and Jackson from Myriad; Castanias and Coruzzi from Jones Day; Katyal and others from the solicitor general's office, as well as a number of observers, began to arrive at the court before nine. The case was scheduled for argument in Courtroom 201, a large chamber on the second floor, accessible via a bank of dingy, rattling elevators.

Upon exiting the elevators, the visitors found themselves in a low-ceilinged waiting area furnished with an assortment of nondescript office chairs, benches, and end tables. A white porcelain drinking fountain adorned one marble wall; beside it stood a glass display case containing patent memorabilia—an antique circuit diagram, a certificate signed by Andrew Jackson, an old leather-bound tome. Without a doubt, this was a court for nerds.

As the different groups of attorneys, clients, and observers arrived, they acknowledged one another with polite nods, then clustered together, speaking softly and avoiding anything that might be overheard. Herman Yue, Judge Sweet's former clerk, had taken the train from New York to DC to hear the argument. He wanted to see how the court treated the decision on which he had spent so much of his time and energy.

At the front of the waiting room, near the elevators, was a brass document stand—the kind in which a street café might display its menu. At precisely 9:00 a.m., a court employee entered this antechamber through a side door. He unlocked the display case and lifted its glass cover. Into it he slipped a sheet of paper bearing the court's seal and titled "Calendar Announcement." As soon as he stepped away, attorneys from each group approached as quickly as decorum would permit. Here they would learn, for the first time, which judges had been assigned to their case.

The announcement indicated that two cases were scheduled for argument that morning. *AMP v. Myriad* was first, at ten. The court's computer had selected "Panel B" for that morning's arguments. Panel B consisted of Judges Kathleen O'Malley, Kimberly Ann Moore, and William Bryson. But the judges who would hear arguments in *AMP v. Myriad* were listed as "Panel B+." Panel B+ was essentially Panel B, but with one substitution. Such substitutions were common and generally occurred when one of the randomly selected judges was disqualified from hearing a particular case due to a conflict of interest. In the *Myriad* case, Judge O'Malley could not sit on the panel because her brother was a partner at Jones Day. So a vacancy existed on Panel B, and a substitution was made to form Panel B+, which would hear the case.

This is where Ravicher's motion to recuse Chief Judge Rader may have come home to roost. When there is a vacancy on a randomly selected panel at the Federal Circuit, under the court's internal operating procedures, that vacancy is filled by the Chief Judge. So, in place of Judge O'Malley, Chief Judge Rader assigned someone else to the panel. That person was Judge Alan Lourie. Short of placing himself on the panel (which would have required the court to rule on Ravicher's motion), Rader could not have picked a judge less sympathetic to the ACLU's case. Lourie had written the court's opinion in *Mayo* just a few months earlier, upholding the patent on a diagnostic blood test that had been challenged as a "law of nature." And he had done so not once, but twice, doubling down on the patent even after the Supreme Court had asked him to reconsider. To make matters worse, the third person randomly selected for this morning's panel, Judge William Bryson, had also been on the panel in *Mayo*, and had voted, along with Lourie and Rader, to uphold the patent. Castanias withheld a smile as he read the assignment sheet. He had been dealt a good hand.

Shortly after 9:30 a.m., the doors to the courtroom opened and the waiting crowd was allowed to enter. The overall gestalt of the courtroom was one of manufactured dignity—faux mahogany paneling, industrial blue and gold carpeting, and mass market hanging lamp fixtures.

The dais on which the judicial bench stood was raised at least five feet above the floor. As a result, the counsel who sat at the tables immediately before the bench were required to crane their necks in order to see the judges peering down at them. There were two high-backed chairs at each of the counsel tables. Hansen and Park took their seats at one of the tables, Castanias and Coruzzi at another. The third counsel table, which had been added that morning, was for the government. Many in the room were surprised to see Neal Katyal, the acting SG himself, confidently take his place at that third table.

There was a quiet buzz of conversation as the courtroom filled with spectators. The gallery of cushioned benches was divided by a central aisle, and the spectators naturally separated along partisan lines, as if at a wedding.

At ten o'clock, a young man in a suit emerged from a door behind the dais and announced that all cell phones should be turned off. The room became quiet. The young man ceremoniously said, "All rise," triggering a susurration of expensive wool and cotton and silk as the assembly stood as a single mass. With this, the three robed judges silently padded into the chamber and filed toward their leather chairs behind the bench.

The most senior judge, Alan Lourie, sat in the center. With steel-gray hair parted neatly on one side, Lourie possessed a beaklike nose, large protruding ears, and thin lips that he occasionally sucked in while thinking.

To Lourie's left sat Judge Kimberly Ann Moore, the most junior member of the panel and, still in her early forties, one of the youngest members of the court. Moore's straight brown hair fell below her shoulders, giving her a youthful air that she counteracted with an unnaturally severe expression. Whereas Judge Lourie was a chemist, Moore was an electrical engineer. She had received her bachelor's and master's degrees from MIT before realizing, while enrolled in the PhD program, that she really wanted to be a law professor. Her husband was a patent lawyer and her engineering background made her a shoo-in for a position at George Mason University, where she was hired to teach intellectual property law. Now that she was a judge, Moore grilled attorneys appearing before her as though they were first-year law students, urging and prodding them toward the right resolution—her resolution—to a thorny issue. But in her private chambers, Moore relied heavily on the advice of senior judges like Lourie, whom she viewed as a mentor.

To Lourie's right sat Judge William Bryson. With his neatly trimmed white beard and wire-rimmed glasses, Bryson presented a professorial image—more so than Moore, who actually had been a professor. Unlike the other members of the panel, Bryson did not have a scientific or technical background, nor had he ever taught at a university. As a Harvard undergraduate he had studied liberal arts, then worked as a magazine reporter before going to law school. Yet Bryson had one of the most inquisitive minds on the court. No matter what the technology, often to the dismay of his clerks, he needed to understand how it worked in excruciating detail.[1]

---

1 Judge Bryson's display of scientific curiosity on the bench is not merely overcompensation for his lack of a technical degree. In his spare time, Judge Bryson is an amateur astronomer. But his interest does not extend merely to observing distant objects through the many large telescopes that he owns. Judge Bryson also builds his own telescopes, which includes grinding and polishing his own mirrors and lenses.

Judge Lourie surveyed the courtroom, then began in a distinctive Bostonian accent (pronouncing "this court" more like "thess cawt") to make a few administrative announcements. A moment later, he announced the case and invited Castanias, representing the appellants, to begin.

Each advocate had twenty minutes to present his case. Much of Castanias's time was spent wrestling with different analogies. Judge Moore got them focused on a hypothetical involving a new mineral embedded in a rock. If it requires significant effort to extract the mineral from the rock—to isolate and purify it—does that make it patentable? Castanias instinctively answered "yes."

Judge Moore challenged him, asking why the ingenuity displayed by Myriad wasn't in the process for extracting the gene, rather than the gene itself. The gene—the DNA—was not created by Myriad. "God made it. Man didn't make it," she asserted.

Castanias looked up at Judge Moore, whom he knew to be a devout Catholic. "God can make a tree," he said, "but man can make a baseball bat." The isolated *BRCA* genes patented by Myriad, he explained, "are the products of molecular biologists. They're not products of nature. They're not the products of God . . ." Moore nodded in approval.

Castanias's twenty minutes sped by without too many bumps in the road. When he concluded, Judge Lourie invited Hansen to the podium. Hansen had scarcely begun when the judge interrupted him. "Mr. Hansen," he said, "this court has held that mere purification of natural material is not patentable subject matter . . . Here, what we have is a gene which in its natural state is covalently bonded with other genetic material, so it's not just purifying the gene. It's isolating. It's not purifying it. Why didn't the district court err when it focused on purification rather than isolating?"

Hansen froze. He didn't know what the judge was talking about. What did "covalently bonded" mean? In all of the hours that he had spent with Mason and Chung and the rest of their technical advisors, that term had never come up. For the first time in forty years of legal practice, Hansen felt like a law student who had missed some key point in the previous night's reading assignment, and now the professor and the entire class were staring at him, waiting for his answer. So Hansen did the only thing that he could do—what squirming law students in the same position have always done—he faked it.

Covalent bonds, Hansen figured, must be special in some way. But since the science team had never discussed covalent bonds, and even Myriad hadn't mentioned them in its brief, they couldn't be *that* important. So Hansen confidently announced to the court that it simply didn't matter whether covalent bonds were broken or not. Then he changed the subject. "Let me give a different metaphor that might be helpful," he said. "A kidney in my body is in its natural state. If a surgeon cuts me open and slices out my kidney and takes it out and holds it in his hand, it's an isolated kidney but it's still a kidney. It's not an invention."

Lourie, not convinced, smiled. "We don't want any kidney problems here."

Polite laughter filled the courtroom. Hansen continued, "The fact that the surgeon used a scalpel instead of a chemical is . . . irrelevant. Just because . . ."

"Well, I'm not sure about that," Judge Bryson cut in. He clearly wasn't satisfied with Hansen's breezy dismissal of the covalent bond point. Hansen valiantly resisted Bryson's attempts to understand why breaking such bonds wouldn't result in a new composition of matter.

Finally, Judge Moore saved Hansen by moving to a different line of questioning. She challenged his assertion, made toward the end of the brief, that Myriad's *BRCA* patents could inhibit whole genome sequencing and other cutting edge technologies.

Now they were back on familiar ground. The rest of Hansen's argument went smoothly. When he concluded, Judge Lourie called Neal Katyal to the lectern.

"General Katyal, welcome to the court," Lourie said, smiling.[2] The solicitor general had never argued in his court before, and Lourie was honored, even though he, and everyone else in the room, knew that Katyal was a lame duck scheduled to be replaced by Don Verrilli as soon

---

2 Unlike the surgeon general, the solicitor general of the United States does not hold a military commission. However, tradition holds that he or she is addressed as General when appearing in federal court. Hence, the honorific General Katyal.

as his confirmation hearings were over. If anyone understood Katyal's awkward position, it was Judge Bryson. Before entering the judiciary, Bryson himself had been a career attorney at the Department of Justice. He worked his way up and served as deputy solicitor general from 1986 to 1994, and twice during that period was called upon to serve as acting SG. Both times someone else was appointed as to the permanent spot above him.[3]

As Katyal stepped up to the podium, he was prepared to argue all of the points made in the government's brief, whether drawn from physics or chemistry or molecular biology. He spoke for less than a minute before Judge Moore asked him to explain why isolated genes should not be patentable. Katyal was ready for her. Here he unveiled a metaphor that he had been working on for weeks, one that he had tossed around endlessly with Mark Freeman at DOJ and Francis Collins and Kathy Hudson at NIH. It was the "magic microscope."

Just imagine, he said, that you had a magic microscope. This microscope could zoom to whatever resolution you wanted. It could see chromosomes, genes, even individual nucleotides. If a patent claimed something that the microscope could see in exactly the form in which it was claimed, then that thing existed in nature, and it could not be patented. But if a patent claimed something that the microscope could not find in nature, it was patentable. So isolated and purified DNA, even if it existed outside the body, could be found within the body in exactly the same form in which it existed outside the body. Thus, it was not patentable. It existed in nature. But cDNA, which included only the coding regions of a gene but excluded the introns interspersed among them, did not exist in nature. No matter how powerful the microscope, you would never find a naturally occurring cDNA within a human cell. Thus, cDNA was patentable.

Many who heard this argument were confused. Castanias had never considered such a metaphor, but it made little sense to him. Ben Jackson,

---

3 One of these appointees was Ken Starr, who was nominated as solicitor general by President George H.W. Bush in 1989. Starr resigned in January 1993 when Bush left office. Bryson served as acting SG until Starr's replacement, Drew Days, was appointed that June.

Myriad's patent counsel, found the metaphor ridiculous. But others were encouraged. Hans Sauer, watching from across the courtroom, saw several people smiling broadly. They thought the microscope was winning the day.[4]

The judges appeared more skeptical. Judge Moore, in fact, seemed upset by the simplistic analogy. "What about [vitamin] B12?" she asked. "What about antibodies? What about, what about, what about? There's about a million things I could thrust out at you right now that . . . exist in nature. Your body will produce them." Are none of those things patentable if manufactured artificially?

No, Katyal responded. While the process for making those substances in the laboratory might indeed be patentable, the substances themselves, if found in nature, are not.

Moore became irate at this. Katyal's theory, she said, would undo decades of patents granted by the Patent Office. It would be highly disruptive.

Katyal tried to assuage her concern. The vast majority of patents issued by the Patent Office, of course, would be unaffected by this theory. But patents on products of nature should not be, and should never have been, allowed.

Here, Moore questioned whether Katyal was speaking on behalf of the government at all. She pointed to the Patent Office's 2001 guidelines on patentable subject matter. "The . . . guidelines conclude that . . . the very thing you're arguing now is *not* patentable subject matter *is* patentable subject matter. I noticed those guidelines have not been pulled down off the website. They continue to be the view of the [Patent Office]. While you're here representing the government, I'm not sure what that really means because it seems like we have a split in the government."

Katyal was not fazed. He explained that he had given the Patent Office ample deference, but their view that naturally occurring DNA, like an

---

4 Kathy Hudson and Francis Collins were clearly pleased with the "magic microscope" metaphor. At a meeting of the NIH directors a few months later, Collins invited Mark Freeman to make a brief presentation about *AMP v. Myriad*. At the conclusion of his presentation, Collins and Hudson presented Freeman with a toy microscope in appreciation of his efforts. On the package, Hudson had taped an illustration of the *BRCA1* gene and words to the effect that "Your genes are owned by you!"

element in the Periodic Table, was patentable was simply wrong. And the government, represented by him, could not write a brief that allowed it.

Moore was not convinced.

In his time allotted for rebuttal, Castanias tried to contest the microscope metaphor. But in a puzzling exchange with Judge Moore over whether the term "magic microscope" was "kitschy" or "catchy," the point that he had hoped to make was utterly lost.

CHAPTER 21

Last Man Standing

AFTER THE FEDERAL Circuit arguments, everyone returned to their respective offices to await the court's decision. While they waited, the big news was *Mayo*. After the Federal Circuit had more or less restated its earlier conclusion, the Supreme Court now granted the Mayo Clinic's petition for *certiorari*, agreeing to hear the case in full.

Many assumed that the only reason the Supreme Court wanted to hear the case was to reverse the Federal Circuit. If so, what did that mean for *AMP v. Myriad*? Speculation was rampant. Blogger Kevin Noonan viewed *Mayo* as a stalking horse for *AMP v. Myriad*, a precursor that would portend which way the winds would blow when it came time for the latter case to be decided. As such, everyone involved in *Myriad* was not only watching *Mayo*, but actively intervening. The hiatus between the oral arguments and the Federal Circuit's decision in *Myriad* gave the ACLU, PubPat and Jones Day attorneys some breathing room to polish their *amicus* briefs for submission to the Supreme Court in *Mayo*.

They heard that the government would also be filing an *amicus* brief in *Mayo*. This time, however, the solicitor general would be supporting the Patent Office, which had granted a patent on the blood test. By now, Don Verrilli's appointment as solicitor general had been confirmed in a relatively smooth 72–16 vote of the Senate. Neal Katyal was again principal deputy SG, though it was rumored that he was on his way out, actively interviewing at private law firms.[1] Mark Freeman, in the Civil Division,

---

1 In September, Katyal accepted a position as a partner in the DC office of the international firm Hogan Lovells.

was beginning to prepare the government's brief. With *Mayo*, the process was far less controversial than it had been with *AMP v. Myriad*. Certainly, the Patent Office and Department of Commerce were pleased that the DOJ was again on their side.

Judge Sweet, too, eagerly awaited the Federal Circuit's decision. No trial judge likes to be reversed, but *AMP v. Myriad* had taken on a special significance for Sweet. In the week after he released his decision, he had his staff print out dozens of copies of the 152-page opinion, which he stuffed into manila envelopes and sent to friends and colleagues across the country. In his silver years, the fame and notoriety that Judge Sweet achieved with that decision was more than invigorating—it was life affirming. So he nervously anticipated what the Federal Circuit would say about his decision.

In the meantime, Sweet was taking his new role as a local spokesman for genetic freedom seriously. He appeared at scientific gatherings and receptions all over town. As in the old days, he would shake hands, slap backs, give hugs and laugh to the tinkle of ice cubes in a tumbler. Laura Coruzzi from Jones Day recalls meeting Sweet at the inauguration of NYU's Center for Genomics and System Biology. The Dorothy Schiff Foundation, run by Sweet's wife, had endowed a chair in genetics at NYU, and was supporting construction of the new Center. But now it was Robert, not Adele, Sweet who was the center of attention. Even Coruzzi's sister Gloria, who chaired NYU's Biology Department, made a point of shaking hands with the judge who had struck down gene patents.

"Someday" Intentions

THE FEDERAL CIRCUIT issued its decision in *AMP v. Myriad* on Friday, July 29, the day before Washington shut down for its August recess. As both sets of lawyers were all too aware, the entire case could have been decided on the basis of procedural standing to sue. And it almost was. Of twenty plaintiffs—researchers, professional associations, genetic coun- selors, advocacy groups, and individual patients—the judges of the

Federal Circuit agreed that only *one* of them had standing to challenge Myriad's patents. That plaintiff was Dr. Harry Ostrer.

Under applicable Supreme Court precedent, someone has standing to challenge a patent if he or she is subject to "a real and immediate injury or threat of future injury" that is caused by the patent holder. Judge Sweet, applying this rule, held that all twenty plaintiffs had standing to challenge Myriad's *BRCA* patents, as each of them was adversely affected or threatened by the patents in some way.

But the Federal Circuit disagreed. The concept of standing, especially in patent cases, had been narrowed over the years. According to the Federal Circuit, the only "injury" that could support standing in a patent suit was being sued for patent infringement. Of the twenty plaintiffs, only three—Ostrer from NYU and Kazazian and Ganguly from Penn—had actually been threatened with a lawsuit by Myriad. The other seventeen plaintiffs—the associations, advocacy groups, and individual patients—may have been apprehensive about Myriad's patents, and may even have suffered health and career consequences as a result of the patents' existence. But because they had not received actual cease-and-desist letters from Myriad, they lacked standing.[2]

And of the three researchers who had received cease-and-desist letters, only one—Ostrer—demonstrated a clear intent to engage in *BRCA* testing if the patents were revoked. That is, in the declaration that Ostrer submitted to the district court, he explained that his laboratory had "all of the personnel, expertise, and facilities necessary to do various types of sequencing of the *BRCA1* and *BRCA2* genes." As a result, he declared that "if the patents were invalidated, I would immediately take steps to begin clinical sequencing of the *BRCA1* and *BRCA2* genes." At a press conference after the Federal Circuit oral arguments, Ostrer, who cut a striking figure—tall and lean with a bushy moustache and commanding voice—reaffirmed that he would "absolutely" begin *BRCA* testing, once the "fear of reprisal" was lifted.

---

2 Ellen Matloff claimed that Yale had also received a cease-and-desist letter, though this fact was contested by Myriad.

This was enough for the court. Judge Lourie acknowledged that Ostrer "has alleged a concrete and actual injury traceable to Myriad's assertion of its patent rights."

Kazazian and Ganguly, on the other hand, had not. Even though their lab at Penn had engaged in *BRCA* testing before Myriad threatened them, they were not as definitive about their future plans. Kazazian, for example, stated that if Myriad's patents were invalidated, and his lab "decided to resume *BRCA* testing," they would be able to do so "within a matter of a few weeks." These statements were more tentative than Ostrer's. Kazazian did not say that he *would* resume *BRCA* testing, only that *if* he decided to resume testing, he would be able to do so rapidly.

Ganguly, whose research had shifted in focus over the last decade, was similarly tentative. She declared that "If Myriad's *BRCA* patents were invalidated, I would immediately *consider* resuming *BRCA* testing in my laboratory." She would consider it, but she did not state that she absolutely would resume *BRCA* testing.

Hansen, of course, had recognized the difficulty that these qualifications might cause. He had gently suggested to both Kazazian and Ganguly that they might want to be more definitive in their statements. But the researchers, trying to be as accurate as possible, could not definitely say that Penn would resume *BRCA* testing more than a decade after they had been forced to discontinue it. The world had changed since 1998. It wasn't clear that a university lab could effectively compete with Myriad in 2009, and they couldn't commit their university to do so. What's more, thanks to the efforts of Barbara Weber, who collaborated with Myriad during the *BRCA2* search, Penn was now a co-owner of some of the *BRCA2* patents with Myriad and the University of Utah. Yet Penn had assigned control over its share of the patents to Myriad. If Penn started to conduct *BRCA* testing, would it be in breach of its contract with Myriad? Kazazian and Ganguly didn't know, nor did they, or Penn's legal counsel, want to find out. So in their declarations, they equivocated.

Unlike Judge Sweet, who had viewed the standing question liberally and wasn't bothered by this waffling, Judge Lourie latched onto it. He wrote that Kazazian's and Ganguly's "'someday' intentions are insufficient to support an 'actual or imminent' injury for standing without any specification of when the 'someday' will be." Thus, Kazazian and Ganguly had no standing to sue, leaving only Ostrer.

In a sense, this result didn't materially hurt the plaintiffs' case. They weren't seeking money damages, only invalidation of the patents. And the patents could just as easily be invalidated at the request of one plaintiff as twenty. Nevertheless, Ganguly was visibly upset by her loss of standing. "Those small words," she mused afterward. They had made all the difference.

And then, in an unexpected turn, the plaintiff's standing became even more tenuous. After twenty-one years at NYU, Harry Ostrer received an offer to move uptown to the Albert Einstein College of Medicine at Yeshiva University and its affiliated Montefiore Hospital. The position was attractive as it would give Ostrer the opportunity to work with several of his longtime collaborators on the genetics of the Jewish people, a topic on which he was writing a book of his own. Ostrer announced his departure from NYU in July—a few weeks *before* the Federal Circuit released its opinion. The announcement was not missed by Jones Day, and its implications were significant.

The way Castanias saw it, back in 1998, Myriad had sent a cease-and-desist letter to Ostrer in his capacity as head of NYU's genetic testing laboratory. Now, in 2011, Ostrer should lose standing on two different theories. First, the "threat" that Myriad made back in 1998 was against NYU and not Ostrer personally, so once Ostrer left NYU, he should be entirely free from that threat. Second, even if the threat were made against Ostrer personally, the threat would be removed when he moved to a hospital, like Montefiore, that did not have a genetic testing lab. Suddenly, Ostrer's standing looked much shakier than it did the week before. And if Ostrer lost standing, the case would be over. Myriad would win.

A month after the Federal Circuit decision, Castanias filed a motion for rehearing at the Federal Circuit. He argued that Ostrer's departure from NYU changed everything. The ACLU responded with a new declaration by Ostrer, in which he asserted that his move to Montefiore did not lessen his intention to begin *BRCA* screening when and if Myriad's patents were eliminated. In fact, Ostrer said that he had already obtained approval from Montefiore to open a genetic testing lab capable of conducting *BRCA* testing when and if the patents were overturned. Apparently satisfied with Ostrer's assertion, the Federal Circuit denied Myriad's new motion without comment.[3] Only one plaintiff of twenty had standing, but the court was reluctant to let that one go.

Three-Way Split

WHY WAS JUDGE Lourie so intent on preserving the thinnest of threads upon which to hang standing? It certainly wasn't out of sympathy for the plaintiffs' cause. To understand his thinking, consider what would have happened if the Federal Circuit had simply given Myriad a victory on standing grounds (something that Jones Day and Myriad fervently hoped it would do). If the Circuit disposed of the case on procedural grounds, it never would have reached the substantive issue—whether or not human genes are patentable. And it was unclear when, if ever, the Circuit would have another chance to rule on the big, juicy patentability issue. So it was important for them to speak on the issue now and, in so doing, overrule Judge Sweet's decision. As Hansen concluded, "The [Federal Circuit] granted standing to [Ostrer] because they so desperately wanted to rule against us on the merits."

On the matter of Myriad's method claims, the Federal Circuit actually found some common ground with Judge Sweet. Under the Federal Circuit's "machine or transformation" test, a patent claiming abstract ideas or mental steps would only pass muster if it caused some transformation in the physical world. Myriad claimed a "method for screening a tumor sample," by "comparing" the *BRCA1* sequence from a tumor sample to a *BRCA1* sequence from a non-tumor sample. According to

---

3 The ACLU also filed a motion for rehearing, arguing that Matloff, who claimed that she received a cease-and-desist letter from Myriad while at Yale, had standing, as did the American College of Medical Genetics, of which Ostrer was a member. Like Myriad's motion, it was denied without comment.

Judge Lourie, this claim recited nothing more than abstract mental steps. There was no transformation of anything, so the claims failed.[4]

Of course, the attentive patent bar, upon reading Judge Lourie's opinion, quickly adapted. In new patent applications, they began to include one or two transformative steps for good measure. So patents would continue to issue, provided that they checked the "transformation" box, and Judge Lourie could rest assured that his invalidation of Myriad's method claims would not harm the future of the industry.

But the crux of the case revolved around Myriad's composition of matter claims—its attempt to patent the *BRCA* genes themselves. Here, the three federal circuit judges diverged. Judge Lourie, who wrote for the majority, focused on the Supreme Court's product of nature test from the *Chakrabarty* bacterium case—was the claimed invention "markedly different" from what already existed in nature? Judge Lourie concluded that isolated from purified genes do, indeed, "have a distinctive chemical identity and nature" from genes occurring in the body. The linchpin of his reasoning, as Hansen had feared during the oral argument, was the need to sever the covalent chemical bonds that attached the gene at its two ends to the rest of the chromosome. This process, which Judge Sweet had largely dismissed as unimportant, was key to Judge Lourie's reasoning. The breaking of these chemical bonds resulted in a new molecule, one that had a "distinctive chemical identity" from what was found in the body. The isolated and purified gene was thus a product of human effort, not a product of nature, and it was patentable.

Judge Lourie rejected Katyal's "magic microscope" test as a misunderstanding of "the difference between science and invention." He also dismissed the many other analogies raised by the parties and *amici*—patents potentially claiming things like lithium, diamonds, atomic particles, leaves and even human kidneys. None of these, Lourie reasoned, is relevant to isolated and purified DNA.

---

4 One Myriad method claim did pass muster—a claim directed to screening potential new cancer drugs by seeing how they affected the growth of cells with a mutated *BRCA* gene. It wasn't clear how or why this method would be useful in screening new chemotherapy drugs, but since it required "growing" the cells that had been treated with a drug candidate, it created enough of a transformation that the claim survived.

Judge Lourie concluded by reminding the reader that the Patent Office had been granting patents on isolated and purified DNA for nearly thirty years. He cited the MIT article estimating that 20 percent of the human genome had been patented—not as a warning, but as a sign that the practice was well established. If such a longstanding practice was going to be reversed, he said, the reversal should be driven by Congress, not by the courts.

Judge Bryson, while agreeing with Judge Lourie with respect to standing, the patentability of cDNA and the non-patentability of Myriad's method claims, disagreed with him on the treatment of Myriad's composition of matter claims. He begins his dissenting opinion by evoking ordinary common sense. "From a common-sense point of view," Bryson writes, "most observers would answer, 'Of course not. Patents are for inventions. A human gene is not an invention.'"

Interestingly, Judge Bryson, the only member of the panel lacking a technical degree, was also the only one who delved into the discovery of the *BRCA* genes. He notes that "Myriad was not the first to map a *BRCA* gene to its chromosomal location. That discovery was made by a team of researchers led by Dr. Mary-Claire King. And Myriad did not invent a new method of nucleotide sequencing. Instead, it applied known sequencing techniques." What Myriad did, he continues, was find the *BRCA* gene (an unpatentable discovery) and then isolate it in the lab. Yet none of these achievements, even though difficult, materially changed the gene. Thus, for its principal (and only known) purpose—encoding a tumor-suppressing protein—it did not differ markedly from the gene existing within the human cell. As a result, the isolated gene should not be patentable any more than the element lithium isolated by scientists for the first time, or the leaf of a newly discovered tree snapped from its branch.

Judge Moore was the swing vote on the panel. Though she agreed with Judge Lourie's result, she wrote separately to make a few additional points. Reading her opinion, one can only surmise that she, an electrical engineer by training, was trying to out-chemistry the chemist. The first part of Judge Moore's opinion reads like an organic chemistry textbook. In no fewer than six pages, she (with the assistance of her law clerk, a chemistry PhD) indulges in an extended exegesis on the chemical structure of DNA. Far surpassing Judge Lourie's chemistry lesson, Moore's opinion includes two detailed diagrams replete with hexagonal molecular stick figures—including bonds covalent and otherwise—and other arcana. All this, it seems, is presented to make the point that DNA is not simply an abstract sequence of letters—A, T, G, and C—but an actual three-dimensional molecule that exists in the real world. As she explains, "Just because the same series of letters appears in both the chromosome and an isolated DNA sequence does not mean they are the same molecule." In this respect, she reaches the same conclusion as Judge Lourie—an isolated and purified gene is markedly different from the gene that exists within a human cell.

Judge Moore also takes time to deride Katyal's "magic microscope" test for its "curb appeal" and "child-like simplicity." Over the course of four pages, Moore explains the many ways in which Katyal's microscope would not, in fact, "see" the same chemical compounds within the cell as in an isolated gene ("the microscope must make some decisions: should the isolated DNA begin and end in a phosphate? a hydrogen? a hydroxyl? a methyl group? an acyl group?").

These critiques are not surprising, given Judge Moore's technical orientation. But then Judge Moore does a remarkable pivot. Showing an unexpected degree of sympathy for the ACLU's position, she writes, "If I were deciding this case on a blank canvas, I might conclude that an isolated DNA sequence . . . is not patentable subject matter. Despite the literal chemical difference, the isolated full length gene does not clearly have a new utility and appears to simply serve the same ends devised by nature."

What? Wasn't this Judge Sweet's reasoning? Despite the literal chemical differences between the cellular and isolated genes, a gene is a gene is a gene. It seems that Judge Moore, despite her painstaking demonstration of the chemical difference between isolated DNA and cellular DNA, might not think that the chemical difference is really that important. Yet in the end she concurs with Judge Lourie and concludes that isolated genes are patentable. Why?

Judge Moore's reasoning here can best be summed up by Tevye's tuneful refrain in *Fiddler on the Roof*—"Tradition!" Like Judge Lourie,

Moore places great weight on the longstanding practice of the Patent Office in granting gene patents, as well as the settled expectations of the patent bar and the biotechnology industry. She writes that these settled expectations "cannot be taken lightly." They deserve deference. She attributes the industry's "outpouring of scientific creativity" to the promise of exclusivity offered by the patent system. As such, it would be exceedingly disruptive to alter the property landscape on which so many had relied over the years.[5]

To Hansen, this logic made no sense. The simple fact that something had been done for a long time didn't make it right. After all, we had segregated schools for forty years, but that was no reason to keep schools segregated. If something was broken, it should be fixed. The fact that a bad practice had been going on for a long time made it even more pressing to end it now.

Nevertheless, Judge Moore cast her vote with her mentor, Judge Lourie. The score was 2–1 against them or, as Hansen liked to put it, 1.51–1.49. In either case, isolated and purified genes could once again be patented.

Like Chessmen Moved Forward

THE WHIPSAW REVERSAL of Judge Sweet's decision by the Federal Circuit had much of the scientific community scratching its collective head. Interest in the issue was exemplified by a special plenary session held at the Annual Meeting of the American Society of Human Genetics in October 2011, a few months after the release of the Federal Circuit's decision. The Society had invited a panel of international patent experts to discuss the status of gene patenting around the world. But the biggest draw at the 7,500 person event was Judge Sweet, himself.

The judge energetically offered his views on the importance of the gene patenting question to society. First, he praised the work of his former clerk, Herman Yue, his trusted translator and guide through the world

---

5 Judge Bryson countered this argument by noting that the existence of large numbers of patents on human genes actually had the potential to hinder scientific research and technology development—a prospect advanced by academic writers including Eisenberg.

of genetics. When Sweet revealed that Yue held a PhD in molecular biology—a fact that few knew at the time—the assembled scientists broke into applause and cheers. Sweet then pronounced that humanity was at the threshold of a revolution in biotechnology, health, and human affairs, quoting from Goethe, "Daring ideas are like chessmen moved forward; they may be beaten, but they may start a winning game."

At the conclusion of Sweet's comments, the traditional question and answer period began. Hands shot up throughout the large hotel ballroom. Following a gesture from the judge, the moderator recognized a tall, casually dressed man with a dark, bushy moustache near the front row. It was Harry Ostrer. In a loud voice, the geneticist and sole surviving plaintiff in the case addressed the packed room. He recalled the first time that Tania Simoncelli contacted him about the lawsuit. He had thought it sounded like a fool's errand. But he joined anyway. It was for the patients, he said. They needed to know what was in their genes. And now, finally, they were close. Ostrer predicted that they would go all the way to the Supreme Court. And, God willing, they would win.

PART III

HIGHEST COURT
IN THE LAND

Patents cannot issue for the discovery of the phenomena
of nature . . . Like the heat of the sun, electricity, or the
qualities of metals, [they] are part of the storehouse of
knowledge of all men. They are manifestations of laws
of nature, free to all men and reserved exclusively to none.

—Justice William O. Douglas, U.S. Supreme Court (1948)

CHAPTER 22

Déjà Vu All Over Again

KATHLEEN MAXIAN WAS smoking pot. Like a rebellious teenager, she was skulking in her own basement with a Tupperware bowl full of weed and a pack of Cheech & Chong rolling papers. It was for the pain, of course. Two years after having her ovaries removed, Maxian's cancer had recurred. Now, in August of 2011, she was back on chemo, and the pain was hard to deal with. Her husband told her that it was OK to smoke in the living room or the bedroom or anywhere else she wanted. But somehow the basement felt right for this illicit medicinal ritual.

Which is why Maxian almost didn't hear the phone when it rang that afternoon. The caller wasn't her mother or one of her friends. It was someone she had never spoken with, but whose call she was expecting.

Maxian's neighbor Mollie Hutton was a genetic counselor at the Roswell Park Cancer Institute in Buffalo, where Maxian was being treated. During her training, Hutton had been a summer fellow in Ellen Matloff's program at Yale. She had recently received an email from Matloff with the subject line "I need stories." Matloff was looking for people who wanted to share their experiences with *BRCA* testing. Hutton immediately thought of her friend Maxian and told her about the call for volunteers. Maxian was ecstatic.

When Matloff called Maxian that afternoon, it didn't take long for the two women to hit it off. Matloff knew all about BART. She had been one of the first people to realize that Myriad's standard BRACAnalysis test wasn't detecting all of the important *BRCA* mutations. Back in 2005 she had asked Myriad whether her lab at Yale could run additional *BRCA* tests on patients who tested negative under the standard BRACAnalysis test, but who had a strong suspicion of hereditary cancer. The company said no. Now, Matloff explained that the ACLU had brought a lawsuit to break up the monopoly that Myriad had on *BRCA* testing by challenging Myriad's patents.

Maxian had heard about the lawsuit. When the news came out a couple of years ago, she had called her local ACLU chapter, but they never called her back. Matloff wasn't surprised. She explained that the litigation was being handled by the National Office, and it was uncharted territory for most of the local ACLU chapters. It was a shame that they hadn't been able to recruit Maxian as one of their plaintiffs back in 2009. Matloff herself was a plaintiff. But there was a way that Maxian could help now.

Matloff explained that the appeals court had just ruled against them, upholding Myriad's patents on the *BRCA* genes. It had also knocked out nineteen of the twenty plaintiffs. Now the ACLU was appealing to the Supreme Court. But the Supreme Court hears only a tiny fraction of the cases that are appealed. So they needed to raise awareness of the issues nationally, to show the court that the case was important enough to hear. They needed to make more noise than ever. Would Maxian help?

Maxian's answer was clear and decisive: "Hell, yes!"

After that, things moved quickly. Maxian's story became *the* story—a new face for the ACLU's PR effort. She hadn't been diagnosed with ovarian cancer until July 2009, two months after the lawsuit was filed, so she wasn't a plaintiff. As a result, Maxian's story would not be tainted by standing and other procedural distractions. She could focus entirely on the impact of Myriad's patents, and how they distorted the *BRCA* testing landscape so badly that her sister had not been offered the BART test that they both needed until it was far too late.

The interviews started almost immediately. Maxian's story was covered by news outlets from the *New York Times* to *The Oncology Nurse*. CNN interviewed not only Maxian, but her oncologist, her sister, and her sister's genetic counselor. Matloff called Myriad "sleazy" on live television. A scrolling headline at the bottom of the screen labeled Myriad's $700 charge for BART "blood money," and Eileen sobbed, "I'm so mad at Myriad. It's even hard just to talk about it."

They flew Maxian to New York where she was interviewed by ABC and visited the ACLU offices. She was impressed by Sandra Park. Like so many others involved in the case, Park was a young woman, but she was making a difference. Maxian resolved that if she made it through chemo and ever became healthy enough to work again, she would do the same. No more installing corporate neon signs on buildings. She would do her part to help.

Two weeks before Christmas, 2011, an ACLU camera crew rolled up to Maxian's home. She was nearing the end of her second six-month chemo regimen, her hair still closely cropped in a style that loudly proclaimed "cancer patient." The four-and-a-half minute video was titled "The Fight to Take Back Our Genes." It featured Maxian sitting beside a small Christmas tree in her living room, as well as stills with her sister and mother. Matloff played the talking head, explaining the science and the patents from a book-filled office. Shots of Myriad's patent and a cease-and-desist letter made appearances, as did a BRACAnalysis kit and a faceless lab technician pipetting fluid—possibly blood—into test tubes. Unlike the ACLU's earlier video, none of the lawyers made an appearance. The only mention of the lawsuit appeared in a series of text screens interspersed without commentary throughout the video:

The U.S. Supreme Court is about to decide
whether a private company can own a piece of you.
This is about WOMEN'S HEALTH
This is about ADVANCING SCIENCE
This is about THE FIGHT AGAINST CANCER
This is about THE RIGHT TO MAKE INFORMED DECISIONS
This is about MORE THAN A COURT CASE
This is about TAKING A STAND

The rhetorical impact of this message was substantial, yet it avoided the legal issues raised in the case. This omission was intentional. Were the video to conclude with the line, "This is about THE PRODUCT OF NATURE DOCTRINE AND STANDING TO BRING SUIT," much of its impact would have been lost. In its appeal to the public, the ACLU needed to frame the case as a matter of women's health and civil liberty. The fine points of patent law would take a back seat in the court of public opinion.

Can We Help?

THE ACLU FILED its *cert.* petition with the Supreme Court on December 8. Though she occasionally spoke with Park and Hansen, Tania Simoncelli did not play an active role in its preparation. Most of the drafting fell to Park, who by now had taken a leading role in the case. Simoncelli was busy with her job at the FDA but, more importantly, was on maternity leave, recuperating following the birth of her son. When she returned to the FDA in the spring, she would take on a new role as a senior advisor in the Office of Medical Products and Tobacco.

Early in January, Simoncelli got a call from her friend Jen Lieb, the DC biotech consultant. Lieb told Simoncelli that she was representing a new Bay Area start-up entering the genetic testing business, and its founders wanted to talk with her about submitting an *amicus* brief in the case. A few days later, Lieb organized a conference call with the start-up's two founders, Sean George and Randy Scott, who explained that their new company, Invitae, would be a spinoff from Genomic Health, a publicly traded cancer diagnostic lab that Scott had founded in 2000. But unlike Genomic Health, Invitae would offer a full battery of genetic tests to doctors and hospitals. The days of one-gene, one-company testing were over, they said. Invitae's goal was to aggregate all the world's genetic tests into a single assay at an affordable price. The problem, they said, was gene patents. Invitae would face the same problem that Affymetrix faced with its gene chips. A company couldn't combine genetic tests that was patented. At best, it could offer a Swiss cheese testing platform, full of holes, each caused by a patent. George, who had previously worked at Affymetrix, had the same aversion to gene patents as Barbara Caulfield, the Affymetrix lawyer who had originally helped convince the ACLU's leadership to greenlight the case.

What Scott and George wanted to know from Simoncelli was whether they could help the case by submitting a brief in support of the plaintiffs and against gene patenting. The New York ruling against Myriad had been encouraging, but the reversal by the Federal Circuit posed a serious problem, not only for them, but for the entire emerging field of personalized medicine. Companies, healthcare providers, and researchers already had the technical ability to run massive, multi-gene tests that could detect thousands of individual mutations. All they lacked was the *legal* right to do so. Lieb was advising them on some possible legislative angles, but if they could support the ACLU's suit in some way, they were interested.

Yes, Simoncelli said, thrilled. Having biotech companies speak out against gene patenting would be tremendously helpful. If nothing else, it would show the Supreme Court that the industry wasn't uniformly in favor of gene patents. Those patents could, in fact, hurt some of the most innovative and promising companies in the field.

Grant–Vacate–Remand

THE SUPREME COURT took nearly four months to issue its decision in *Mayo*. When it did so in March 2012, it surprised the biotech community. Instead of supporting the Patent Office's position, ably argued by Don Verrilli, the Court struck down the blood test patent. In a unanimous decision written by Justice Breyer, the Court held that the challenged patent claims simply recited a law of nature—the relationship between metabolite levels in the blood and a patient's ability to metabolize a drug—and a mental step taken by a physician upon seeing the test result. The claimed "invention" was no more an invention than Newton's law of gravity or $E = mc^2$. None of these fundamental scientific laws merited patent protection.

The *Mayo* decision sent shock waves through the patent community. It was one thing for the Supreme Court to refuse a patent on a flimsy idea like the utility bill formula in *Bilski*, but the blood testing method in *Mayo* was a real invention—a medical procedure that was being administered to patients all over the country and possibly saving lives. The Federal Circuit had twice upheld the patent in *Mayo*, but it was overruled. Instead, the Supreme Court held that in order to patent an invention based on a "law of nature," the claims must add something of significance, or cover an "application" of that law of nature.

The *Mayo* decision would clearly have major implications for *AMP v. Myriad*. But before the parties could analyze them fully, the Supreme Court acted. On Monday, March 26, *AMP v. Myriad* appeared on the Court's weekly roster of summary dispositions with the following terse order:

> The petition for a writ of certiorari is granted. The judgment is vacated, and the case is remanded to the United States Court of Appeals for the Federal Circuit for further consideration in light of *Mayo Collaborative Services v. Prometheus Laboratories, Inc.*

It was another GVR (grant-vacate-remand), the same approach that the Court had used in *Mayo* after *Bilski* was decided. They would be heading back to the Patent Court.

The Supreme Court's remand of the Federal Circuit decision in *AMP v. Myriad* was curious. *Mayo* was a case about method claims, and the Federal Circuit had already invalidated most of Myriad's method claims. The only one that survived was an obscure claim for screening therapeutic compounds. Everyone knew that claim was just a throw-in, added by Myriad's lawyer because he thought it might be useful someday. But it had nothing to do with Myriad's diagnostics business. Could that be what the Supreme Court was concerned with? Or did the high court think that its decision in *Mayo* also affected Myriad's composition of matter claims? Anything was possible, but as Hansen and Castanias both knew, trying to discern the intention of the Supreme Court was like reading tea leaves.

The Federal Circuit, also unsure of the high court's intent, requested that the parties brief both issues. The briefs dutifully paid homage to the Supreme Court's decision in *Mayo*. Castanias again challenged Harry

Ostrer's standing to bring suit following his departure from NYU, and the ACLU again argued that, in addition to Ostrer, Matloff, Reich, the American College of Medical Genetics and the other plaintiffs all had standing. Everyone again rolled out their favorite metaphors—baseball bats, gold ore, kidneys—and filed their briefs in June.

Back to Main Justice

MYRIAD'S ARGUMENTS IN its second Federal Circuit brief were largely the same as they were in its first. But one thing that Myriad tried to do the second time around was persuade the solicitor general to change his position in the case, either by supporting Myriad's patents or not intervening at all.

Ray Chen, the Patent Office Solicitor, and the rest of his staff also saw the return to the Federal Circuit as an opportunity to convince the solicitor general to abandon Neal Katyal's cDNA compromise and support all of Myriad's patent claims. This hope wasn't as outlandish as it might seem. Katyal was now out of the picture, as were many of the other leading players who had opposed gene patents in 2010. That was Washington. Political appointees came and went quickly, but the Patent Office remained steadfast.

So when Castanias requested a meeting with the DOJ, the Patent Office lawyers were more than happy to attend. Rick Marsh flew to DC for the meeting, and invited Hans Sauer and BIO's outside counsel to attend as well.

When they entered the large conference room on the fifth floor of Main Justice, with its iconic view of the National Archive, Marsh was taken aback by the number of government lawyers in the room. Every seat at the large conference table was occupied, and he recognized almost no one.

Castanias did most of the talking. He did his best to present the case the same way he had with the Federal Circuit. The lawyers around the table asked a few questions, scribbled notes on yellow legal pads, and

Don Verrilli inherited the case from Neal Katyal when he was appointed solicitor general.

tapped noisily on laptops. After an hour, the Myriad team was politely excused. As they filed out of the conference room, they saw the ACLU team—Hansen, Park, and Ravicher—waiting to enter. Apparently the DOJ wished to give equal time to each side. There was little eye contact between the two groups as they passed each other in the hall.

Despite the efforts of Myriad and the Patent Office to sway Verrilli, it was not to be. The DOJ lawyers were not convinced, and had no desire to undo the Solomonic compromise fashioned by Freeman and Katyal. And despite the other departures and staff changes within the administration, Francis Collins was still the Director of NIH. Somehow the lanky, plain-spoken geneticist had survived three presidential administrations at the highest levels of the agency's leadership—no small feat on the Washington, DC, carousel. And Collins felt strongly about the matter. So the memoranda that Verrilli received from his staff recommended staying the course. Verrilli, who liked and respected Katyal, found no compelling reason to deviate from the position laid out by his ambitious, young predecessor. The government's brief thus advanced the same compromise that Freeman had forged the last time: genomic DNA is not patentable, but human-made cDNA is.

The Patent Court, Redux

ORAL ARGUMENTS FOR the second run at the Federal Circuit were scheduled for late July 2012, almost a year after Judge Lourie's first opinion was released. Castanias was vacationing with his family in Colorado, escaping the unbearably hot and muggy DC summer. He knew that the issues, the metaphors, and the pushback from the judges would all be the same as they were the first time around, so there was no pressing need to cancel his vacation. He later joked with a reporter that "being out in the mountains is good for mental clarity."

Castanias flew back to DC a few days before the arguments and reviewed the issues with Coruzzi and Brian Poissant, who had come down from New York. A new senior associate, Jennifer Swize, had also been assigned to the case. She helped prepare the brief that they submitted to the Federal Circuit, and stayed in DC to prepare Castanias for the argument.

In *Mayo*, the Federal Circuit hadn't bothered to schedule a second round of oral arguments when it got the case back from the Supreme Court on GVR. Castanias wasn't sure why the court had summoned them all back to argue this time, except that the Federal Circuit judges might still be smarting from their reversal in *Mayo*. Maybe they wanted to make absolutely sure that they dotted their i's and crossed their t's. They didn't want to be reversed again.

Hansen also prepared for his return appearance at the Federal Circuit. If nothing else, he learned about covalent chemical bonds. Other than that, not much had changed. He was ready.

THE FEDERAL CIRCUIT panel on rehearing again consisted of Judges Lourie, Moore, and Bryson, who seemed more or less of the same mind as they had the first time around. The question on everyone's mind was whether Judge Moore would "flip" and vote with Judge Bryson, this time rejecting Myriad's composition of matter patents.

Because Harry Ostrer was now the sole plaintiff in the case, he felt an even greater personal commitment to it. He attended the July arguments at the Federal Circuit with his wife, his daughter, and a friend, all of whom had come down from New York for the day.

Though the lawyers, the judges, and the arguments were more or less identical, giving some commentators a sense of *déjà vu* in the courtroom, at least the government's representation had changed significantly. Mark Freeman, who had been so instrumental in fashioning the government's brief in 2010, was now on a one-year temporary assignment to the solicitor general's office.[1] He had worked with Verrilli to prepare the government's Supreme Court brief in *Mayo*, among other things. So when *AMP v. Myriad* came back to the Civil Division, Freeman was not on deck to handle it. Instead, the case went to one of his colleagues, Melissa Patterson. Scott McIntosh again supervised and Beth Brinkman, head of the appellate staff, oversaw the work.

Once the decision was made to abide by the government's prior position, Verrilli found little reason to devote much energy to the case. Katyal's personal appearance at the Federal Circuit had been unorthodox, to say the least. Verrilli was too busy to argue in any court other than the Supreme Court. What's more, he didn't feel that a do-over at the appellate court was worth the time of one of his deputies. Instead, Verrilli allowed the Civil Division to handle the second Federal Circuit argument. The person assigned to make the oral argument was the most junior person on the case, Melissa Patterson.

At the oral argument, Patterson did an excellent job explaining and justifying the government's cDNA compromise. At one point, Judge Lourie asked her whether the government had abandoned the "magic microscope" test, which had not appeared in its revised brief. Patterson smiled. Though the government still found the microscope to be a useful metaphor, she explained, it also understood that the metaphor had not been particularly "helpful" to the court. The magic microscope was dead.

---

1 It is not uncommon for attorneys in the DOJ to be assigned or detailed on a temporary basis to a different unit, usually for a period of one or two years. Freeman, during his stint at the SG's office, was busy on cases that involved issues ranging from Indian law to sentencing guidelines for drug offenders.

"I Decline the Opportunity to Act"

DURING AUGUST THE entire capital shuts down, its residents flee the baking sidewalks and stultifying atmosphere and are replaced by busloads of tourists sporting fanny packs, selfie sticks, and plastic water bottles. Castanias was with his family at Rehoboth Beach on the Delaware shore. Despite the slow summer season, he checked the Federal Circuit's disposition sheet on his Blackberry every morning at eleven, just to see if anything required his attention.

One scorching day in mid-August, with his kids playing in the sand and his wife reading under a beach umbrella, Castanias noticed an entry for *Myriad* on the court's website. That was quick, he thought. He went back to the air-conditioned hotel room and accessed the decision over the sluggish Wi-Fi network. As he scanned the pages, he was struck by its similarity to the last decision issued by the court. Again, Judge Lourie wrote the decision, with a concurrence by Judge Moore and a dissent by Judge Bryson.

Castanias called Swize at the office in DC. He asked her to run a redline of the decision, a comparison of the two documents that would highlight added and deleted text. The result was not surprising.

Judge Lourie's new opinion was essentially a cut-and-paste of his earlier opinion, with a few added references to the Supreme Court's *Mayo* decision. Lourie reasoned that the Supreme Court's holding in *Mayo*, which pertained to method claims and laws of nature, was not particularly relevant to the analysis of Myriad's composition of matter claims.

Judge Bryson's dissenting opinion was also similar to his earlier opinion, in which he did not find the chemical differences between cellular and isolated DNA to be meaningful. He then added two paragraphs to address the Supreme Court's decision in *Mayo*. Unlike Judge Lourie, Judge Bryson felt that *Mayo* supported the conclusion that isolated and purified *BRCA* genes are unpatentable products of nature. In the end, Bryson came out just as he had a year before.

The only one of the three Federal Circuit opinions to change significantly between 2011 and 2012 was Judge Moore's. Whereas her earlier opinion focused on the structural chemical differences between naturally occurring DNA and isolated DNA, that entire discussion was cast aside, textbook diagrams and all. In her revised opinion, Judge Moore took to heart the Supreme Court's challenge to find a way to reconcile *Mayo*, a case involving method claims, with the law regarding compositions of matter.

She first reasoned that the Supreme Court's observations about patenting laws of nature *should* apply to compositions of matter. She criticized as "untenable" Myriad's (and perhaps Judge Lourie's) attempt to limit the holding of *Mayo* to methods incorporating laws of nature. Trying to reconcile the older cases involving products of nature with the Supreme Court's holding in *Mayo*, she reasoned that a composition of matter should be patentable if it has "markedly different characteristics" from what is found in nature *and* is a useful improvement over nature. As a result, Judge Moore concluded that isolated DNA *is* patent eligible, but not because of its chemical structure alone. Instead, the structural changes exhibited by the isolated DNA, coupled with its different and beneficial utility, make the isolated DNA patentable.

Judge Moore also reiterated her concern with upsetting the settled expectations of the industry. Her struggle with the issues presented by Myriad's patents is evident when she writes, "It is tempting to use our judicial power [to invalidate Myriad's patents], especially when the patents in question raise substantial moral and ethical issues related to awarding a property right to isolated portions of human DNA—the very thing that makes us humans, and not chimpanzees. The invitation is tempting, but I decline the opportunity to act where Congress has chosen not to . . . The patents in this case might well deserve to be excluded from the patent system, but that is a debate for Congress to resolve. I will not strip an entire industry of the property rights it has invested in, earned, and owned for decades unchallenged under the facts of this case."

Judge Moore's second opinion was thoughtful and represented a genuine attempt to reconcile the different strands of Supreme Court precedent with the underlying goals of the patent system. Yet she was unable to convince her brethren to join her. Instead, her synthesis of *Mayo* with the older product of nature cases represented the view of a single judge, and Lourie's chemistry-based analysis, which Moore concurred in, became the holding of the court. Neither would last long.

In Castanias's view, the revised Federal Circuit decision was a win. Myriad still had most of its patents, and the coming appeal to the Supreme Court had been pushed off by half a year. During that period, Myriad had earned roughly $250 million from *BRCA* testing. Thank you, Supreme Court.

CHAPTER 23

Air Force 1

A MONTH AFTER the Federal Circuit released its revised opinion, it was time for the plaintiffs to submit a new *cert.* petition to the Supreme Court. Park took the lead drafting this crucial document. There were three grounds on which they could seek reversal of the Federal Circuit's decision against them: the Federal Circuit's denial of standing to nineteen of the twenty plaintiffs, its affirmation of Myriad's method claim for screening a cancer therapeutic, and its affirmation of Myriad's composition of matter claims covering the *BRCA* genes themselves.

The Supreme Court had recently exhibited an interest in standing issues, and the conservative Justice Scalia, many felt, was obsessed with standing (mostly as a mechanism for preventing conservation groups from bringing claims under the Federal environmental laws). At this point, the legal standing of the nineteen plaintiffs other than Ostrer was not critical to the ACLU's case. The Federal Circuit had allowed Ostrer to remain, which was technically all they needed to appeal. But to Hansen, the standing issue was more than a procedural detail, it was a matter of principle. He didn't want the case to become known as "*Ostrer v. Myriad Genetics.*" He wanted every one of the original twenty plaintiffs, especially the individual women, to remain listed in the case's name. If nothing else, this would remind the Court what was really at stake.

Hansen also didn't care much about Myriad's only surviving method claim—the one for screening a cancer therapeutic, which nobody was actually using. But given that the Supreme Court had already vacated and remanded the case once in view of *Mayo*, he figured that the Court might have something more to say about method claims. So they offered the Court the opportunity to grant *cert.* to invalidate Myriad's remaining method claim, if it so chose.

The big issue, the one that everybody was thinking about, was whether the *BRCA* genes themselves were patentable. When submitting a *certiorari* petition to the Supreme Court, one phrases the appeal in terms of the "questions presented" by the case and the decisions of the lower courts. To attract the Court's attention, these questions must be drafted in a way that conveys their importance, as well as their timeliness and the urgency of the matter at hand. Most lawyers, even skilled Supreme Court advocates, try to cram as much information as possible into these one-sentence questions, essentially stating their arguments, Jeopardy-like, in the form of a question. In a case like *AMP v. Myriad*, the third question presented to the Court would typically have been phrased along these lines:

> Did the Federal Circuit err by holding that isolated and purified DNA does not constitute a product of nature as defined by this Court in numerous prior precedents, thereby rendering such isolated and purified DNA patentable subject matter under Section 101 of the Patent Act?

Ravicher and Park had worked up several variants of this question for Hansen to consider. But to Hansen, they were too much. Too many dependent clauses, double negatives, and cross-references. He was convinced that the Court needed to understand the issue the way he had understood it when Simoncelli walked into his office eight years ago. It needed to be a common sense question. Not a patent question, not a genetics question. Just a common sense question, as Judge Bryson had perceived it. "Are human genes patentable?" That was it. Only four words. Hansen thought it was probably the shortest question presented for *certiorari* in the history of the Court. But he liked it.

Myriad, of course, opposed the petition on all three grounds. Even though the Federal Circuit had not ruled 100 percent in Myriad's favor and had invalidated most of its method claims, there was no reason to believe that the Supreme Court would treat the company's patents more favorably than the Patent Court had. Especially in view of the *Mayo* decision, the Court didn't seem particularly enamored of patents that veered too close to laws of nature. But given that the Supreme Court heard only a tiny fraction of the cases submitted to it, and that it had just ruled in *Mayo* a few months ago, many people thought it was unlikely that the Court would grant *cert.* on another patent case so soon. Right now, the odds were with Myriad.

Doing the Math

OTHER THAN A tiny number of cases that the Supreme Court is constitutionally obligated to hear, such as suits between two or more states and cases involving ambassadors and other public ministers, the members of the Court decide which cases they will accept. There are no firm rules regarding the criteria that they must use to decide which cases to choose—the justices can apply whatever criteria they like. The only real rule is the "rule of four," which says that a grant of *certiorari* requires the affirmative vote of at least four of the nine Supreme Court justices.

During the Court's 2012 term (running from October 2012 to June 2013), it received approximately 7,500 *cert.* petitions, 6,000 of which were filed by self-represented (*pro se*) criminal defendants, usually from their prison cells, and 1,500 of which were "paid" cases, in which the parties were represented by counsel. Of these, the Court granted *cert.* in only ten *pro se* cases and eighty-two paid cases. Setting aside the *pro se* cases, that resulted in a grant rate of 5.5 percent.

Despite these long odds, there was one factor weighing in the plaintiffs' favor: in recent years, the Supreme Court seemed to be interested in patent cases. After all, in the past three years, the Court had weighed in on both the *Bilski* and *Mayo* cases, plus a smattering of other patent-related disputes. The ACLU and PubPat filed their second *cert.* petition on September 25. Then they hoped for the best.

Park, who had worked long hours on the petition, was looking forward to a short break. In mid-October, she was invited to a conference in Orlando organized by FORCE, the advocacy group for *BRCA* mutation carriers, to receive the organization's Spirit of Empowerment award on behalf of the ACLU. (Ellen Matloff, the Yale genetic counselor and advocate, had won it the year before, along with Francis Collins, the NIH director; Joanna Rudnick, the producer of the documentary film *In the Family*, had won it the year before that.) This year, the award recognized the ACLU's advocacy in "the successful challenge of Myriad's gene patents"—an honor that Park hoped would not be premature.

In her acceptance speech, Park reflected on how far the ACLU had come with FORCE over the past four years. When the group had invited her to speak at one of their meetings back in 2009, almost no one knew about the patents, and many were nervous about a lawsuit that might endanger their only source of *BRCA* tests. But now, FORCE was a significant ally. The prolonged round of applause after Park's speech proved it.

Weathering the Storm

ON OCTOBER 22, 2012, a tropical storm named Sandy was gaining strength in the Caribbean. By October 28, now a full-fledged hurricane, Sandy's first rain drops fell on the Jersey Shore. The full brunt of the storm hit New York City on October 29, its fourteen-foot surge overflowing the East River and flooding tunnels and streets; its eighty mile per hour winds downing power lines and bringing the city to a standstill.

The ACLU offices in Lower Manhattan closed, the surrounding streets and building lobby waist-deep in water. Nevertheless, work continued. The law firm Skadden, Arps allowed the ACLU to use overflow office space in one of its Times Square buildings while the storm damage was remediated downtown.

They were still camped out in Times Square a month later when the Supreme Court released its weekly decisions on *certiorari*. Crammed into a small meeting room with a computer, Park and a dozen other lawyers were eagerly monitoring the Supreme Court's website to see which cases had been accepted by the Court that week. In addition to *AMP v. Myriad*, they had filed a *cert.* petition in *Windsor v. United States*, a case challenging the federal Defense of Marriage Act.

When the electronic page showed that *cert.* had been granted in *AMP v. Myriad*, Park and the others cheered. They were on their way to the Supreme Court—this time for real.[1] But the Court had decided to hear only one of the three questions presented, the big one: "Are human genes patentable?"

Moving On and Coming Back

CHRIS HANSEN WAS not part of the celebration in the ACLU's borrowed offices. Three weeks earlier, after forty years at the ACLU, Hansen had officially retired.

There had been a small reception, then drinks with some of his long-time colleagues. He promised that he would come back to finish the *Myriad* case if *cert.* were granted, but that was all. The rest of his cases had been transferred to others. At the ACLU headquarters on Broad Street, his office lights were out and the shelves were cleared. The eerie Tanguy print was gone from its place of pride beside the window. At the age of sixty-five, having spent his entire career at the ACLU, Hansen was ready to go home.

But now, at least for a short time, he would be back. The ACLU team's next task was to submit briefs to the Supreme Court explaining in more detail why it should overrule the Federal Circuit on the composition of matter claims. The bulk of this work fell to Park, who drafted the majority of their brief with Hansen's input. At this point, Steve Shapiro, the ACLU's Legal Director, began to take a more active role in the case. The ACLU filed twenty Supreme Court briefs each year, and Shapiro was involved in all of them. He knew each Justice and his or her leanings on various issues, he knew what was on their docket, what they had agreed to hear and what they had passed on. Shapiro's knowledge became an invaluable asset to the team as they advanced toward the final stage of the case.

---

1 The Court granted cert. in *Windsor* the following week. The case, a significant victory for the ACLU and for marriage equality, was decided two weeks after *AMP v. Myriad*.

Just as Hansen came out of a well-deserved retirement to finish the case, Tania Simoncelli took time off from her job at the FDA to assist with this final appeal. Though she was not a lawyer, no one else on the team had been involved in the gene patenting debate for as long as Simoncelli. She helped to refine the scientific arguments—distinguishing DNA patents from patents on small molecule drugs, insulin, and other biotech products; recruiting additional *amici*; and consulting with Chris Mason and other members of the scientific team.

Dan Ravicher also re-engaged with the effort once *cert.* was granted. Following the hearing before Judge Sweet, Ravicher's role had diminished. He had helped to analyze the patents and to develop the technical arguments that won Judge Sweet over. But once Hansen and Park knew the patents and the arguments backward and forward, Ravicher moved on to other pursuits—fundraising for PubPat, teaching at Cardozo, and challenging Monsanto's patents on genetically modified seeds. Another factor was the ACLU itself. Despite everyone's good intentions, Ravicher had found collaborating with the civil rights giant to be challenging. The ACLU was great at publicizing its own people, but Ravicher observed that it seemed to have a hard time sharing credit with others. Nevertheless, Ravicher had started down this road, and he was committed to seeing it through. And, to be perfectly honest, few patent lawyers would ever see the inside of the Supreme Court; Ravicher wasn't going to miss that.

The Fate of the Industry

THE SUPREME COURT's decision to hear *AMP v. Myriad* caused concern not only among the Myriad legal team, but in the biotech community at large. In the past three years, the Court had heard two major patentability cases—*Bilski* and *Mayo*. And now it wanted to hear a third one? This would be more patentability cases than the Court had considered over the last century. What was going on?

Clearly, four of the justices felt that the patent law was still not being applied correctly, or at least not according to *their* peculiar view of it.

In *Mayo,* Justice Breyer, writing for the Court, didn't defer to the views of the Patent Office or the Federal Circuit, though both had expertise in this complex area of law. Back in 2006, in his dissent in *Metabolite,* Breyer seemed to be itching to say more about the patentability of natural products. And now, odds were that the high court had decided to hear *AMP v. Myriad* so that it could reverse the Federal Circuit. As one commentator noted, "The Supreme Court does not take cases to offer congratulations to the lower court." The result could be bad news for Myriad and its *BRCA* patents.

This worried Hans Sauer at BIO and other observers, as the direction in which the Supreme Court was headed seemed to presage less patent protection for all types of inventions that depended on "natural" phenomena. This made Sauer ask why Myriad was heading to the Supreme Court in the first place. Except for Myriad, Athena Diagnostics and a few other companies (none of which were members of BIO), patents on isolated human genes were not particularly important to the biotech industry, especially after the entire human genome was made publicly available in the early 2000s. But in trying to do the right thing by the women affected by Myriad's business practices, the Court might wreak all kinds of unanticipated havoc on *other* biotech inventions that depended on some natural law or product of nature. It was exactly as Judge Moore had feared. Patents on everything from vitamin B12 to synthetic antibodies were at risk. The fate of the entire industry was riding on Myriad's shoulders.

Clearly, input from the industry was needed now more than ever. So as the new year rolled in, with President Obama beginning his second term in office, Sauer and others began to urge every company, trade group, and bar association with an interest in biotech or patent law to speak up and submit an *amicus* brief to the Supreme Court. Briefs would be due in mid-March, so there was little time to spare.

But then, something unexpected happened. On January 9, 2013, the Supreme Court handed down a decision in a case that hardly anyone in the biotech industry had been following. It involved a trademark dispute over knockoff athletic shoes, something completely unrelated to genetics or patents. Yet the legal question that it raised struck a chord. Could this new decision make *AMP v. Myriad* conveniently go away before it reached the Supreme Court? If so, an iconic Nike shoe might save gene patenting, not to mention the rest of the biotech industry.

If the Shoe Fits

IN 1982, NIKE introduced a thick-soled, air-cushioned sports shoe branded "Air Force 1." The shoe became one of the industry's top sellers, proudly worn by NBA stars, hip hop legends, and tens of millions of teenagers around the world. Like many such fashion icons, the Air Force 1 attracted countless knockoffs, counterfeits, and imitations. To deal with these competitors, Nike had amassed an arsenal of trademarks covering not only the Nike and Air Force 1 names and the famous Nike "swoosh" but also the designs of the shoes themselves. Thus, one 2008 trademark registration held by Nike purported to protect "the design of the stitching on the exterior of the shoe, the design of the material panels that form the exterior body of the shoe, the design of the wavy panel on the top of the shoe that encompasses the eyelets for the shoe laces, the design of the vertical ridge pattern on the sides of the sole of the shoe, and the relative position of these elements to each other."

In 2009 Nike sued a company known as Yums, which was selling a line of candy-colored sports shoes whose designs closely resembled the Air Force 1. Among other things, Nike alleged that Yums was infringing its trademark on the Air Force 1 design. But unlike most low-budget counterfeiters, Yums, a sizable company in its own right, fought back. It claimed that Nike couldn't trademark the design of a shoe, and as a result Nike's Air Force 1 trademark was invalid. This allegation terrified Nike. Nike depended on the Air Force 1 trademark to go after dozens of counterfeiters. So far, Yums was the only one that had been bold enough to challenge it in court. But if Yums were successful and Nike lost the trademark, the sportswear giant's broader enforcement program would be crippled.

In an effort to avoid a credible challenge to its valuable Air Force 1 trademark, Nike did something that was clever but risky. It unilaterally authorized Yums to use the Air Force 1 trademark by sending it a written "covenant not to sue"—an irrevocable commitment that Nike would never bring suit against Yums for infringing the Air Force 1 trademark. At that point, Nike argued, Yums was no longer being threatened. As a result, there was no longer a "case or controversy" for the court to adjudicate. If Yums wasn't being injured, the Constitution said that courts shouldn't spend taxpayer dollars to hear the case. The court should thus dismiss Yums's challenge to Nike's trademark. It was an argument about standing and it was based on the same rules that resulted in nineteen of the twenty *AMP v. Myriad* plaintiffs being dismissed from their case.

The trial judge in the Southern District of New York, sitting in the same courthouse as Judge Sweet, agreed with Nike, as did the Court of Appeals for the Second Circuit. To the surprise of many, Yums appealed to the Supreme Court, which agreed to hear the case. And then, in a unanimous decision written by Chief Justice Roberts, the Court affirmed the rulings of the lower courts. Because Nike had effectively eliminated its ability to sue Yums for trademark infringement, there was no longer a dispute between the parties that warranted the use of scarce judicial resources. Yums no longer had standing to challenge Nike's trademark, even if it wanted to. The case was dismissed.

Going for Broke

WHAT IF, MYRIAD's lawyers wondered, they followed Nike's example? There was only one plaintiff left in their lawsuit—Harry Ostrer. What would happen if Myriad issued Ostrer an ironclad, no-cost covenant not to sue? Myriad could promise never to assert a *BRCA* patent against Ostrer again. Wouldn't that eliminate any controversy between the parties? Shouldn't that lead to a dismissal of the litigation, as it had in *Nike*? And if it did, wouldn't that save Myriad, not to mention the rest of the industry, from the potentially disastrous Supreme Court decision that was barreling toward them?

It might. The *Nike* gambit had supporters within both Myriad and Jones Day, as well as many of Myriad's closest allies. Two weeks after the *Nike* decision was published, Kevin Noonan, Myriad's longtime supporter, wrote a blog post titled "Is It Time for Myriad to Concede . . . for the Good of the Biotechnology Industry?" In it he wrote, "The question must be asked whether the prudent thing for Myriad to do is grant Dr. Ostrer a covenant not to sue on all the patents and claims involved in this litigation. The result would be to render the issues before the Court moot . . ." Noonan reasoned that Myriad had little to gain by winning the lawsuit: it had many other patents, and the ones in the case were set to expire in just a few years. Accordingly, Noonan concluded that Myriad had no reason to continue the litigation other than "a desire to be vindicated." And in Noonan's view, Myriad's desire for vindication was far outweighed by the risk that another wildcard Supreme Court patent decision posed to the industry.

But the *Nike* strategy had risks, too. What if the Supreme Court didn't find Myriad's covenant to be strong or broad enough? Maybe they would think differently about covenants relating to patents on breast cancer genes versus trademarks on athletic shoes. After all, there was half a century of precedent recognizing a public interest in clearing the market of invalid patents. If the Court refused to dismiss the case, Myriad could look like it was trying to do something sneaky to avoid a ruling on the merits. That could poison Myriad's legitimate arguments in the eyes of the Court.

And what if the *Nike* gambit *did* work and the Supreme Court dismissed the case as moot? What would happen then? Ostrer would have carte blanche to run a *BRCA* testing operation without paying Myriad. And not just him, but presumably his employer, too. What if Ostrer left Montefiore and went to a major commercial laboratory like LabCorp or Ambry? What if he decided to consult for every lab in the country? For Myriad, that outcome could be just as bad as having the patents invalidated in court.

And, from a procedural standpoint, what would happen to the lower courts' rulings if the Supreme Court now declined to hear the case? Under an obscure legal doctrine called judicial *vacatur*, if an appeal is rendered moot by the unilateral action of the party who prevailed in the lower court, then the lower court ruling must be vacated or set aside. So if Myriad, which largely prevailed at the Federal Circuit, issued Ostrer a covenant-not-to-sue, effectively mooting the Supreme Court appeal, would Judge Lourie's favorable decision be vacated? And what about Judge Sweet's ruling at the district court? Would that remain in effect, eliminating every one of Myriad's patent claims? Or would they simply go back to square one in the Southern District, but this time with twenty plaintiffs who knew how to draft a declaration to preserve standing?

The risks and uncertainty of the *Nike* gambit were just too great. Myriad convened a meeting of its Board of Directors, and the decision was made. The Federal Circuit had upheld their patents. They were winning. So why throw the game now? This was no time to go on the defensive with an elaborate lawyers' trick. They would go to the Supreme Court and offer their best arguments, but they would not authorize Ostrer or anyone else to operate under their patents. It may have been careful legal strategizing, or professional hubris, but whatever the reason, Myriad was going for broke.

CHAPTER 24

With Friends like These

FROM THE BEGINNING, *AMP v. Myriad* attracted significant public attention. This was due in part to Myriad's longstanding feud with the academic medical and genetics communities, as well as the ACLU's efforts to spread the word about the case. As a result, even at the district court level the case attracted a number of *amicus* briefs filed by "friends of the court." *Amicus* briefs at the district court level are rare, but *AMP v. Myriad* had fourteen of them, from groups like the American Medical Association and the Canavan Foundation supporting the ACLU, and a range of biotech companies and patent-oriented organizations supporting Myriad.

By the first appeal to the Federal Circuit, the biotechnology, medical, and disease advocacy communities were fully mobilized, leading to an even larger number of *amicus* filings. What's more, the Federal Circuit is particularly receptive to *amicus* briefs, even to the extent that it reportedly maintains an *amicus* "invite list" including bar associations and other groups whose views it values. In *AMP v. Myriad*, the Federal Circuit received more than thirty *amicus* briefs on behalf of more than sixty different parties. These numbers far exceeded the Federal Circuit's average of four *amicus* briefs per patent case.

And at the Supreme Court, the *amici* turned out in record numbers. Over the last fifty years, *amicus* filings have become common at the Supreme Court. Professor Allison Orr Larsen reports that, during the Court's October 2012 term, *amicus* briefs were filed in sixty-one of the seventy-nine cases heard by the court. For example, the groundbreaking affirmative action case *University of California v. Bakke* (1978) attracted fifty-four *amicus* filings, and the controversial abortion case *Webster v. Reproductive Health Services* (1989) attracted seventy-eight. *Bilski*, the 2010 patent case, drew sixty-five *amicus* briefs, more than any previous patent case. And *AMP v. Myriad* was close behind, with nearly fifty *amicus* briefs filed on behalf of more than eighty different parties. It would not be unfair to say that virtually everyone with an interest in biotechnology or patents weighed in on the case.

Some *amicus* briefs simply rehash the parties' legal or policy arguments and do little more than create extra work for the law clerks who are forced to review them. Some raise novel legal arguments that the parties did not make, which can help the Court consider perspectives that they otherwise might not have. For example, in *AMP v. Myriad* one pair of law professors submitted a brief arguing that patenting parts of the human body violated the Thirteenth Amendment, which abolished slavery in 1865, and the Southern Baptist Convention submitted a brief contending that such patents upset "the fundamental relationship between human beings and God." Some *amicus* briefs present the Court with the perspectives of organizations that will be affected by a decision. Thus, genetic testing companies like Invitae and GeneDX submitted briefs supporting the ACLU's position, and Genentech, Gilead, and CropLife International filed briefs in support of Myriad. Myriad even marshaled the support of patient groups to its cause. One such group, Lynch Syndrome International, argued that eliminating the patent incentive that encourages companies to develop tests for their particularly deadly syndrome would inevitably harm sufferers. Specifically, they wrote, "genetic diagnostic testing can help individuals intelligently decide whether to form a family; whom to form a family with; whether or not to have children; whether or not to adopt in lieu of naturally conceiving children; and whether and how long to use contraception"—all of which would not be possible without a market for genetic diagnostic tests backed by patents.

But the most controversial *amicus* briefs are those that present supplemental "facts" for the court's consideration—data about industries, markets, technology, science, medicine, public opinions, and more. The problem with these briefs is that they are not subject to the strict rules and procedures that govern the introduction of evidence at trial—objections, cross-examination of witnesses, challenges on appeal. These facts are simply presented to the court without any verification. They are, in the words of Professor Larsen, "eleventh-hour, untested, advocacy." Yet the Court relies on them, sometimes heavily. Larsen found that from 2008 to 2013, the Supreme Court cited facts presented in *amicus* briefs on 124 separate occasions.

One of the most influential *amicus* briefs in *AMP v. Myriad* was submitted by President Obama's science advisor, Eric Lander. Lander, who participated in some discussions of the case at NIH and DOJ, was not nearly as involved as Francis Collins or others at NIH and OSTP. Despite his prominent role in the Obama administration, Lander preferred Cambridge to Washington. He spent much of his time there at the towering glass and steel headquarters of the Harvard–MIT genomics powerhouse known as the Broad Institute, of which he was the director.

One afternoon, Lander attended one of the regular scientific seminars held at the Broad. Some recent work was being presented by a researcher looking at Aicardi-Goutières syndrome, a rare genetic disorder of the brain and nervous system. The syndrome affected DNA repair, and resulted in the accumulation of DNA fragments within a patient's cells.

That evening, Lander took a car to Boston's Logan Airport. As always, he downloaded some reading material to his iPad to make the most of the otherwise unproductive journey. He had previously read Judge Sweet's opinion in *AMP v. Myriad* and was sorry that it had been overturned by the Federal Circuit. Now, given the pending Supreme Court arguments, which he was arranging to attend, he thought he should read the Federal Circuit's opinion in full.

As he began to read, he thought that Judge Lourie was doing well. The opinion contained the obligatory lesson in elementary molecular biology, figures showing double helical strands of DNA and the mechanisms of RNA transcription. A, T, C, G—Lander covered all of this in the freshman biology class he taught at MIT. So far, so good.

But then something caught Lander's eye. Judge Lourie, in describing how Myriad's "isolated and purified" *BRCA* genes differed from DNA found in nature, wrote that "the isolated DNA molecules before us are not found in nature . . . They have to be chemically cleaved from their

native chemical combination with other genetic materials." Lourie's conclusion was that the only way that isolated genes could be created was
through a human intervention that broke the covalent chemical bonds
linking those genes to the rest of the chromosome.

But wait, Lander thought. Hadn't he just seen slides that morning that
illustrated DNA fragments floating around the cells of a patient with
Aicardi-Goutières syndrome? Those DNA fragments weren't cleaved from
their chromosomes in a lab. The covalent bonds tethering them to the rest
of the chromosome were broken through some natural process triggered
by the disease. Lander looked for a citation to support Judge Lourie's
statement. There was none. Had the judge made it up, or perhaps taken
Myriad's word for it? Whatever the reason, Lander concluded that Judge
Lourie was wrong. And his mistake was one of the fundamental bases for
the court's decision. Because, as Lander knew, isolated DNA fragments
could be products of nature, too. And as such, they should not be
patented.

The next morning, the discrepancy in the Federal Circuit's opinion
continued to nag at Lander. He was in California, but called his office
anyway. He discussed his observation with a few colleagues and they
decided to find all the scientific literature that existed on cellular DNA
fragments. Forty years of journal articles were waiting for him when he
got back. As Lander made his way through them, he became incensed.
This wasn't a new discovery. Researchers had known for *decades* that DNA
fragments broke away from chromosomes without external chemical intervention. Judge Lourie was wrong. The Federal Circuit was wrong. Lander
felt morally obligated to bring this to the Supreme Court's attention.

It never occurred to Lander that his own interests might be affected
by the outcome of *AMP v. Myriad*. Alongside his many academic positions, Lander was also a co-founder of the biotech firm Millennium
Pharmaceuticals. Millennium, which was formed in 1993 during the
heyday of gene discovery, planned to develop drugs targeted to disease-related genes.[1] Like Myriad, Millennium derived much of its value from
patents. As described in a 1999 *Fortune* magazine profile, "Millennium's

---

1 Millennium was acquired in 2008 by Takeda Pharmaceuticals for approximately $8.8 billion.

brain trust realized early that a broad genomics effort would produce an abundance of intellectual property. Not only would it reveal thousands of disease-related genes . . . but also every single one might become a little patent factory." Lander was a direct beneficiary of these little patent factories.

But Lander seemed to derive perverse enjoyment from arguing against the interests of his own companies. At conferences, he was known to embrace positions opposite those taken by his companies' lawyers. And as for their reaction to his intervention in *AMP v. Myriad*, he couldn't have cared less. "I didn't ask them. It didn't matter," he says nonchalantly.

So that afternoon, Lander started to sketch the outline of an *amicus* brief for the Supreme Court. While he knew exactly what scientific argument he needed to make, he had never filed such a brief. He would clearly need some legal assistance, and he couldn't ask the in-house counsel at Millennium or the Broad to help.

Kendall Square in Cambridge, Massachusetts, is, according to some, the epicenter of the biotech industry, and across the river in Boston are dozens of law firms that cater to the legal needs of the companies crowding the corridor between Harvard and MIT. As the head of one of the most prominent genomics research institutions in the world, not to mention a high-flying biotech company of his own, Lander had never had a problem finding a lawyer. In fact, law firms were usually begging for his business. But not this time. Not one of the firms that Lander had used in the past was willing to assist with a brief opposing gene patenting. It was the same problem that Lori Andrews had faced twenty years earlier when she litigated the Canavan case. No law firm wanted to be responsible for killing the goose that laid the golden egg.

Eventually Lander recruited Glenn Cohen, who ran the health law program at Harvard, to help with the brief. Cohen, in turn, was able to recruit a couple of lawyers from the West Coast. Lander also consulted with some of the people within the Obama administration who knew the details of the case—which was not difficult given Lander's position as one of the president's top science advisors.

When the brief was done, Lander was pleased. It began modestly, with the intention of providing "information and perspective concerning several scientific issues at the center of the case." But once Lander got going, he held no punches. He cited thirty-two peer-reviewed scientific articles published between 1974 and 2012 in support of his conclusion that the Federal Circuit's "foundational assumption" that DNA fragments do not occur in nature was "demonstrably incorrect." Of all the *amicus* briefs filed in the case, Lander's would assume a prominence second only to the solicitor general's.

Moot, Moot

ARGUING BEFORE THE Supreme Court of the United States is one of the greatest honors that an American lawyer can achieve. Of the three oralists in *AMP v. Myriad*, Verrilli had appeared most often before the Court. Both as SG and in private practice before that, he had handled some of the most important intellectual property and technology cases of the past two decades. This would be his twenty-seventh Supreme Court argument. Castanias had previously argued three times before the court, and Hansen only once, twenty years ago, in a school segregation case from suburban Atlanta.

But no matter how skilled or seasoned a lawyer is, one can never be too prepared to appear before the nine justices of the Supreme Court. Thus, there is a time-honored tradition among Supreme Court advocates to practice—or moot—their arguments in advance of their court date.

Mooting a case is very different from practicing a speech in front of a mirror. Oral argument, especially at the Supreme Court, is the legal equivalent of a contact sport. The nine justices—arguably nine of the smartest and most opinionated people in the country—constantly interrupt the speaker, and one another, with questions real, hypothetical, and purely rhetorical. They try to guide the speaker toward their own views of the facts and the law, or away from the views of their colleagues. They speculate, theorize, and pontificate. Though they generally do not harass the speaker out of meanness, this is sometimes the impression that comes across.

Arguing before the Supreme Court is part theater, part science, part law school exam. It is essential that the advocate know not only the facts and legal theories of the case, but the questioning style of each justice and his or her particular perspectives on, and experience with, other cases. There is a veritable cottage industry that analyzes the voting patterns, questioning styles, and policy leanings of the nine justices of the Supreme Court. For example, one study reports that between 2009 and 2011, Justice Scalia, the most vocal justice on the bench, asked about a thousand questions, an average of nineteen per case. Chief Justice Roberts asked eight hundred, Justice Samuel Alito asked four hundred, and Justice Clarence Thomas, infamous for his apparent somnolence during arguments, asked none at all. In terms of the types of questions asked, Justice Alito seemed the most interested in policy issues, which represented 46 percent of his questions from the bench, while the three female justices, Ginsburg, Sotomayor, and Kagan, devoted most of their questions (over 60 percent) to legal doctrine.

Castanias mooted his argument at Jones Day, in an expensive, state of the art mock-up of an actual courtroom. Brian Poissant and Laura Coruzzi came down from New York to participate, along with a variety of other partners and associates. Their job was to pose questions that the justices might ask and to help Castanias think on his feet about the case and its interrelated parts. Castanias had the advantage of mooting with several partners who had themselves argued before the Supreme Court, as well as numerous former clerks who knew how the justices approached different issues. Jones Day paid a premium to hire these former Supreme clerks, and there was a reason why.

Hansen mooted twice, once at the ACLU and once at Cardozo School of Law. At the ACLU session, Steve Shapiro played the role of Chief Justice Roberts, convening the proceedings. Lori Andrews was there with Park and other ACLU lawyers asking a variety of questions. Harry Ostrer played himself, as plaintiff. The questions, which he knew were being posed to test Hansen, maddened Ostrer nonetheless. He furiously gesticulated and scribbled notes to Hansen about how to answer. Afterward, Hansen told him that at the Supreme Court, Ostrer had to behave. He could show no emotion whatsoever, and notes would not be possible, as only counsel were permitted to sit near the speaker.

Hansen's second mooting session was held at Cardozo, where Dan

Ravicher and the Public Patent Foundation had their offices. Ravicher hosted the event, and recruited a range of law professors, local practitioners and students to quiz Hansen during his argument. This was good practice because, unlike the judges on the Federal Circuit, the justices of the Supreme Court were generalists who knew little about patent law or science. This would be Hansen's fourth time arguing *AMP v. Myriad* in court—he felt prepared.

It was Verrilli, though, arguing for the cDNA compromise, who had the most accomplished panel of moot court judges. He had first been briefed on the scientific issues by Eric Lander, who flew from Boston to DC to coach the solicitor general in the intricacies of human genetics. At the mooting session, they placed a podium on one end of the long table in the solicitor general's conference room at Main Justice. Through the large window opposite the table there was a clear view of the National Archive, whose ornate rotunda displayed original copies of the Declaration of Independence, the Constitution, and the Magna Carta. The large red, white, and blue flag flapping in front of the Archive's massive Corinthian columns was a picture straight out of a patriotic video—known to insiders as "the best view money can't buy."

About sixteen people were seated around the long table, with an additional row of seats around the outer perimeter of the conference room. There must have been thirty or so people in the room. They included the assistant and deputy SGs working on the case, as well as Mark Freeman and Melissa Patterson from the appellate staff. These lawyers, who knew the legal arguments and precedents backward and forward, bombarded Verrilli with hard questions. They tried to trip him up, to trap him the way Justice Scalia might, or to guide him down fallacious paths, as the two Harvard professors—Breyer and Kagan— likely would. According to Verrilli, the DOJ lawyers were brutal. "They figure out all the hardest questions. They don't hold back."

In addition to the large cast of lawyers, Lander, as well as a contingent of scientists from NIH, participated in Verrilli's mooting sessions. Francis Collins was there, along with Eric Green, who led NIH's Genome Institute, and Larry Brody, a senior geneticist who had studied the *BRCA* genes.

Interestingly, Hansen and Park also came to Washington at Verrilli's invitation. They sat along the back wall of the conference room during one of his mooting sessions. The SG often invited the party on whose side it was arguing to participate in a moot. Though in this case the solicitor general's brief stated that it was submitted on behalf of "neither party," the government's position was clearly more closely aligned with the plaintiffs' position than Myriad's. Thus, the ACLU team was invited to provide input in Washington, and some of the DOJ attorneys attended one of Hansen's moots in New York.

While Verrilli stood at the podium at the far end of the long conference table, the researchers quizzed him on the scientific details, made sure he was correctly distinguishing cDNA from genomic gDNA and why it mattered. Lander, of course, mentioned his own *amicus* brief and what he perceived as the fatal flaw in the Federal Circuit's reasoning. Verrilli soaked it all in. He had received intensive genetics tutorial sessions from a team of NIH researchers, and had become surprisingly adept with the scientific concepts at hand. What's more, unlike the judges of the Federal Circuit, the Supreme Court justices were not chemists or engineers. Verrilli likely wouldn't face any off-the-wall scientific questions. For good measure, though, Hansen warned Verrilli about covalent bonds.

One thing that everybody agreed on was the need to reduce each legal argument to its bare essentials. The Court itself offers this advice in its instructions for potential oralists, noting, in charmingly anachronistic terms, that "preparing for oral argument at the Supreme Court is like packing your clothes for an ocean cruise. You should lay out all the clothes you think you will need, and then return half of them to the closet." A few paragraphs later, the Court's manual shifts to more practical advice, including the important tip that "a legal sized pad does not fit on the lectern properly. Turning pages in a notebook appears more professional than flipping pages of a legal pad."

The Court urges advocates to make their presentations as interesting as possible, adding that they "should speak in a clear, distinct manner, and try to avoid a monotone delivery." As one distinguished Supreme Court advocate warned thirty years earlier, "the greatest barrier to an effective argument is boredom."

Of course, respect and protocol are paramount as well, and the Court makes it clear that an advocate should never interrupt a Justice. "Give your full time and attention to that Justice—do not look down at your notes, and do not look at your watch or at the clock . . . If you are speaking and a Justice interrupts you, cease talking immediately and listen."

And, perhaps the most useful—and the most frequently disregarded—advice offered by the Court concerns the content of the arguments presented: "The Supreme Court is not a jury. A trial lawyer tries to persuade a jury with facts and emotion. Counsel should try to persuade this Court by arguing points of law."

Saving Lives

As THE ATTORNEYS prepared for their April 15 arguments, the ACLU continued to generate public interest in the case. In February, *SCOTUSBlog*, the leading outlet for news about the Supreme Court, ran a special online symposium about gene patenting featuring contributions from leading patent scholars on both sides of the debate. Feature stories about gene patenting appeared in the *New York Review of Books* (March 7), *The New Yorker* (April 1), and *WIRED* (April 13).

Myriad was not silent during this period either. In February, the company celebrated a significant milestone—it conducted its one millionth *BRCA* test. Then, a few days before the oral arguments, its CEO Pete Meldrum issued a self-congratulatory press release entitled "Myriad Genetics, the Supreme Court, Gene Patents, and Saving Lives."

As the court date drew near, television crews began to produce segments on the case. CBS Evening News filmed Kathleen Maxian at her home in Buffalo. She was in recovery. That February, she had been admitted to Memorial Sloane Kettering in New York to have her lymph nodes stripped—a complex and delicate procedure to remove potentially cancerous tissue. The procedure had gone well, and she was now back on chemo. Nevertheless, she mustered the strength to appear on camera. She and her husband had even managed to look relatively at ease taking their little black dog for a walk. His name was Bart.

CHAPTER 25

Oyez, Oyez, Oyez!

THE SUPREME COURT of the United States occupies a grandiose marble edifice of neoclassical design, its stately columns evoking the Parthenon of Athens and its thirteen-ton bronze doors depicting jurisprudential milestones like the Code of Justinian and the Magna Carta. Since 1935 the justices of the Supreme Court have presided in this stately building, across the street from the sprawling Capitol complex, a couple miles down Pennsylvania Avenue from the White House. From its chambers have emanated judicial rulings that have shaped and reshaped American law, defining the contours of every citizen's rights under the Constitution and the powers and limits of government.

Hearings are generally open to the public, but seats in the small gallery behind the reserved sections are distributed on a first-come, first-served basis. In important cases, getting a seat means arriving early, sometimes days before the argument. For the cases scheduled to be heard on April 15, 2013, the line began to form the night before.

A light drizzle fell over the sleeping capital, obscuring the moon that occasionally peeked through the low hanging clouds. Myriad attorneys Rick Marsh and Ben Jackson, who had flown in from Salt Lake City earlier that day, had dinner with the Jones Day team, then took a leisurely stroll around the Capitol, discussing the case and how far it had come. As they passed the Supreme Court building, they saw the line forming in the dark, people camping out in a queue stretching into the plaza from the public entrance to the court. Interesting, they thought. We'll have an audience.

Those waiting outside huddled under umbrellas, some crouching on plastic bags spread over the wet pavement. By dawn the line had grown until it snaked across the plaza and onto the adjoining sidewalk, glistening

Some spectators waited overnight in line to hear
the Supreme Court arguments on April 15, 2013.

in the morning sun. Latecomers would not get a seat, but might be admitted to the rear of the gallery to observe the proceedings in rotating three-minute shifts.

The rain subsided as the sun brightened over Washington, bathing its white marble edifices in imperial glory. This week, the capital's massive government offices and commercial blocks were further brightened by the outpouring of pink and white cherry blossoms that blessed the city every spring. The blooming of the delicately flowering trees, gifts of the Japanese government more than a century ago, is eagerly awaited by both residents and visitors to the capital. The peak of the bloom had arrived a week ago, so by now many of the trees had dropped their blossoms, creating a damp, fragrant carpet of petals across the city's streets, sidewalks, and plazas.

BRCActivists

GIVEN THE CONTROVERSY and passion that often surrounds Supreme Court cases, the marble steps of the court building have been no strangers to protest. Over the years these steps have witnessed rallies and marches, candlelight vigils, the singing of hymns, the burning of draft cards, and

*Donna Kaufman of Breast
Cancer Action at rally outside
the Supreme Court.*

the desecration of flags. All of these exhibitions have been carried out by a vocal and engaged citizenry making their voices heard before the highest court of the land. They are the angry, the concerned, the bereaved, the hopeful. The arguments in *AMP v. Myriad* were no exception.

Demonstrators began to arrive as the sun rose. They held signs and placards; a few chanted into megaphones. DON'T PATENT OUR GENES! the signs read. NO ONE SHOULD OWN OUR DNA! OUTLAW HUMAN GENE PATENTS! They were mostly women, young and old, mothers and daughters, several wearing pink ribbons and other paraphernalia of the breast cancer movement. They called themselves "BRCActivists." Observing the pink cherry blossoms throughout the plaza, some were inspired; some thought it was a sign.

The rally, as they called it, had been organized by Breast Cancer Action, one of the ACLU's organizational plaintiffs, and FORCE, the "previvor" advocacy group that helped to recruit some of the individual plaintiffs. Many of the participants at the rally had never been to the Supreme Court before. Donna Kaufman, a military veteran and breast cancer survivor who volunteered with BCA, had done much of the legwork to plan the event. She carried a poster showing photos of women who were too ill to attend in person. Lisa Schlager, FORCE's Vice President of Community Affairs and Public Policy, was there with her daughter

Rachel. The had spent the previous evening making signs and coming up with catchy slogans. None of them were real political activists, so being on the steps of the Supreme Court, chanting into a megaphone, watching important-looking people get out of black SUVs and hurry toward the Court building was a surreal experience.

Soon, news vans bristling with satellite dishes and antennas rolled up to the sidewalk. Camera crews and correspondents arranged themselves in advantageous positions, the Court building—partially obscured by scaffolding and plastic bunting—still looming grandly in the background. No cameras were permitted inside, so the reporting would be done from the curb. Tents had been set up to protect the equipment from the rain. Along the perimeter of the plaza, the Capitol police were outfitted in Kevlar, dark glasses, and visible weapons. It seemed like they came expecting Abbie Hoffman and ACT UP. Instead they got moms and tweens.

A little before nine the ponderous doors to the Supreme Court building opened and the spectators filed in. Security was tight. As in all federal buildings, visitors are greeted with obligatory metal detectors and x-ray machines. But at the Court the guards go a step further and confiscate all electronics—phones, laptops, recorders—as well as all notebooks,

Demonstrators outside the Supreme Court.

briefcases, portfolios, and anything else beyond a simple notepad and pen. Nothing posing even the slightest risk may enter the Court's hearing chamber.

Some of the wet and bedraggled crowd who had waited all night began to trade places with new arrivals who emerged from taxis and sedans. Those departing were line-standers, enterprising scalpers who queued up at concerts, art exhibitions, and court hearings for a fee. Once their patrons arrived, the standers took their money and traded places with those who had hired them.

The legal teams entered through side doors. A medium-sized chamber on the first floor of the building is set aside for counsel to wait and prepare before their arguments. The doors to this Lawyer's Lounge generally open at nine o'clock, but Hansen and Simoncelli arrived a few minutes early. He asked her to sit with him. Stiff-backed wooden chairs were arranged in five rows facing the front of the room. They each took a seat in the back row.

"This case never would have happened without you," he said quietly.

"Or without you," she replied.

They spoke for a few minutes, reminiscing about the incredible journey they had taken together. When others started to file into the room, Simoncelli stood.

"Break a leg," she said. He smiled, and she left to go find her place in the ticketholder line.

All the King's Horses

THE COURT'S GRAND hearing chamber, its floor covered in thick, crimson carpet, filled quickly. Between the gallery and the justices' high bench are a set of polished mahogany tables, behind each of which stand four high-backed chairs. On the table, in front of each seat, lay a pair of neatly crossed goose quill pens—a tradition begun by Chief Justice John Marshall in the 1800s. The Court's protocol manual explains, "The quill pens at counsel table are gifts to you—a souvenir of your having argued before the highest Court in the land. Take them with you. They are handcrafted and usable as writing quills." Chris Hansen, Sandra Park, Dan Ravicher, and Lenora Lapidus, head of the ACLU's Women's Rights Project, occupied the counsel table to the right. Myriad's counsel from Jones Day—Greg Castanias, Laura Coruzzi, Jennifer Swize, and Dennis Murashko, a junior lawyer on the team—occupied the table to the left.

A few weeks earlier, there had been a small disagreement between the ACLU and the Department of Justice over who would sit at the plaintiffs' counsel table. Traditionally, when the government supports a party, its attorneys occupy one or two of that party's seats at the table. But here, the government was still trying to broker a compromise, adopting neither party's side completely. As a result, neither Myriad nor the ACLU would cede any of its four seats to the solicitor general or his staff. So the Court squeezed a third counsel table into the space between the two parties' tables. Here sat Don Verrilli, his gray moustache neatly trimmed, wearing the long Edwardian morning coat, gray vest, and striped trousers required of the nation's chief litigation counsel.[1] Beside Verrilli were two of his deputies, along with Mark Freeman, who had now returned to the Civil Division after his one-year stint with the SG. Though Freeman only ranked seventh among the eight names listed on the SG's brief, nobody had done more to formulate the government's position in the case. He would go home with his goose quill today.

Behind the counsel tables stood three rows of chairs reserved for attorneys who were admitted to practice before the Supreme Court. A waist-high brass railing separated them from the main gallery, in which the general public could observe the proceedings from five rows of long, pew-like benches. Behind these benches, interspersed among the massive marble columns at the back of the courtroom, were additional chairs for members of the public who were not lucky enough to get reserved seats. They would cycle through in three-minute increments, getting a quick glimpse of the Court in action. On the left wing of the courtroom, nestled among similar columns, were seats for the press corps, and on the right were two rows of upholstered benches for the justices' personal guests, and less comfortable chairs for their law clerks.

---

1 The only female SG to that date, Elena Kagan, received a special dispensation from the Court to wear a conservative pants suit during her arguments.

Toward the front of the gallery sat Pete Meldrum, Rick Marsh, and Ben Jackson, from Myriad. Brian Poissant, the Jones Day partner who argued the case in Judge Sweet's courtroom, was with them, pointing out the more interesting architectural features of the grandiose courtroom.

Several other members of the plaintiffs' legal team were also in attendance. Steve Shapiro and Aden Fine from the ACLU were members of the Supreme Court bar, and sat in front of the brass railing. Simoncelli, now back among the observers, had a ticket for the general gallery, where she sat beside Anthony Romero, the ACLU's Executive Director. The Court gives six such seating tickets to each advocate who argues before it. Hansen gave one to Simoncelli and the other five to members of his family who traveled to Washington to watch his performance—his two sisters and their husbands, and a niece who was herself undergoing treatment for breast cancer. At least four other relatives waited in the line outside, but they didn't arrive early enough to get gallery seats. Instead, they would be ushered through the rear of the courtroom in three-minute increments to see what they could.

Several of the twenty original plaintiffs in the case also attended the oral arguments. Haig Kazazian and Arupa Ganguly from Penn had paid line standers to stake out places in the queue the night before the argument. Kazazian and Ganguly arrived around 7:00 am to replace their fill-ins. Ganguly's line-stander had secured place number twenty-two, which was sure to earn her a seat, but Kazazian's less diligent stand-in was only fifty-ninth in line. Once again, however, Haig Kazazian got lucky, as sixty members of the public were admitted to the Court that day. Wendy Chung, the geneticist from Columbia hadn't taken any chances. She had waited patiently in the rain all night, and her perseverance had paid off—she got a seat in the gallery. Ellen Matloff, the genetic counselor from Yale, also got lucky. She had been prepared to wait in line outside the Court, but at the last minute Park found a ticket for her—as Matloff recalls, the ticket had been reserved for a VIP who, for some reason, gave it up.[2]

Mary Williams, executive director of AMP, the lead plaintiff, and

---

2 The extra ticket may have belonged to John Holdren, the president's science advisor and head of OSTP, who had planned to attend the oral arguments but was unexpectedly called away on other White House business.

Karuna Jagger from Breast Cancer Action were also in attendance. Jagger had helped to organize the rally outside, but she made sure to get a seat inside while her co-organizer, Donna Kaufman, led the chanting on the courthouse steps. Lisbeth Ceriani, now in remission and largely recovered from her difficult radiation and chemo, traveled from Boston for the event. It was Ceriani's first time in Washington.

Harry Ostrer, the sole plaintiff with standing in the case, brought his fifteen-year-old son Nathaniel along to witness the historic argument. Ostrer took a few deep breaths once he sat. As Hansen had warned him after the mooting session, Ostrer needed to rein in any excitement and remain composed in the courtroom.

To Ostrer's other side was someone who introduced himself as being from the Patent Office, there to glean what he could about any rulings that might affect its procedures. But other than this designated note-taker, most of the Patent Office attorneys and officials who had worked on the case stayed away. Ray Chen and David Kappos, in particular, had no stomach for what was about to occur. Likewise, Mark Skolnick, Myriad's co-founder, had little interest in hearing the ACLU denigrate his discovery or his company. Hans Sauer from BIO also stayed home, even though his office was just a few blocks from the Court. "I'm not going to that circus," he said, feeling queasy about the damage that the case could wreak on the biotech industry well beyond Myriad.

An array of invited dignitaries occupied the reserved seats for guests of the Court. They formed a veritable who's who of American science. James Watson, the aged but energetic co-discoverer of the famous "double helix," sat toward the front. Watson had filed an *amicus* brief pointing out that *he* hadn't seen fit to patent the structure of DNA itself, and didn't see why Myriad should be entitled to patent its far less significant discovery.

Near Watson was another Nobelist, Harold Varmus. Varmus had recently resigned as president of the prestigious Memorial Sloan Kettering Cancer Center in New York to return to NIH as director of the National Cancer Institute, NIH's largest funding body. Nearby was the government's genomics dream team, led by Francis Collins, the director of NIH. Collins's involvement with the *BRCA* genes predated that of anyone else in the room. Twenty-three years ago Mary-Claire King called him to propose a collaboration to find the gene that she had mapped to the long arm of chromosome 17. Since then he had led the Human Genome Project and NIH itself. Collins marveled at how far they had all come. Next to Collins was Eric Green, his longtime lieutenant and the current director of NIH's Genome Institute.

Eric Lander sat near Simoncelli. Lander was accompanied by his daughter Jess, now a high school teacher. Two of Lander's staff from the Broad Institute were also in attendance. They had helped Lander to prepare his *amicus* brief and were excited to see how the case turned out. It was no small feat for Lander to get four coveted tickets to the arguments, but he had approached this task with the same dogged determination as everything else and, after making multiple requests through different channels, had ensured that his team could be there.

Judge Robert Sweet, now ninety-one, also occupied a seat of honor. He had obtained his ticket through the office of Justice Sotomayor, his former colleague in the Southern District of New York. To a large degree, Judge Sweet had laid the groundwork for the case four years ago. Now he wanted to see what would happen to his baby. Seated beside the judge was a young woman who knew absolutely nothing about the case. She was there because her husband was scheduled to argue the second case that morning, and she had been advised to arrive early to get a good seat.

In addition to the dignitaries, the parties and their counsel, scattered throughout the packed courtroom were a variety of other interested observers, most of whom had obtained their tickets through some back-channel connection or the infamous line outside. Dan Kevles, a prominent historian of science from Yale, sat beside Lander and his daughter. Chris Mason avoided the line with a reserved ticket from his friend in Senator Reid's office and sat next to Ellen Matloff. Bob Cook-Deegan and Arti Rai from Duke attended along with a half-dozen Duke law students who had driven all night and camped out in front of the court building to secure their places in line. Lori Andrews was there, dressed impeccably, as always, accompanied by two colleagues who had worked with her on the Canavan case nearly twenty years earlier. For Andrews, the fact that she was sitting in the Supreme Court after so many years was vindication enough, no matter how the case turned out.

Supreme Court oral arguments, especially in important cases, are like NFL playoff games to law professors. In addition to Rai and Andrews, Josh Sarnoff, who had offered his advice to the ACLU team early on, was there. Sarnoff, a Supreme Court junkie (at least where patents were concerned), had attended the oral arguments in both *Bilski* and *Mayo*, and was now eager to see the third installment of the Court's patentability saga. John Whealan from George Washington University, his silver ponytail standing out against his dark suit, was also in attendance.

The press section was full, with reporters brandishing notebooks and pencils, all without the benefit of electronic recording devices. Long-time Supreme Court correspondents like Nina Totenberg from NPR were there, seated by seniority at the front of the press section, as well as science and women's health journalists who seldom ventured into court.

Finally, by sheer dint of luck, a group of high school students visiting Washington were seated in the gallery. Places had been reserved for them months in advance, long before the hearing schedule was published. Whether they knew it or not, they would witness history being made that morning.

How the Cookie Crumbles

A LOUD BUZZ of conversation reverberated within the marble walls of the Supreme Court's hearing chamber. Every seat was filled, and the first wave of public observers was ushered to the seats at the rear. A minute before 10:00 a.m., a tall woman with shoulder-length gray hair entered and walked to one of the small desks adjoining the Court's high bench. Like Verrilli and his deputies, she wore formal attire: a dove gray butler's coat, a charcoal vest and silk tie and carried in her hand a heavy wooden gavel. She was Pamela Talkin, the Court's long-time marshal, and her entrance signaled the commencement of the proceedings. At precisely 10:00, a high-pitched chime sounded and the room quickly became quiet.

All eyes focused on the long mahogany bench, behind which nine black leather chairs were precisely arrayed. It was raised several feet above the floor, causing everyone in the chamber to crane their necks to view it. About twenty feet above the center of the bench hung a large bronze-cased clock. Above it a gigantic marble frieze spanned the room, depicting in its center two Zeus-like figures representing the Majesty of Law and the Power of Government. Behind the bench, four massive Ionic columns stood out against a heavy red curtain like set pieces in a Cecil B. DeMille film.

A moment later the marshal brought down her gavel with a reverberating clack, and called out, "The Honorable, the Chief Justice, and the Associate Justices of the Supreme Court of the United States!"

At that point, the nine justices, each robed in black, silently emerged though three openings in the curtain and took their places behind the bench. The marshal stood erect and called out in stentorian tones, "Oyez! Oyez! Oyez![3] All persons having business before the Honorable, the Supreme Court of the United States, are admonished to draw near and give their attention, for the Court is now sitting. God save the United States and this Honorable Court!"

These centuries old trappings were new and wholly unexpected to many of the observers. Several of the scientists, especially, stared with disbelief at the pomp and spectacle unfolding before them. Was this truly where the fate of gene patenting in America would be decided? By an institution that seemed to have been plucked straight from the set of *Harry Potter*?

The gavel fell again, and the justices took their seats, followed by counsel and the spectators. The chief justice, seated at the absolute center of the bench, began with some routine administrative announcements, then got under way. "We'll hear argument first this morning in Case 12-398, Association for Molecular Pathology versus Myriad Genetics." He paused and looked down at Hansen, already standing at the podium. "Mr. Hansen?"

---

3 The word "oyez" (pronounced "o-yay") is an archaic throwback to the Norman-French judicial tradition. It means "hear ye" or, in modern terms, "your attention, please."

Though Hansen had been here before, it was still a thrill to stand at the small podium, just a few feet from the nine justices. From this very podium legal heavyweights like Erwin Griswold had argued against the publication of Richard Nixon's Pentagon Papers, Archibald Cox had defended affirmative action in *University of California v. Bakke,* and Thurgood Marshall struck the decisive blow against school segregation.

Now it was Hansen's turn, and he began. "Mr. Chief Justice, and may it please the Court." He had only twenty-five minutes to speak. Of the thirty minutes usually allotted to the plaintiffs, five had been ceded to the solicitor general. Hansen spoke crisply and efficiently. "One way to address the question presented by this case," he said, "is what exactly did Myriad invent? And the answer is nothing." Hansen looked up at the justices. He could tell that they were interested. This case fascinated them.

"Myriad unlocked the secrets of two human genes," he continued. "These are genes that correlate with an increased risk of breast or ovarian cancer. But the genes themselves—where they start and stop, what they do, what they are made of, and what happens when they go wrong—are all decisions that were made by nature, not by Myriad."

"Now, Myriad deserves credit for having unlocked these secrets," Hansen paused for effect. "Myriad does not deserve a patent for it."

This is as far as Hansen got before he was interrupted. He was pleased that they let him make it all the way through his opening statement. It was good, and he knew it.

The first to question him was Justice Ginsburg. At eighty, she was the oldest member of the Court. She appeared tiny, perched behind the bench and peering at him through thick, owlish glasses. Ginsburg, of course, knew the ACLU well. She had served as its general counsel from 1973 to 1980 and was the founder of its Women's Rights Project. This was well before Hansen arrived, but it was no coincidence that the ACLU seated Lenora Lapidus, the current head of the Women's Rights Project, at the counsel table in clear view of the justices. Ginsburg was also intimately familiar with cancer. She had twice been diagnosed—with colon cancer in 1999 and pancreatic cancer in 2009—and still mourned the loss of her beloved husband Marty to the disease in 2010. Now, Justice

Ginsburg was asking about patents, her husky Brooklyn accent under-scoring a tough, no-nonsense approach to the law.

The questions were basic, and Hansen had prepared well for them. Chief Justice Roberts and Justices Kennedy, Alito, Scalia, and Sotomayor all joined the discussion, trying to tease out when a natural product—a leaf broken from a branch, gold mined from the ground—was sufficiently transformed from its native state to merit a patent.

Justice Kagan, the former Harvard Law dean, then asked about financial incentives, as she often did. Without patents, what would motivate researchers to discover useful things like the *BRCA* genes? Kagan, who had been solicitor general when the case came to the Department of Justice, appreciated what her successor Neal Katyal had done—she knew how difficult it must have been to go against the Patent Office. Yet it seemed right that she should air some of the arguments being made by the Patent Office and others who supported the patents.

Hansen replied that academic researchers were already close to finding the genes when Myriad discovered them. They weren't after patents, though—they were paid by the taxpayers to make discoveries, they were curious about nature, they wanted Nobel prizes. This got a chuckle from the audience, especially from those who did have Nobel prizes. But Scalia and Kennedy weren't satisfied, and Kagan slipped into law professor mode, leading Hansen along for the benefit of her colleagues: "I thought you were going to say something else, Mr. Hansen . . . which is that, notwithstanding that you can't get a patent on this gene, there are still various things that you could get a patent on that would make this kind of investment worthwhile . . . But if that's the case, I want to know what those things are rather than you're just saying . . . we're supposed to leave it to scientists who want Nobel Prizes. And I agree that there are those scientists, but there are also companies that do investments in these kinds of things that you hope won't just shut down." Scalia and Kennedy didn't seem entirely satisfied, but the discussion moved on.

Justice Sotomayor, who saw her old friend Robert Sweet sitting in the special guest section, raised the issue of cDNA. She asked Hansen why cDNA, if constructed in the lab, should not be patentable. In other words, she wanted to know why the ACLU had not adopted the government's compromise position. Sotomayor had no training in science, but she had clearly been thinking about cancer genetics. In fact, as long ago as 2010 she surprised acting SG Katyal during an oral argument in a seemingly unrelated case about employment law. She asked him whether an employer could legally ask an employee, "What's your genetic makeup, because we don't want people with a gene that is predisposed to cancer?" Katyal, flummoxed by the off-the-wall question, sidestepped it. Had Justice Sotomayor seen one of the ACLU's videos, or the *60 Minutes* segment, or one of the many ads that Myriad had run in New York? Or maybe the justice had a particular interest in genetics. Though not a scientist herself, she had been married to one for seven years. Sotomayor and her ex-husband, the biotech patent attorney and blogger Kevin Noonan, had remained on good terms after their split. And Noonan was certainly steeped in, and highly vocal about, the controversy surrounding Myriad and its cancer screening test.

The discussion prompted by Justice Sotomayor's cDNA question lasted a few minutes, with several of the other justices chiming in. Justice Breyer, especially, seemed eager to dive into the language of the patent claims. And with that, Hansen's twenty-five minutes were up. He was pleased. None of the questions had been too damaging, and he emerged relatively unscathed from the crucible.

Chief Justice Roberts next invited the solicitor general to the podium. Verrilli rose and confidently approached. As solicitor general, he appeared before the Court regularly. He knew each of the justices, their quirks and tics, and they knew him. Justice Kagan had been Verrilli's predecessor in the SG's office—disregarding Neal Katyal's stint as acting SG—and had started them down this road. Both Chief Justice Roberts and Justice Alito had also spent part of their careers in the SG's office. Of all the advocates in the spacious chamber, Verrilli was the most in his element.

Verrilli had been allotted only ten minutes to speak, yet during that ten minutes he was articulate and forceful. Standing tall in his formal morning coat, he evoked a more genteel era of the law, an era that most of the justices seemed to relish. The SG defended the government's compromise position rejecting the patentability of genomic DNA, but

conceding the patentability of cDNA. He assured the Court that permit-
ting patents on cDNA would not unduly burden the research enterprise,
then rebutted the "settled expectations" arguments that had so moved
Judges Lourie and Moore at the Federal Circuit. In doing so, he referred
to the Court's recent decision in *Mayo*. He reminded the justices that
the patentee there had also argued that rejecting its patents would disrupt
the biotech industry. Yet avoiding disruption to a particular industry,
even if it were likely, did not justify misapplication of the law. This
argument was all the more forceful, given that Verrilli himself had argued
to uphold the patents in *Mayo*. He showed that, notwithstanding his
position in that case, he had taken the Court's holding to heart. The
justices seemed to appreciate Verrilli's candor.

Hypomania

GREG CASTANIAS SPOKE last. He was allotted a full thirty minutes.
While the justices questioned both Hansen and Verrilli, they had clearly
reserved their worst for Castanias. It was what one observer would later
describe as a bloodbath.

Castanias began by invoking history—first, the Court's decision to
allow a patent on the oil-eating bacterium in *Chakrabarty* thirty-three
years earlier, then the sixteen years since Myriad's first gene patents
began to issue. But he barely finished his prepared introduction before
Justice Sotomayor interrupted him. She wanted to know how a human
gene could satisfy the test for patentability. "I always thought that to have
a patent you had to take something and add to what nature does," she
said, echoing the Court's holding in *Mayo*. Castanias responded obscurely,
questioning what a "product of nature" would be in this scenario—the
genome? A chromosome? A gene?

Justice Sotomayor did not let him make the point that he was driving
at. Instead, she unleashed the first in a flood of hypotheticals. Hypothecial
scenarios—"hypos"—are standard features of legal argumentation. They
are "what if" scenarios, sometimes implausible, sometimes outlandish,
but generally intended to illustrate the limits of some legal doctrine or

argument by comparing it to something more familiar. Law professors love to pose hypos in class, and their students, who go on to become lawyers and judges, can't resist using them after they graduate. On the Supreme Court, Justice Breyer was famous for regularly deploying a menagerie of mind-bending puzzle challenges to test the sometimes blurry boundaries of legal doctrines: tomato children, green-eyed turkeys, a hairbrush in the shape of a grape, King Tut's abacus-wielding accountant, rabbit-ducks, and raccoons with an appetite for garage door sensors have all made appearances in the former professor's questions from the bench.

Justice Sotomayor was also skilled at coming up with unusual hypothetical scenarios. "Look," she said to Castanias, "I can bake a chocolate chip cookie using natural ingredients—salt, flour, eggs, butter—and I create my chocolate chip cookie. And if I combine those in some new way, I can get a patent on that. But I can't imagine getting a patent simply on the basic items of salt, flour, and eggs, simply because I've created a new use or a new product from those ingredients." It was classic Sotomayor—combining down-home folk wisdom with incisive legal analysis, a skill she claims to have learned from her Abuelita while growing up in the Bronx. Castanias tried to interject, but the justice continued, "Explain to me . . . why gene sequences are not those basic products that you can't patent."

Castanias did not get the chance to explain, only to protest that "simplistic analogies" like cookies were not particularly useful in this case. Justice Alito then asked him to explain again. In attempting to do so, Castanias said, "When you look at those particular [DNA] sequences, there was invention in the decision of where to begin the gene and where to end the gene. That was not given by nature. In fact—"

Justice Scalia loudly interrupted him, displaying his characteristic bombast. "Well, well, well, well, this is something I was going to ask you," he said, looking pleased with himself. "I assume that it's true that those abridged genes, whatever you want to call them, do exist in the body . . . You haven't created a type of gene that does not exist in the body naturally."

Castanias took Scalia's bait, responding with yet another analogy. "A baseball bat," Castanias explained, "doesn't exist until it's isolated from a tree. But that's still the product of human invention to decide where to begin the bat and where to end the bat."

Now Justice Breyer sprang into action. He was tall, gaunt in appearance, with a hawk-like nose and penetrating gaze. With his colleagues and clerks he could be warm and charming, but when he considered an advocate before him to be less than forthcoming, he scrutinized the unfortunate lawyer like a frog laid open on a dissecting table. "Well, that's true," Breyer began, "but then you were saying something that I just didn't understand because I thought the scientists who had filed briefs here, as I read it, said it's quite true that the chromosome has the *BRCA* gene in the middle of it and it's attached to two ends. But also in the body, perhaps because cells die, there is isolated DNA. And that means that the DNA strand, the chromosome strand, is cut when a cell dies, and then isolated bits get around, and there may be very few of them in the world, but there are some, by the laws of probability, that will in fact match precisely the *BRCA1* gene. Now, have I misread what the scientists told us, or are you saying that the scientists are wrong?"

Castanias frowned. "Well, I will tell you that—"

Breyer added, as an afterthought, "I probably misread it. There's a better chance that I've misread it." This triggered muted chuckles from the gallery.

Castanias tried to respond without giving too much away. "Well, no, I think you may have read some of the submissions correctly, Justice Breyer. I think that's a question—"

"Well, which one have I not read?" Breyer asked.

Castanias answered, "I think that's a question of some dispute in this record."

"So in other words, you're saying that the Lander brief is wrong," Breyer said. Now it was clear where Breyer was going. He was referring to Eric Lander's brief, the one that tried to discredit the Federal Circuit by arguing that isolated genes *did* occur naturally in the human body, albeit under very rare circumstances.

Breyer continued, not letting up. "I want to know because I have to admit that I read it and I did assume that, as a matter of science, it was correct. So I would like to know whether you agree, as a matter of science, that it is correct—not of law, but of science—or if you are disagreeing with it, as a matter of science."

Castanias seemed to be taken aback. Lander's brief had raised scientific concepts that had never been introduced at the lower courts. It was eleventh-hour evidence, submitted well after any opposing expert could be found to contradict it. It was unheard of for the arguments in an *amicus* brief to be raised by one of the justices from the bench. There were too many points raised by the parties themselves to spend time on the uncorroborated views of friends of the Court.

But this was Breyer. He and Lander were practically neighbors on one of the exclusive streets near Harvard. So of course Breyer had read Lander's brief. Now it was taking on a life of its own.

Castanias did what he could. "What I will tell you is that what are called pseudogenes—" Castanias was referring to a concept that had been raised in Robert Nussbaum's original declaration to the district court. In a 1984 paper, Nussbaum and colleagues reported that fourteen cDNA copies of the coding region of a particular gene were found naturally throughout the genome. The plaintiffs tried to use this fact to prove that cDNA wasn't necessarily human-made, but could occur naturally in the human body. It was the same point later made by Lander in his *amicus* brief. The Federal Circuit dismissed the pseudogene argument because Nussbaum's results were not related to the *BRCA* genes, and there was no evidence that *BRCA* pseudogenes naturally appeared elsewhere in the genome. But both Lander and Justice Breyer seem to have overlooked this prior line of argumentation.

Breyer, focusing on the Lander brief, cut in, as though addressing an evasive witness. "I'd like a yes or no answer."

Castanias hesitated. "So the answer—I would say the answer is no because there is no evidence—"

Breyer would not let go. "Was the answer no, you do not disagree with it?"

"I do disagree with it with the following—"

"As a matter of science," Breyer clarified.

Castanias swallowed hard. "As a matter of science with the following—"

Breyer continued, "OK. Very well. If you are saying it is wrong, as a matter of science, since neither of us are scientists, I would like you to tell me what I should read that will, from a scientist, tell me that it's wrong."

"You want me to tell you something from a scientist that you should read that tells you that it is wrong?"

"Yes," said Breyer. "I need to know."

Castanias was backed against a wall. Was Lander, the president's own scientific advisor, wrong? Of course not. But Lander's arguments didn't make legal sense. Castanias flipped through the papers that he had brought with him to the podium. "I think you could look at the declaration . . . for Dr. Kay, for example . . . You'll find an extensive discussion in there of the technology here and of the genetics." Castanias looked up from the podium toward Breyer. "But, Justice Breyer, just to explain the finishing thought, what Dr. Lander says in his brief is that these pseudogenes, which are undifferentiated fragments, exist in the body. What hasn't been brought to the forefront is something that is new and useful and available to the public for allowing women to determine whether they have breast or ovarian mutations that are likely to result in cancer."

At this point, Castanias was foundering on rough seas. Dr. Kay's declaration did not actually address Lander's point, though the reply brief that Myriad filed with the Court *did* discuss the DNA fragments identified by Lander. That discussion, which was probably Coruzzi's careful work, was buried on pages fifty-one to fifty-two of the brief. It explained that even if genes and other random pieces of DNA were found in the human bloodstream, these were not deliberately "isolated" as that term was used in the patent. But, under the intense pressure of the clock and Breyer's aggressive questioning, Castanias didn't quite connect the dots.

Seeing what had happened, Chief Justice Roberts stepped in. Breyer had totally derailed the conversation, and the chief justice appeared eager to preserve some semblance of dignity in the proceeding. "Can I get back to your baseball bat example?" Roberts asked.

"Sure," Castanias said.

Roberts, a sports fan, seemed intrigued by the baseball analogy. He drew Castanias out—when was the baseball bat transformed from a product of nature (wood) into a patentable item (a bat)? Before Castanias could answer, Justice Kagan posed a hypothetical about the patentability of a leaf broken from a tree branch in the Amazon. Castanias did his best to juggle these competing analogies.

But Justice Ginsburg was still stuck on Lander's brief. "Do you concede . . ." she asked Castanias, "do you concede, at least . . . that Judge Lourie did make an incorrect assumption, or is the Lander brief inaccurate with respect to that, too?"

Castanias could not seem to bring himself to concede the point. He looked up at Justice Ginsburg, peering down at him from behind the high bench. "Judge Lourie was exactly correct to say that there is nothing in this record that says that isolated DNA fragments of *BRCA1* exist in the body," he said. But it was a lawyer's answer. Too contingent, too cagey.

Justice Alito who was, with Chief Justice Roberts, a George W. Bush appointee, was also a baseball fan. He pulled them back to the bat analogy with a new and even more bizarre hypothetical. "Suppose that in—I don't know how many millions of years trees have been around, but in all of that time—possibly someplace a branch has fallen off a tree and it's fallen into the ocean and it's been manipulated by the waves, and then something's been washed up on the shore, and what do you know, it's a baseball bat!" There was laughter from the gallery. Alito concluded, "Is that what Dr. Lander is talking about?"

The analogies were hardly making sense any more. A tree branch carved by the *ocean* into a baseball bat? Castanias gamely forged ahead. "That's pretty much the same as what he's talking about," he answered before trying to steer the conversation back toward more solid ground— the incentives built into the patent system.

But Breyer took this as an opportunity to repeat a favorite speech of his. "I'd be interested in your view," he said, "that the patent law is filled with uneasy compromises because on the one hand, we do want people to invent. On the other hand, we're very worried about them tying up some kind of whatever it is, particularly a thing that itself could be used for further advances." But rather than waiting for Castanias to express his view, Breyer went on at length, speculating about the leaf in the Amazon, the entire notion of "uneasy" compromises echoing his famous 1970 *Harvard Law Review* article "The Uneasy Case for Copyright"—the article that got him tenure at Harvard.

There was no fitting response to Breyer's lengthy dissertation, so Justice Kagan brought them back to isolated DNA. She asked, "Do you think that the first person who isolated chromosomes could have gotten a patent?"

Castanias hemmed and hawed, but Kagan, also reverting to professor mode, would not give up. Castanias objected that something needed to be useful in order to justify a patent. Kagan, who had developed a reputation as the Court's humorist, quipped, "Chromosomes are very useful." Eventually, Castanias conceded that, in his view, the first person to find a full chromosome could indeed patent it.

At that, Kagan's eyes lit up, and she sprang her trap. What about the first person who found a liver?

Castanias resisted mightily, but Kagan was soon joined by Breyer and Sotomayor, all pushing him toward the fatal admission that he must have seen coming. Isolated DNA was not the same as a human organ, he argued, because it's not just removed from the body, it is part of a full chromosome, and to excise it requires great human effort and scientific ingenuity. It was Sotomayor who finally locked him within his own labyrinth of words. "If you cut off a piece of the whole in the kidney or liver, you're saying that's not patentable, but you take a gene and snip off a piece, that is? What's the difference between the two?"

Castanias had no choice. He admitted that there was no difference. Under his theory, apparently, you could patent a kidney. The arguments continued for a few more minutes, but the fatal admission had been made. After that, it was over.

The marshal's gavel loudly closed the proceedings as the chief justice, looking up, quietly announced, "The case is submitted."

Sandra Park and Chris Hansen after the Supreme Court argument.

Everyone rose as the justices disappeared into the tall velvet curtains behind them. As the lawyers collected their things, Park noticed that Hansen had left his white goose quill on the counsel table. "Chris," she said, picking up the prized memento and handing it to him. "Don't forget this."

He looked at it and shook his head. He was retired. This was his last legal argument. "I don't need it," he said. Then, after a pause, "Give it to Tania."

Debrief

FOLLOWING THE ORAL arguments, the Myriad team debriefed in Jones Day's offices. The mood was somber. They knew that the argument had not gone well. The justices, especially Breyer, Kagan, and Sotomayor, seemed outright hostile. They didn't care a whit for Patent Office practice—Kagan had called the agency "patent happy." Other justices were harder to read. Roberts seemed to give them the benefit of the doubt. Ginsburg was stuck on Lander's *amicus* brief. Kennedy's questions didn't give much indication of where he was. Alito's hypothetical about the ocean and the baseball bat still had them scratching their heads. Scalia, usually one of the most outspoken justices, hadn't said much at all. And Thomas, as always, was a cipher. He had said nothing, asked no questions, spent most of the argument swiveling in his chair and gazing up at the ceiling. How this mixed bag of reactions would translate into a decision was anybody's guess.

The plaintiffs' team was more upbeat. They exited the Supreme Court building and joined the rally on the steps. Many of them posed for photos and spoke with reporters. There were hugs all around. Eric Lander and Bob Cook-Deegan gave each other a big, NBA-style high five on the courthouse plaza. Steve Shapiro from the ACLU asked Park how she thought the arguments went. "We're going to win," she said.

Afterward, the legal team, some of the plaintiffs, and a few of Hansen's visiting relatives took cabs to the ACLU's DC office on McPherson Square. They got box lunches from the restaurant downstairs and Romero, Simoncelli, and Hansen gave impromptu speeches about the day and the journey that had brought them there. Spirits in the room were high. Nobody wanted to jinx it, but all the signs from the Court were positive. When they finished their lunches, Park, Matloff, and others dispersed to the various news studios around town to tape interviews for the evening broadcasts.[4]

The press reactions to the oral arguments were mixed. Most recognized the significance of the case. Nina Totenberg said on NPR's Morning Edition that "there is no way to overstate the importance of this case to the future of science and medicine." But even sympathetic observers were confused by what had happened at the Supreme Court. In *Science*, veteran reporter Eliot Marshall titled his story, "In a Flurry of Metaphors, Justices Debate a Limit on Gene Patents."

Many, particularly in the patent community, were frustrated by the justices' questions and the lawyers' responses. Kevin Noonan, in particular, called out the Court and counsel for their excessive use of bad analogies, writing that "Isolated human DNA is not a tree (Justice Kagan); sap from a tree (Justice Breyer); a chocolate chip cookie (Justice Sotomayor); a baseball bat (Greg Castanias); or a liver or kidney (Chris Hansen)."

Instead, Noonan suggested that the Court should have considered

---

4 Later that day, two bombs were detonated near the finish line of the Boston Marathon, killing three people and injuring hundreds more. Much of the news coverage of the oral arguments in *AMP v. Myriad* was superseded or delayed as a result.

*Tania Simoncelli
and Eric Lander in
a celebratory mood.*

whether any of the following real-world substances isolated from nature should continue to be patent-eligible: a chemical compound isolated from crude oil that is useful as a lubricant, a chemical compound from a plant that is useful as a drug, an isolated antibiotic produced by bacteria, an isolated cucumber gene that extends freshness or an isolated human gene that stimulates red blood cell production. Noonan's questions might have been less colorful, but more relevant to the issues at hand.

In the end, Noonan spoke for many in the patent community who seemed unhappy with all of the players in the case. He specifically bemoaned the "ACLU's [perfidy] (in advancing arguments having little resemblance to the law or the facts), the government's duplicity (asserting that genomic DNA should not be patent-eligible after more than thirty years of granting, and continuing to grant, such patents) . . . incompetence by Myriad's representatives (for being unable to do the near impossible in answering the Court's sometimes incoherent questions) [and] Myriad's irresponsibility (for continuing a suit on patents that do not protect its core business and putting at risk patents [necessary to others] for commercialization)." In Noonan's view, everyone shared the blame for what many patent lawyers envisioned as the train wreck to come.

The Angelina Effect

THOUGH THE ACLU's public relations machine continued to run at full steam after the oral arguments, the most significant piece of publicity during this critical period came from a source that was wholly unexpected—Hollywood superstar Angelina Jolie.

Jolie rose to box office stardom through her portrayal of strong heroines like Lara Croft (*Tomb Raider*). As the highest-paid actress in Hollywood, Jolie was one of the most recognizable celebrities in the world. She had also tested positive for a deleterious *BRCA1* mutation.

On May 14, 2013, the Monday after Mother's Day, Jolie, a mother of six, published a *New York Times* op-ed titled "My Medical Choice." In it, she discussed her family's history of breast cancer, as well as the detection of her own *BRCA1* mutation. She then shocked the world with the revelation that, three months earlier, she had undergone bilateral mastectomy surgery to reduce her risk of cancer. "I am writing about it now," she said, "because I hope that other women can benefit from my experience. Cancer is still a word that strikes fear into people's hearts, producing a deep sense of powerlessness. But today it is possible to find out through a blood test whether you are highly susceptible to breast and ovarian cancer, and then take action."

After describing her procedure and painful recovery, Jolie acknowledged the barriers that still exist to broad *BRCA* testing: "It has got to be a priority to ensure that more women can access gene testing and lifesaving preventive treatment, whatever their means and background, wherever they live. The cost of testing for *BRCA1* and *BRCA2*, at more than $3,000 in the United States, remains an obstacle for many women."

It was a direct indictment of Myriad's pricing practices and, indirectly, the exclusivity afforded by its patents. Jolie did not name the case or the company, but what she was referring to was crystal clear. More than a few people noticed. *TIME* called Jolie's announcement "a cultural and medical earthquake" and featured her on its cover two weeks later. Requests for *BRCA* testing surged across the United States, Canada, the United Kingdom, Australia, New Zealand, and elsewhere. The

Angelina Jolie, who revealed her BRCA *test status in a* New York Times *op-ed, was featured on the cover of* TIME, *May 27, 2013.*

phenomenon became known as the "Angelina Effect," and the timing couldn't have been better.

But which side did Jolie's announcement help most? Myriad's stock rose sharply after the op-ed, as demand for its *BRCA* tests surged. Andrew Cohen, writing in *The Atlantic*, predicted that "it is more likely than not that the justices will uphold Myriad's patent—its monopoly—over research for these genes. This is that type of Court." But the ACLU team was heartened as well. Jolie criticized Myriad, without naming the company, for its pricing and patenting. That was definitely something that the justices might notice.

CHAPTER 26

9–0

Justice Clarence Thomas is an enigma. The only African American justice on the Supreme Court, the putative successor to civil rights icon Thurgood Marshall, Thomas is fundamentally opposed to all forms of affirmative action. Known for his sphinxlike silence at the Court's oral arguments, within chambers he is gregarious and voluble. But Thomas's ultra-conservative views are in many cases so extreme that he finds it difficult to persuade even fellow conservatives to join the opinions that he writes. As a result, he is often reduced to writing solo concurring or dissenting opinions in cases involving civil rights, abortion, free speech, gun control, criminal procedure, and other politically charged issues. But every justice has an obligation to share the burden of writing the Court's opinions. So Thomas is often left with cases that are uncontroversial or highly technical in nature. For example, patent cases.

Over the past decade and a half, Clarence Thomas has written some of the most important Supreme Court decisions concerning patent law, most of which have been decided by unanimous (9–0) votes. Consistent with this trend, on June 13, 2013, Justice Thomas was announced as the author of a unanimous 9–0 decision in *AMP v. Myriad*.

For all of the criticism that Justice Thomas attracts, at least his opinions are short. On a Court increasingly known for its verbosity, Thomas often gets his point across in a few words. *AMP v. Myriad* is no exception. The opinion is a mere eighteen pages—only 12 percent of the length of Judge Sweet's opinion in the district court.

And the winner was? *The U.S. government*, but not the Patent Office. In effect, Justice Thomas and the Court adopted the solicitor

Justice Clarence Thomas delivering the Supreme Court's opinion
in AMP v. Myriad.

general's arguments almost entirely. Genomic DNA in the form in which it exists in the human body is an unpatentable product of nature. Complementary DNA that is created in the laboratory is a patentable construct of humankind.

The opinion closely echoes many of the statements made by Verrilli and Hansen at the oral arguments:

"Myriad did not create anything. To be sure, it found an important and useful gene, but separating that gene from its surrounding genetic material is not an act of invention."

"Groundbreaking, innovative, or even brilliant discovery does not by itself satisfy the §101 inquiry."

And, in a direct rebuke to Judge Lourie, the chemist of the Federal Circuit, Justice Thomas wrote, "Nor are Myriad's claims saved by the fact that isolating DNA from the human genome severs chemical bonds and thereby creates a nonnaturally occurring molecule. Myriad's claims are simply not expressed in terms of chemical composition, nor do they rely in any way on the chemical changes that result from the isolation of a particular section of DNA. Instead, the claims understandably focus on the genetic information encoded in the *BRCA1* and *BRCA2* genes."

This was, in the end, a validation of Judge Sweet's DNA-as-information theory.

And not to leave Judge Moore out of his crosshairs, Justice Thomas also rejected the argument that thirty-five years of gene patents justified their continuation. Without offering much explanation beyond an observation that the solicitor general did not buy this argument, he concludes that, on balance, the Patent Office's practice of issuing gene patents should not stand.

Though Justice Thomas wrote the opinion for the Court, that did not stop Justice Scalia from getting in the last word. Appended to Thomas's opinion is a peculiar three-sentence concurrence by Justice Scalia. It reads, in full:

> I join the judgment of the Court, and all of its opinion except Part I—A and some portions of the rest of the opinion going into fine details of molecular biology. I am unable to affirm those details on my own knowledge or even my own belief. It suffices for me to affirm, having studied the opinions below and the expert briefs presented here, that the portion of DNA isolated from its natural state sought to be patented is identical to that portion of the DNA in its natural state; and that complementary DNA (cDNA) is a synthetic creation not normally present in nature.

The offending "Part I-A" is merely a summary of the scientific background of the case—the obligatory molecular biology primer that explains to the reader what the case is about. What did Justice Scalia find objectionable about this material and the other "fine details of molecular biology" included in the opinion? And why did he feel compelled to disavow it so completely?

Maybe Justice Scalia was taking a jab at Justice Breyer and the other amateur scientists on the Court. Or maybe he was getting at a deeper, and more fundamental, question about the nature of judicial lawmaking in a world increasingly dominated by complex technologies that judges (and juries) can never hope to understand. We may never know, as Justice Scalia passed away during a hunting trip in 2016.

Getting the News

DECISIONS OF THE Supreme Court are generally announced on Mondays, but in the final weeks of the Court's term, additional days may be designated to accommodate the large number of opinions. Thursday, June 13, was such a day. Park was working on a brief in another case, keeping one eye on the Supreme Court's website, occasionally hitting the refresh key. Several other cases were announced that morning. Then, finally, Park saw an announcement on her screen. Case 12-398, *Association for Molecular Pathology v. Myriad Genetics*. She held her breath as she read. It was a unanimous opinion written by Justice Thomas. They had won!

Park called Hansen, but he didn't pick up. He was at home, surfing the web and reading email, when he saw a message on which he was cc'd. It was about the case. Somebody was telling a colleague that they had won the case, assuming that Hansen already knew. Hansen sat back and smiled, satisfied.

After trying Hansen, Park texted Simoncelli, who was attending a meeting at FDA. When Simoncelli saw the text, she jumped out of her chair and told her surprised colleagues that she had to go. She ran back to her office and called Park. They laughed and congratulated each other, marveling at how they had gotten here, and a little sad that the three of them—Park, Simoncelli, and Hansen— were miles apart during this sweet moment of triumph.

Next, Park called Lisbeth Ceriani. Still in her pajamas, Ceriani turned on her computer to listen to the audio recording of the announcement from the Court.

Justice Thomas, whose voice Ceriani had never heard before, given his silence during the oral arguments, began in a slow, flat monotone, "This case comes to us on a writ of *certiorari* to the United States Court of Appeals for the Federal Circuit . . ." She listened, literally at the edge of her seat. After about three minutes, Thomas got to the good part: "Myriad's principal contribution was uncovering the precise location and sequence of the *BRCA1* and *BRCA2* genes. Myriad did not create or alter the genetic information encoded in the two genes or the genetic structure of the DNA itself. It identified important and useful genes, but this discovery, groundbreaking as it was, does not by itself satisfy the Section 101 inquiry. We, therefore, conclude that merely finding the location of the *BRCA1* and *BRCA2* genes does not render the genes patent eligible as new compositions of matter . . . Consequently Myriad's patents on the *BRCA* genes are invalid."

Ceriani fell back in her chair. Did she understand that right? Was it possible? As she later told Park, "I was euphoric. I was just really happy, and I felt validated . . . It's not just some abstract scientific concept out there. It impacts people's daily lives. It impacts people's mothers and sisters and wives and their healthcare decisions. Being able to simplify it down to that. It's a basic human right to be able to see your own blood, to see your own genes, and to know what that means. I felt really validated by the decision."

CHRIS MASON WAS teaching a class on genomic medicine at Weill Cornell, where he was now an assistant professor, when his phone vibrated in his pocket. Knowing that the Court was announcing decisions that morning, he surreptitiously glanced at the flashing screen and saw a text from Hansen.

He looked up at his morning classroom of bleary-eyed medical students. "We've won!" he said, meeting their looks of incomprehension with a huge smile. "Your genes are free!"

ELLEN MATLOFF HADN'T heard that the Court would be announcing decisions that Thursday, so she wasn't expecting the news. Instead, she was singing children's songs and eating cake on the last day of her three-year-old's preschool. Outside, it was raining—a Connecticut summer downpour. Around eleven, Matloff and her daughter ran back to the car, getting soaked in the process. They jumped in and Matloff glanced at her phone, which she had left on the passenger seat. It was buzzing, and she could see the message counter displaying double digits.

As Matloff recalls, "When I played the first message, it was one of my closest friends saying, 'You won! You won!' At first I didn't know what she meant, because she was kind of hysterical. Then I got it. I started crying. Then my daughter started crying. She was like, 'What's wrong.' I was like, 'Honey, it's OK.' There were all of these reporters to get back to. It was just complete chaos. I remember driving home in the rain, bawling and saying to her 'We won. We won. It's a good thing. Mommy won. We won our case.' It was just overwhelming."

UPON HEARING THE news, Francis Collins was elated. It had been a decades-long road from his *BRCA1* collaboration with Mary-Claire King, but it was finally over. He sent a Tweet to his thousands of followers: "SC: 'A naturally occurring DNA segment is a product of nature and is not patent eligible merely because it has been isolated' Woo Hoo!!!"

SALT LAKE CITY is in the Mountain Time zone, two hours behind the east coast. It was a hot summer day, and by 10:30 a.m. the temperature was already pushing ninety degrees. By the time Ben Jackson arrived at Myriad's office, news of the decision had been out for several hours. Another attorney met him in the lobby. "We got the decision," he said.

From the look on the attorney's face, Jackson gathered that the news wasn't good. "Well?" he asked.

"It's bad," said the attorney. "We lost on everything but cDNA."

Jackson exhaled a long breath. "OK," he said. "This is going to be a long day."

Jackson was right. Before the day was over, three different diagnostics companies—Ambry Genetics, GeneDx, and DNATraits—announced that they would now offer *BRCA* testing at prices as low as $995. Ambry's website displayed a majestic image of the Supreme Court building under the banner "Your Genes Have Been Freed." And on a second screen: "*BRCA1* & *BRCA2*—Now Accepting Samples."

Another company adopted a more tongue-in-cheek approach. It posted an online ad depicting a pointy black hat with two stockinged legs emerging from beneath its wide brim—a clever allusion to *The Wizard of Oz*. The caption read, "DING DONG . . . In a unanimous decision, the U.S. Supreme Court rules gene patents invalid, finally opening the door for more complete and cost effective testing options."

Victory!

NOT SURPRISINGLY, THE ACLU was quick to declare victory. In a Churchillian blog post titled "VICTORY! Supreme Court Decides: Our Genes Belong to Us, Not Companies," Park wrote, "We celebrate the Court's ruling as a victory for civil liberties, scientific freedom, patients, and the future of personalized medicine."

In the "if you get lemons, make lemonade" department, Myriad also claimed the decision as a victory. Its press release brazenly announced: "Supreme Court Upholds Myriad's cDNA Patent Claims" and went on to remind the market that "following today's decision, Myriad has more than five hundred valid and enforceable claims in twenty-four different patents conferring strong patent protection for its BRACAnalysis test." Myriad's stock rose 13 percent following this news, but closed down by 6 percent when the markets fully grasped what had happened.

The Patent Office responded immediately, issuing a matter-of-fact memorandum to all patent examiners stating: "As of today, naturally occurring nucleic acids are not patent eligible merely because they have been isolated. Examiners should now reject product claims drawn solely to naturally occurring nucleic acids or fragments thereof, whether isolated or not, as being ineligible subject matter under 35 U.S.C. §101."

Others were less reticent about their views of the decision. BIO called it "a troubling departure from decades of judicial and [Patent Office] precedent." The American Intellectual Property Law Association wrote that it would "throw into question patent protection for important technology that is critical to improving health for the public." Many criticized not the Court's legal reasoning, but its understanding of science. One commentator in the *New Republic* ridiculed the Court's "bizarre" distinction between gDNA and cDNA in an article titled, "The Supreme Court Reveals its Ignorance of Genetics." Not surprisingly, the indomitable Gene Quinn of *IP Watchdog* spoke for many in the patent community, deriding the decision as a loss for science, for the patent system, and for people dependent on medical technologies. He predicted that the decision would result in "a near complete cessation in many areas of personalized medicine" and heralded it as "a big win for those who wish to copy innovators," ending his exclamation-filled tirade with the ersatz rallying cry, "The war on patents continues!"

But perhaps the greatest feat of spin-doctoring was achieved by Jones Day. Despite the fact that the Court's ruling on the patentability of cDNA would have virtually no effect on Myriad's business, which did not use cDNA, the law firm released its own press release that not only touted its role in Myriad's "victory," but attempted to portray the decision as a history-making win for all patent holders:

> In a closely followed and highly publicized case, Jones Day secured an important Supreme Court victory for Myriad Genetics, Inc. on the patent-eligibility of its composition-of-matter claims directed to complementary DNA, or cDNA—synthetic DNA molecules created by scientists in a laboratory . . . Unlike other recent decisions in which the Supreme Court struck down all patent claims at issue in those cases, the Myriad decision marks the first time since 2001 that the Court has upheld patent claims as patent-eligible subject matter under Section 101 of the Patent Act.

As P. T. Barnum reputedly said, "Any publicity is good publicity, as long as they spell your name right."

CHAPTER 27

Aftermath

IT HAS NOW been more than a decade since the ACLU and PubPat filed their complaint in *AMP v. Myriad*, and the case has had a lasting impact on the many people involved, as well as the broader industry and legal landscape.

In December 2013, the scientific journal *Nature* named Tania Simoncelli one of its "ten people who mattered" for fighting "to keep genes open to all." After three years at the FDA, Simoncelli moved to OSTP, the White House agency that had been so instrumental in marshaling support to overturn gene patenting within the administration. After a stint working with Eric Lander at the Broad Institute in Boston, she moved back to California to take a position as director of science policy at the Chan Zuckerberg Initiative (funded in large part by Mark Zuckerberg of Facebook fame), where she seeks to accelerate biomedical research and change the health research ecosystem to better serve the interests and needs of patient communities.

Simoncelli was never replaced in the science advisor role that she created at ACLU. But Sandra Park has emerged as a leader in the area of genetics advocacy and policy. She has filed *amicus* briefs in subsequent patent cases, including *Alice v. CLS Bank*, the fourth in the Supreme Court's quartet of patent eligibility decisions, and worked with Simoncelli, Bob Cook-Deegan, and others to file a first-of-its-kind complaint on behalf of four patients (including Runi Limary, one of the plaintiffs in *AMP v. Myriad*) who were denied access to their genomic data after being tested by Myriad.

Chris Hansen is enjoying his well-deserved retirement in Mount

Vernon, New York, reading, teaching the occasional law class, and continuing to frown at the national news. During Hansen's forty years at the ACLU, he saw the organization change dramatically. It is now more institutional and more respected than it was before. But it is also harder to win cases, as the increasingly conservative courts have imposed higher and higher procedural hurdles to the successful litigation of civil rights cases. During his long and distinguished career Hansen scored many major victories in addition to *AMP v. Myriad*—cases like *Willowbrook, Brown v. Board,* and *Reno v. ACLU.* Which of these history will rank as the most significant is hard to guess. But from Hansen's standpoint, no case involved issues that were more interesting, more novel, and more intellectually challenging than *AMP v. Myriad.*

Lenora Lapidus, the head of the ACLU's Women's Rights Project who championed the gene patenting challenge from the moment she heard about it, died from breast cancer in 2019 at the age of fifty-five. Barbara Caulfield, the general counsel of Affymetrix who helped to persuade Steve Shapiro that the biotech industry would not fail because of the case, also died from cancer. The deaths of these and many more women make it clear that, despite advances in genetic testing and immune system therapies, the war on cancer is still far from over.

Dan Ravicher, the only patent lawyer on the ACLU team, left New York and the Public Patent Foundation, which has more or less shut down. He now practices law amid palm trees and hurricanes in Miami. Lori Andrews, whose early work was so instrumental in kicking off *AMP v. Myriad*, now writes about personal privacy, mobile apps, and social media. She has been listed as one of Chicago's Best Lawyers in the field of Biotechnology Law every year since 2012.

Judge Robert Sweet, whose district court opinion laid the groundwork for the invalidation of Myriad's patents, passed away quietly at the age of ninety-six at his vacation home outside of Sun Valley, Idaho. He remained active on the bench and enjoyed ice skating at the Sky Rink until the very end. His former law clerk, Herman Yue, now practices patent law in New York.

Wendy Chung, Harry Ostrer, Haig Kazazian, and Arupa Ganguly all continue to work in genetics, though none of them is offering *BRCA* testing to patients. "There's just no need," Kazazian says, "when companies will do it these days for so little." Even Harry Ostrer, whose written declaration insisted that he would resume *BRCA* testing if Myriad's patents were overturned, thereby giving him standing in the case, shakes his head. His work, too, has moved on.

Chris Mason continues to swab the world's subway stations, hospital floors, and public benches for microscopic denizens—most recently in the fight against COVID-19. Despite his work with the ACLU, Mason today considers himself to be "very pro-patent." He has started four companies and is an inventor of several patented technologies, though none, he insists, claims products of nature or natural laws. The official website for his lab in New York says that he is "working on a ten-phase, 500-year plan for the survival of the human species on Earth, in space, and on other planets." What that means is anybody's guess. But knowing Mason, it will likely be important.

Mary-Claire King, who first discovered the neighborhood in which *BRCA1* resided on chromosome 17, is still conducting research at the University of Washington. She has become one of the biggest public supporters of *BRCA* testing, and has proposed that every woman in America be tested. Yet, even now, more than a quarter century after *BRCA1* was sequenced, there is not widespread agreement about who should be tested, or what measures should be taken in response to a positive result.

David Kappos, who served as Director of the Patent Office during *AMP v. Myriad* and Don Verrilli, who argued the U.S. Government's case before the Supreme Court, are now both in private practice. Ray Chen, the Patent Office solicitor who argued *Bilski* at the Federal Circuit (twice), was nominated by President Obama to the Federal Circuit in 2013. He now sits as a judge on the Patent Court. Kimberly Moore, who was one of the most junior judges on the Federal Circuit when *AMP v. Myriad* was heard, became the chief judge of that court in May 2021. At a recent bar association event at which Judge Moore was the keynote speaker, one audience member volunteered that her concurring opinion in *AMP v. Myriad* was one of the greatest legal opinions ever written. Her role in history is surely not over.

Francis Collins, who successfully led the monumental Human Genome Project, continues to serve as director of the National Institutes of Health. Having thrived under five very different presidential administrations (Clinton, Bush, Obama, Trump, and Biden), he has announced no plans to step down. Mark Freeman is still in the Civil Division of the Department of Justice, where he now heads its elite appellate staff.

The White House's interest in patents did not end with *AMP v. Myriad*. OSTP and NEC helped to shape the administration's position in *Alice v. CLS Bank*—which concerned the patentability of software and business methods. That time around, President Obama himself got involved, and in 2014, for the first time ever, or since, the president mentioned patent reform in his State of the Union address.

The lawyers who represented Myriad continue to do well. Rick Marsh retired as general counsel in 2019 and is now serving a three-year term as president of the Mormon Church mission in Arequipa, Peru. Ben Jackson, who began his career as a student intern at Myriad, is now its general counsel and leads the company's legal department.

Brian Poissant, who argued *AMP v. Myriad* at the district court in New York, retired from Jones Day in 2014, and Laura Coruzzi, Jones Day's molecular biology expert, has joined a private biotechnology company. Greg Castanias and Jennifer Swize, however, are still with the firm. In 2018, Castanias was back at the Supreme Court, this time opposing the Patent Office. In the case, *SAS Institute v. Iancu*, he argued that the Patent Office should have *less* discretion to reject a company's challenges to its competitor's patent claims. And this time, despite significant grilling by Justices Sotomayor, Ginsburg, and Breyer, five of the nine justices agreed with him. Though the *SAS* case was nothing like the sweeping assault on gene patents that Castanias had so mightily resisted in *AMP v. Myriad*, the Court's decision did make it marginally easier to challenge otherwise invalid patent claims, aligning Castanias, in an ironic turn, with his former adversaries at ACLU and PubPat.

In 2020, the marketing group Acritas ranked Jones Day as having the strongest U.S. law firm brand in the country (for the fourth year in a row). And in 2018, the firm broke a new record by hiring eleven of the Supreme Court's thirty-eight outgoing law clerks—more than any other law firm in history. Each of the new recruits reportedly received a signing bonus of $400,000.

For carriers of *BRCA* mutations and their families, *AMP v. Myriad* has been life-changing. *BRCA* testing is now available from some vendors for as little as $199, and even direct-to-consumer genetic testing companies like 23andMe offer limited *BRCA* testing. Myriad has reduced its pricing in many cases by half, and BART testing is now a standard part of Myriad's *BRCA* offerings. Today, nearly every woman who wishes to be tested can be. Lisbeth Ceriani's daughter has been tested, but won't be told her results until she is eighteen. Kathleen Maxian, who became a vocal spokesperson for the case in its later stages, is in remission from her ovarian cancer and now leads the nonprofit Ovarian Cancer Project.

DESPITE THE LOSS of its patents, Myriad continues to lead the *BRCA* testing market. There are many reasons that Myriad has held onto its market share—its tests still have an excellent reputation for accuracy among clinicians and genetic counselors, they are now covered by most insurance carriers, whereas some of its competitors' tests are not, and insurers—led by Medicare—are still willing to pay relatively high prices for the tests. Moreover, Myriad has added numerous additional genes and disease indications to its test offerings, expanding well beyond *BRCA*. As a result, the company's hereditary cancer testing revenues have again begun to rise, reaching $480 million in its 2019 fiscal year (on total revenues of $851 million).

Less than thirty days after the Supreme Court's decision was issued, Myriad (joined by the University of Utah and the University of Pennsylvania, among others) sued LabCorp, Ambry, and several other diagnostics companies that entered the *BRCA* testing market on the heels of the Supreme Court's ruling. Apparently, Myriad was attempting to show its investors that it retained some valuable patent assets, informing the Court that it still held 515 patent claims covering the *BRCA* genes that the Supreme Court had *not* specifically invalidated. This time around, Myriad asserted patent claims covering *BRCA1* and *BRCA2* primers (short segments of DNA only fifteen to thirty bases in length) as well as methods of analyzing *BRCA1* and *BRCA2* sequences.[1] Myriad's theory was that because primers are "man-made" compounds, they were spared by the Supreme Court's 2013 decision. They were wrong.

A trial judge in the Federal District of Utah, Myriad's own backyard, ruled against the company. Because the primers covered by Myriad's patent claims duplicated the DNA sequences of the *BRCA1* and *BRCA2* genes, they were ineligible for patent protection per the Supreme Court's new precedent. Likewise, Myriad's method claims contained no inventive elements other than the primers, which were themselves invalid. Thus, all ten patent claims asserted by Myriad against the competing labs were struck down.

Somewhat surprisingly, the Federal Circuit—the Patent Court itself—affirmed the Utah court's rejection of the patents in a unanimous opinion written by Judge Dyk and joined by Judge Clevenger and Chief Judge Prost (who replaced Randall Rader in that position). After that, Myriad gave up. It settled all remaining *BRCA* cases in February 2015.

Mark Skolnick, Myriad's co-founder, remains happily retired, spending time with his wife, children, and grandchildren at their homes in La Jolla, California, and Parma, Italy. Pete Meldrum, the driving force behind Myriad's business strategy, retired in 2015 to run his charitable foundation. He passed away in 2018 after sustaining a head injury while playing with his grandchildren a few days before Christmas. Meldrum left behind a significant legacy for charitable and educational causes in Utah.

Myriad continues to capitalize on the valuable data that it gathered from patients during the two decades that its patents were in force. It maintains a proprietary database of an estimated twenty-thousand or more different mutations in the *BRCA* genes, though large public efforts have arisen to replicate the information that the company keeps to itself.

As the ACLU intended, the impact of *AMP v. Myriad* reached far beyond Myriad Genetics and the *BRCA* testing market. No company today can monopolize the use of a gene in the United States. As a result,

---

1 The 2014 case was something of a reunion, or perhaps a grudge match. Sandra Park, Lenora Lapidus, and Dan Ravicher all appeared on *amicus* briefs, which were filed by ACLU, PubPat, Breast Cancer Action, and AARP. Myriad, however, was not represented by Jones Day.

patents claiming the genes linked with the hereditary cardiac disorder known as Long QT syndrome, Alzheimer's, diabetes, asthma, hereditary deafness, and a host of other health conditions are things of the past. As technology has advanced, the era of single-gene testing has largely ended—most labs these days test a wide variety of genes simultaneously, and soon whole genome sequencing will be routine. Isolated genes—those "compositions of matter" so beloved by the patent bar during the 1990s— are no longer important in the world of genetic diagnostics. Companies that once bet their fortunes on single-gene diagnostic tests are disappearing. But, notwithstanding the predictions of doom by trade associations and bloggers, the U.S. biotech industry did not collapse under the weight of the decision, nor did biomedical innovation grind to a halt.

That doesn't mean, however, that the outcome of *AMP v. Myriad* has been universally accepted. There have been repeated efforts in Congress to reverse its effects. In May 2019, Senator Thom Tillis (R-NC) and Senator Chris Coons (D-Del.) introduced a draft bill to undo the Supreme Court's decisions in *Bilski, Mayo, Myriad,* and *Alice.* As explained by Senator Tillis, "Section 101 of the Patent Act is foundational to the patent system, but recent court cases have upset what should be solid ground." The intent of this bill was to eliminate or "abrogate," in a single fell swoop, the entirety of the Supreme Court's patent eligibility exclusions, from $E = mc^2$ to BRCA.

Not surprisingly, the ACLU and its allies mobilized to resist the proposed legislation. Sandra Park, Ellen Matloff, Chris Mason, Tania Simoncelli, Josh Sarnoff, Lisa Schlager, and other veterans of *AMP v. Myriad* again charged to the front lines, testifying before congressional committees, drafting letters to Congress, and developing briefing papers seeking to preserve the Supreme Court's rulings. Familiar faces from the other side of the aisle also emerged to support the legislation—David Kappos, Hans Sauer, Sherry Knowles, and many others. Nevertheless, efforts in this area had subsided by late 2019, and the onset of the COVID-19 pandemic in early 2020, with its own overwhelming set of legal issues and controversies, effectively ended discussion of the Tillis–Coons proposal. There is little doubt, however, that it will re-emerge in the future.

A Lasting Legacy

BUT THE IMPACT of *AMP v. Myriad* extends beyond the realm of genetic diagnostics and even patent law. In December 2010, shortly after Neal Katyal filed his "magic microscope" brief at the Federal Circuit, OSTP director John Holdren attended the annual White House holiday party. Because Holdren's wife stayed home, he was accompanied by his forty-one-year old daughter, Jill. Jill Holdren, a brilliant and witty conversationalist who had just returned from an extended stay in Argentina, soon charmed both Barack and Michelle Obama.

With a degree in environmental science from Berkeley, Jill also had training in epidemiology. For years, she had known that her paternal grandmother's death from breast cancer at the age of forty-seven could have had a genetic link, as could her great-grandmother's death from an unknown form of cancer. Yet Jill did not qualify for *BRCA* testing, given that these casualties arose on her father's side of the family and her father, an only child, had never been tested. The year after the White House party, Jill asked her father to get tested because if he were *BRCA* positive, she herself would qualify for testing. But he demurred, not wishing to dredge up memories of his mother's death. Jill persisted. He was, after all, the president's science advisor. He had lunch every week with Eric Lander and Harold Varmus, two of the most famous geneticists in the world. What did they think, she asked? Holdren didn't raise it with them.

In May 2013, Jill read Angelina Jolie's *New York Times* op-ed describing her *BRCA* diagnosis and subsequent surgeries. Her jaw dropped. She immediately sent the article to her father, again begging him to talk with Lander and Varmus. This time, he did, and the two scientists told him that he should absolutely get tested. So Holdren drove to Walter Reed, the giant government hospital across the street from NIH, and asked his physician to order a *BRCA* test for him. It came back positive.

With her father's result in hand, Jill asked her own physician to order a test for her. He did. She was also *BRCA* positive. Not only that, she had Stage 2C ovarian cancer. Her oncologist at Mass General told her, "I want your ovaries out five years ago." Shortly after the Supreme Court's decision in *AMP v. Myriad*, Jill underwent a complete removal of her ovaries and fallopian tubes, as well as a double mastectomy, followed by a course of chemo.

During Jill's treatment and recovery, Holdren and his wife flew regularly from DC to Boston to spend time with her. Each time, President Obama asked after Jill, often sending along a note or a small gift.

On one of these visits, Jill asked her father whether the president and his family had been tested. After all, it was common knowledge that his mother had died from cancer at the age of fifty-three. But Holdren never raised the subject with the president. "It was too personal," he said.

Instead, Holdren arranged a White House visit for Jill and her family. Three weeks after her mastectomy, barely able to move her upper body, Jill, her husband, and their two children, nine and thirteen, appeared with Holdren in the Oval Office. The visit had been arranged as a typical White House photo op, but Jill ensured that the meeting quickly turned substantive. She told the president that if she had been able to get *BRCA* testing earlier, she could have avoided cancer, and that now, after the Supreme Court case, testing was broadly available at an affordable price. Obama listened with interest. Of course, he had known about the gene patenting lawsuit being pursued by his solicitor general, but he had not connected it with breast and ovarian cancer.

A few days earlier Obama had met with Anne Wojcicki, the CEO of the direct-to-consumer genetic testing firm 23andMe. When Jill explained *BRCA* testing to Obama, that prior conversation clicked. "Oh, that's what Anne was talking about," he reflected. Jill urged Obama and his sister to get tested, and concluded by telling him that he really ought to think about the possibilities in genomics and big data. If we developed a better understanding of the genetic origins of certain diseases, she explained, it could make a big difference in people's lives. The president thanked her warmly and ended the meeting.

Two days later, Holdren was driving to Dulles Airport, where he would board a flight to London to meet with the chief science advisors of the G8+5 countries. Halfway to the airport, his cell phone rang. It was the president. "I've been thinking about the conversation I had with your daughter," he told Holdren, his voice crisp and businesslike. "I want you to pull together an initiative that looks at the intersection of genomics, big data, analytics, and public health." Holdren acknowledged the request. Then the president added, "I want this by next Tuesday." It was Thursday.

Holdren immediately called Eric Lander and his associate director for science at OSTP, Jo Handelsman. While Holdren was in London, the three of them coordinated with their staffs to assemble a memo outlining the blueprint for such a program. It would be modeled on ideas that Francis Collins and others had proposed over the last decade and called for a massive effort to understand how genes and the environment shape human health.

Soon, the plan took on a life of its own as Collins, Harold Varmus, and a host of other government leaders rallied behind the president's new initiative. In his 2015 State of the Union Address, President Obama announced, "Tonight, I'm launching a new Precision Medicine Initiative to bring us closer to curing diseases like cancer and diabetes — and to give all of us access to the personalized information we need to keep ourselves and our families healthier." Ten days later, at a White House ceremony, the president revealed that $215 million would be allocated to the new Precision Medicine Initiative—"one of the biggest opportunities for breakthroughs in medicine that we have ever seen." It would assemble the largest-ever cohort of individual DNA samples and demographic data—collected from one million American volunteers—to expand knowledge of the complex interactions among genes, environment, medicine, and health.

Not surprisingly, among the dozens of government experts and policy analysts who brought the president's ambitious new program to fruition was Tania Simoncelli, then at OSTP. Working with Holdren and Handelsman, Simoncelli saw the ultimate vindication of the work that she had begun more than a decade earlier at the ACLU. She drafted the privacy and trust framework for the new initiative's million-person cohort and got the opportunity to present it personally to the president. The victory that Simoncelli helped to achieve in *AMP v. Myriad*—making the information contained in our genes freely available to all, and owned by none—was one of the cornerstones needed to advance us all toward a healthier and more just world.

APPENDIX
The (Legal) Meaning of *Myriad*

THE BOOK THAT you have just read offers a narrative account of the unlikely legal case that ended gene patenting in America. In this appendix, I offer the reader my own thoughts and conclusions about *AMP v. Myriad* and its implications for the law, for innovation, and for healthcare.

As the reader has seen, *AMP v. Myriad* elicited strong, often visceral, reactions—not only among lawyers, but among scientists, healthcare advocates, cancer patients, and their families. The case continues to fascinate and infuriate because it is "about" so many different things. It has layers upon layers of meaning, like the skins of an onion that can be peeled back, one by one, to reveal what lies at its core.

On its outer, most superficial, layer is a question of patent doctrine—a new twist on the old "product of nature" rule. This is the type of question often posed by technical legal cases, and it can be answered, one presumes, through the diligent, if not mechanistic, application of existing legal rules. In such cases, one can ask whether the court got it "right," and, if not, what flaws in its reasoning led it astray. But that is a technical question only. Had that been the only thing at stake in *AMP v. Myriad*, its result would have been of interest to a handful of patent lawyers and biotechnology analysts, but there would not have been demonstrators on the steps of the Supreme Court.

The next layer gets to the *effect* of the patents in question—on healthcare, on the market, and on the individual women who desperately needed *BRCA* testing. This human element underlies many patent cases, especially in the pharmaceutical area. Patents give their owners the exclusive right to control, and price, products that are often essential to life itself. On the other hand, in our market economy, new drugs will not spring into existence of their own accord. Achieving the right balance between access and innovation is perhaps the greatest challenge faced by the patent system. This layer of the onion elicits impassioned advocacy—far more than technical questions of law.

But there is, within *AMP v. Myriad*, a deeper layer still. That is its connection to human genes. Genes hold a privileged place in our collective imagination and our notions of what it means to be human. They are the messengers of heredity—determining our physical features, our mental quirks, our susceptibility to disease. They link us to our families, our ancestors, broader social, ethnic, and regional groups, and, at the deepest level, to all living creatures on earth. And, as something with which we are born, and that we can never change (at least, not yet), we consider genes to be integral to our persona and our identities as human beings. Thinking about genes as different, as special, has been (somewhat derisively) called "genetic exceptionalism." But genes *are* exceptional in ways that other cellular systems and organs in our bodies are not. And for this reason, the notion of *owning* genes somehow shocks the conscience, or at least gives one pause. A case about owning genes was bound to be trouble.

Discussions of *AMP v. Myriad* often reach an impasse due to the number of issues swirling around it. In the next few pages, I try to tease apart the different strands of this complex landscape and offer some thoughts, though hardly definitive answers, on a few of these questions.

Does This Story Have a Villain?

IN AN OFT-QUOTED 2013 article, Bob Cook-Deegan and Lane Baldwin observe that the press and the academy cast Myriad as a "ruthless mercenary" and "a villain in the evolving narrative of biotechnology." Is this characterization fair?

The first thing to note here is that Myriad didn't do anything underhanded when it obtained patents covering the *BRCA1* and *BRCA2* genes. Myriad, the University of Utah, and their lawyers were following well-established rules and practices of the Patent Office, which, by 1994, had been in place for at least a decade. Human-derived substances like insulin, human growth hormone, and erythropoietin had long been patented with little objection from anyone. Likewise, the Patent Office was faithfully, if somewhat myopically, following guiding precedent from the Supreme Court's 1980 decision in *Chakrabarty.* "Anything under the sun made by man" was patentable, so long as it had "markedly different characteristics" from anything found in nature. According to the Patent Office, genes that were isolated and purified met this threshold. As a result, the issuance of patents on isolated human DNA had become routine and accepted practice.

But the Patent Office wasn't simply a rubber stamp agency, approving whatever patent applications came down the pike according to some rote procedure. After all, they rejected the energy traders' flimsy patent application in the *Bilski* case and many others. But with gene patents, many in the Patent Office genuinely believed that they were facilitating innovation and the American economy. It is perhaps for this reason that the controversy was so bitter. Both sides believed that justice was on their side.

AMP v. Myriad was the first case in which the very notion of patenting human genes was challenged. Before that, there had been significant litigation over DNA-based patents, but those disputes all involved challenges to particular patents on technical grounds—novelty, obviousness, and the like. Nobody in the industry seemed interested in bringing down the entire house of cards by challenging the existence of gene patents as a general principle. It was for this reason that nearly every patent expert that Hansen and Simoncelli consulted thought that they didn't stand a chance.

The second thing to recall is that the ACLU chose Myriad as its target for reasons that went well beyond the characteristics of its patents. These considerations are discussed at length in the text, but suffice it to say that they were largely political and strategic rather than legal. In fact, even at the time the lawsuit was filed, most of Myriad's *BRCA* patents were old. By now, almost all of them would have expired. And following the public release of the full sequence of the human genome in the early 2000s, gene patents, as they were issued in the heyday of 1990s gene hunting, were artifacts of the past. Whether or not human genes were

APPENDIX

eligible for patent protection, any patent application that tried to claim a gene whose sequence was already disclosed to the public was doomed to fail on more conventional grounds. So Myriad was the poster child for a legal campaign against an (admittedly bad) patenting practice, but Myriad was by no means the worst or only beneficiary of that practice (recall Athena Diagnostics, Miami Children's Hospital, and the bankrupt holder of the Long-QT gene patents).

This is not to say, though, that Myriad was a particularly altruistic company. Pete Meldrum ran a tight ship, and Myriad was both aggressive and calculating when it came to its competitors. But this is no different from most successful companies, and corporate tactics rarely seem to influence public perception of a company.[1] As Gold and Carbone note, "Myriad pursued a fairly traditional commercialization strategy," and that strategy—for Myriad and for every other company in the biotechnology and pharmaceutical sectors—involved patents.

The final important thing to understand about this story is that Myriad's patents never covered genes as they naturally exist *inside* the human body. As broad as Myriad's patent claims were (and they were much too broad), they only covered "isolated and purified" DNA, which, no matter how expansively those terms are interpreted, refer to forms of DNA that are produced in the laboratory. Many well-intentioned people believed, and were understandably upset, that Myriad owned an integral, invisible part of their bodies. But that simply wasn't the case. And while I firmly believe that Myriad's patents on the *BRCA* genes should never have been granted, it is important to understand what they actually covered.

Baby-Splitting Generally Isn't Good for the Baby

IN ASSESSING THE significance of *AMP v. Myriad*, the first layer of the onion is doctrinal: did the Supreme Court get it right? Was the

---

1 For example, companies whose products are beloved by the public—Disney, Apple, Mattel, McDonald's—are also known to be highly litigious and ruthless in their business dealings.

Department of Justice's and Justice Thomas's interpretation of the product of nature doctrine reasonable, and was it sufficient to overturn the Patent Office's longstanding practice of granting patents on human genes?[2]

To understand what's going on, it is useful to set out the three key contrasts that Justice Thomas establishes when he talks about DNA and genes. Though he doesn't actually lay these out very clearly, and somewhat blurs the distinctions among them, they can help to elucidate the basis for what is and is not patent-eligible:

CELLULAR GENES VERSUS ISOLATED[3] GENES—Cellular genes are what exist within human cells. Hundreds of genes are strung together in long, double-helical strands called chromosomes. Lots of other DNA is interspersed between the genes, and all sorts of other molecules hang off them, like holiday lights on a wire.[4] Along the chromosome, the genes and other DNA are attached end-to-end with covalent chemical bonds— the strong "glue" that Judge Lourie at the Federal Circuit found so important. An isolated gene, on the other hand, is just the gene, detached from the rest of a chromosome, and lacking most of its hanger-on molecules. Generally speaking, and notwithstanding Eric Lander's observations, isolated genes do not exist in the body in any useful form.

GENOMIC DNA (gDNA) VERSUS COMPLEMENTARY DNA (cDNA)—in contrast to the distinction between cellular and isolated genes—a distinction that relates to the positioning of a gene among other surrounding molecules—the distinction between gDNA and cDNA relates to what is "inside" a particular gene. Every human gene contains both exons (DNA

---

2 I won't address the standing issues raised by the case, as the procedures for challenging the validity of patents have changed significantly since *AMP v. Myriad* was brought in 2009. Today, virtually anyone can challenge a patent at the Patent Trial and Appeal Board (PTAB) under a procedure called "*inter partes* review" or "IPR." Yet the IPR procedure is itself under attack, both from those who would prefer to limit the avenues for challenging issued patents at all, and those (like Dan Ravicher) who find that IPRs are more expensive and difficult to maintain than the old *inter partes* reexamination procedures that he employed to great effect at PubPat.

3 For the sake of discussion, I'll refer to "isolated and purified" genes, as referenced throughout the case, as "isolated."

4 We are slowly learning more about the function of the molecules that ride along with DNA. This is the study of epigenetics, which is still in its infancy. See Mukherjee (2016, pp. 393–410).

that codes proteins) and far more numerous introns (DNA that doesn't code proteins). The fundamental difference between gDNA and cDNA relates to the presence of introns—gDNA has both exons and introns, while cDNA has only exons.[5] Genes as they exist in the body are gDNA, but gDNA also exists in the lab—the isolated-purified genes mentioned above are often gDNA. But it is also possible in the lab to construct DNA molecules that consist only of exons.[6] This is cDNA, and cDNA does not exist naturally within the body.

NATURALLY OCCURRING VERSUS SYNTHETIC—this distinction relates to the way in which something is made. Was it created through natural means (i.e., by the body), or in the lab? Though this distinction is often raised in discussions of the "product of nature" doctrine, unlike the other two distinctions noted above (cellular versus isolated genes, gDNA versus cDNA), the distinction between naturally and synthetically created DNA, to the surprise of many, did not factor into Justice Thomas's analysis of patentability.

With these three bases for comparison firmly in mind, we can analyze the Supreme Court's approach to patent eligibility. Interestingly, the Court's opinion, which attempted to follow the government's lead in splitting a particularly colicky baby, has been criticized in equal measure both by supporters and opponents of gene patenting.

GENES ARE NOT PATENTABLE JUST BECAUSE THEY ARE ISOLATED— First, Justice Thomas tackled the significance of *isolating* genes. In the

---

5 It's important to remember that there are LOTS of introns in every gene. The *BRCA1* gene has a total of about 110,000 bases. Only 8,000 of those are in exons that code proteins and they are spread out across the full length of the gene. The other 102,000 bases are non-coding and located within introns. Suppose that a particular protein (a receptor that is activated by a chemical found in catnip) is coded by a sequence of the three DNA nucleotides: C-A-T. In the gene, these three coding nucleotides (exons) are likely spread out and interspersed with non-coding nucleotides (introns), like this: AAT**CA**GGGACC**T**GAA. Without those extraneous introns, the sequence consists of just the exons: **CAT**.

6 The body also makes a chemical that codes for proteins without the intervening introns, but this chemical is RNA, not DNA, and RNA is slightly different than DNA for patent purposes and otherwise.

go-go 1980s, the Patent Office adopted a hyper-technical interpretation of the law, accepting the argument that isolating DNA from a human cell transformed it into a new composition of matter. As such, it was patentable, like a new type of steel alloy or polymer. As Rebecca Eisenberg has pointed out, the first issued patents on human DNA covered the production of artificial proteins, which resembled drugs more than anything else. Because patents on traditional, small-molecule drugs are a mainstay of chemical patent practice, it was a small conceptual leap to accept the same framework for protecting biological compounds like synthetic insulin. It was only later, when companies like Myriad emerged, that patents started to claim complete human genes for no other purpose than diagnosing particular diseases and stopping others from doing so. And by then, it was too late—the floodgates had been opened.

Justice Thomas begins his analysis with the test developed by the Court in *Chakrabarty*: is the substance under consideration "markedly different" from anything that occurs in nature?[7] He first observes that the nucleotide sequences of isolated genes (the precise order of the As, Ts, Gs, and Cs) are identical to the sequences of cellular genes. And given that these sequences are the same, isolated genes aren't "markedly different" from cellular genes. Therefore, isolated genes are not patentable.

Under this theory, it does not matter that isolated genes are detached from the larger chromosome or lack the other molecules that hang onto cellular genes in the body. Those chemical differences are what Judge Lourie at the Federal Circuit viewed as paramount. But Justice Thomas dismisses them as unimportant or, at least, not rising to the level of a marked difference. To him, the real purpose of genes is to encode proteins, and in that regard only the nucleotide sequence matters.

Many of those who favored gene patenting criticized this reasoning. Some remained fixated on the chemical differences between isolated and cellular genes. To them, the breaking of covalent bonds is key—it creates

---

7 Eisenberg and other scholars have wondered why the existence of a substance in nature should matter to the patentability of a synthetic compound that is clearly made by man. After all, how do we know what exists in nature throughout the universe? When scientists develop what they believe to be an entirely new polymer compound, should it matter that the compound is later discovered frozen beneath the ice cap of some extrasolar planet? If so, why?

new and different chemical compounds. Yet as Judge Sweet noted in his opinion, "scientists in the fields of molecular biology and genomics have considered this practice a lawyer's trick." The chemical differences between a gene in the body and the same gene isolated in the lab are irrelevant to the "informational" purpose of genes. I think that the Supreme Court was right to disregard these differences and eliminate the decades-old legal fiction that allowed patenting of human genes as new compositions of matter simply because they are chemically different from their counterparts in the body.

A SYNTHETIC COPY OF A NATURAL SUBSTANCE IS NOT PATENTABLE EITHER—A different objection to Justice Thomas's reasoning was raised by critics like Gene Quinn. In a blog post on the day of the Supreme Court's decision, Quinn argued that synthetic substances *should* be patentable, even if they are identical to substances that exist in nature. Take, for example, an artificial kidney grown in a test tube. It needs to be "indistinguishable from what appears in nature" in order to resist rejection by the host body into which it is transplanted. Shouldn't the artificial but otherwise-found-in-nature kidney be patentable? The Supreme Court said no.

A real-life example of Quinn's hypothetical example arose in Myriad's patent infringement suit against Ambry and other diagnostic labs that started to perform *BRCA* testing immediately after the Supreme Court's decision. This time around, Myriad asserted patent claims covering short DNA primers that are necessary to perform diagnostic testing. Though the primers were clearly synthetic, both the district judge and the Federal Circuit held that they were unpatentable because they duplicated the exact DNA sequence of the cellular genes that they targeted. The fact that the primers were synthetic was irrelevant. Because they were the same as something occurring in nature, they were not patentable.

Objecting to this result, critics argued that society will lose out on lifesaving innovations if we don't offer patent protection to their developers. Yet the Court recognized that patents should be available for innovative *methods* of creating even otherwise-found-in-nature substances. Thus, if Myriad had developed a novel method of locating mutations in the *BRCA* genes, or of using them in a diagnostic test (which it did—see below), then a patent might have been available. But that is a different matter from patenting the biological product itself. Ultimately, the Court ruled that if the product is the same as something found in nature (no matter how innovative the method of producing it), then no patent is available for the thing itself.

Accordingly, if Quinn's lab-grown kidney is identical to a human kidney, then no patent is available, even though a novel method of *growing* a synthetic kidney would be patent-eligible.[8] I think that this result is both reasonable and consistent with practice in many other industries. For example, there are thousands of patents covering improved methods for manufacturing unpatented articles—everything from rubber tires to salt pellets to electronic components, most of which have been around long enough that their patents, if they ever existed, are now expired. Nevertheless, manufacturers have ample incentives to keep developing better methods of production, whether these are superior in terms of speed, cost, sustainability, worker safety, or anything else. And if a competitor copies a patented method of manufacture, the patent holder can assert its rights in court. We even have a law that allows a patent owner to prevent the importation into the United States of an *unpatented* article (a tire, an electronic component, a kidney) that is manufactured outside the United States using a process that is patented in the United States.

Likewise, there are thousands of patents covering improved agricultural methods—methods of growing crops faster, eradicating pests and weeds, yielding bigger and healthier tomatoes, figs, cattle, and salmon. All of these inventive methods are patentable, but the resulting fruits, vegetables, and livestock, so long as they are genetically the same as their naturally occurring cousins, are not. And like manufacturers, agricultural

8 This argument was made, and the same result reached, by the Federal Circuit in a 2014 case involving Dolly the sheep. The Scottish Roslin Institute that created Dolly, the first viable cloned mammal, sought U.S. patents on both the method of cloning a mammal as well as the cloned offspring themselves. The Federal Circuit held that while the Institute's method claims were allowable, the claims covering the sheep were not, given that, under the Supreme Court's ruling in *AMP v. Myriad*, the cloned sheep were identical (by definition) to sheep found in nature (i.e., the original sheep from which they were cloned). *In re: Roslin Institute (Edinburgh)*, 750 F.3d 1333 (Fed. Cir. 2014).

scientists have ample incentives to keep improving even without patents on the fruits, literally, of their labor.

But what about more esoteric compounds, like synthetic antibodies? Shouldn't we grant a patent to the scientists who have developed a new lifesaving antibiotic that is identical to that produced by healthy immune systems? The answer, and the logic behind it, are the same as they are for kidneys, salt tablets, and avocados—no. If the synthetic antibodies are identical to the antibodies produced by the human immune system, then they should not be patentable. If the process or method used to develop, create or manufacture them is novel, then it should be patentable. But if the method is simply a routine process used throughout the field, or a novel method that is patented by someone else, then the developers of the synthetic antibody should not be entitled to any patent at all.[9]

But won't this defeat the incentive system—meaning that nobody will then develop those lifesaving antibodies? I don't think so. As discussed in chapter 6, Genentech faced this very issue when it considered how to protect the synthetic insulin that it developed. Precisely because synthetic insulin was identical to natural insulin, Genentech did not pursue a composition of matter patent on it. Yet Genentech did obtain a patent on the process for making synthetic insulin, a patent that earned it substantial sums over its lifetime, and which resulted in some of the largest patent battles of the day.

But if the method of creating some synthetic substance—insulin, antibodies or kidneys—is truly routine, then anybody should be able to use that method to create the same synthetic substances and more. In that case, the absence of patents should result in more, and cheaper, antibodies, just as generic drugs are produced at a fraction of the cost of drugs that are still protected by patents. Likewise, the difficulty of discovering a naturally occurring substance should have no bearing on whether the substance itself is patentable. That is like saying that the Amazonian explorer who spent six months hacking through the jungle

---

9 It is true that detecting an infringing manufacturing process is often more difficult than detecting an infringing article of manufacture. Yet this practical evidentiary difficulty should not drive the substantive outcome of the law. Rather, procedures could be considered to inspect manufacturing operations that are suspected of infringing legitimate patents.

with machetes deserves a patent on the mushroom that he discovers deep in the jungle, simply because it was hard to find. It is a product of nature, and is not the kind of "invention" that patents were intended to cover.

Finally, some critics have observed that, in practice, it can be difficult, if not impossible, to prove that a patented method is being used to create a synthetic product because factories and production plants are often locked down and their employees are under strict contractual obligations not to reveal the processes used within. It would be far easier, they argue, to point to the resulting synthetic kidney, antibody, or other substance and prove that it infringes a patent. This point is well taken, and it is clearly easier to prove infringement by a product that one has in hand than by a manufacturing process conducted at some secret location. But the patent laws are not designed for the convenience of litigants, and they should not be bent simply to overcome evidentiary inconvenience. There are ample discovery mechanisms within the civil litigation system to adduce evidence of infringement by a manufacturing process, and potential difficulties in that area do not justify changing the law to allow patenting of synthetic substances that are identical to those found in nature.

BUT cDNA IS PATENTABLE—So much for those who would have preserved the patentability of human genes. What about those on the other side of the aisle—the ACLU allies who criticized Justice Thomas for allowing patents claiming cDNA? In a single paragraph, Justice Thomas concludes that "cDNA does not present the same obstacles to patentability as naturally occurring, isolated DNA segments." Why not? Because, he reasons, cDNA lacks the introns found in naturally occurring gDNA. Without the introns, cDNA is "markedly different" from gDNA. Hence, it is patentable.

This particular leap of logic has attracted the ire of many scholars who have analyzed Justice Thomas's opinion. What was wrong with his reasoning? Just cast your eyes up a few paragraphs and you will see where Justice Thomas belittles the chemical differences between isolated genes and cellular genes. He finds these differences immaterial to the information content of the genes, so these two substances are not mark edly different. Hence, no patent on isolated genes.

But if differences like covalent bonds and attached molecules don't make isolated DNA markedly different from cellular DNA, then why does the absence of introns make cDNA markedly different from gDNA? After all, just like the missing covalent bonds in isolated DNA, introns are not relevant to the coding function of a gene. The same exons occur in both gDNA and cDNA in the same order, and both will code the same proteins. It's just that the cDNA version lacks the introns interspersed among the exons. But the cellular mechanisms that code proteins splice out the introns and use only the information from the exons. So if we ignore covalent bonds to reason that isolated gDNA isn't patentable, then why don't we also ignore the absence of introns in cDNA?

In other words, if the important thing about DNA is its informational content—the order of As, Ts, Cs, and Gs—and we ignore other chemical differences that are unrelated to this information, then shouldn't we also ignore introns? Because if we do, *both* gDNA and cDNA should be unpatentable.

The magic microscope metaphor, as charming as it is, just doesn't work here. Or, if it does, it's not being properly focused, because *neither* isolated gDNA nor cDNA is chemically identical to what exists inside the body. Yet the Court treats one as patentable (cDNA) and the other as unpatentable (gDNA). There is really no principled, scientific reason to distinguish between molecules attached to cellular DNA, on one hand, and introns that are interspersed among the coding exons of a gene, on the other hand. Both are equally irrelevant to protein coding, and cDNA should thus not be eligible for patent protection.[10]

---

10 Some commentators have argued that the real difference between gDNA and cDNA is not a scientific one, but a commercial one. Whereas the primary use for isolated gDNA is in genetic diagnostic tests (a comparatively small market), cDNA may be useful in a wide range of therapeutic applications. See Rai and Cook-Deegan (2013), Rai (2014, p. 7). While this is likely true, it seems that there ought to be a better and more reasoned basis for a legal rule than its potential commercial impact. After all, opinions will differ regarding the commercial impact of one rule or another, and commercial facts are generally even more difficult to establish than scientific ones.

Reining in "Invention"

I AGREE WITH the Supreme Court's decision in *AMP v. Myriad* inasmuch as it prohibits the patenting of isolated human genes (though, as mentioned above, I would have extended the ruling to cover cDNA as well). But even as to gDNA, I would have gone further than Justice Thomas, whose entire analysis rests on the "markedly different" test established in *Chakrabarty*. For reasons both rhetorical and analytical, I would have returned to the roots of patent law—the U.S. Constitution.

The Constitution authorizes Congress to grant patents to "inventors" for their "discoveries." The Patent Act says that patents will be granted to "whoever invents or discovers any new and useful . . . composition of matter." Invention is at the root of the patent law. Did Myriad "invent" the *BRCA* genes? No—not under any reasonable interpretation of the word. Though Myriad was the first to isolate and sequence the *BRCA* genes, it did not *invent* the genes themselves. In fact, at the time that Myriad sequenced the *BRCA* genes, neither Myriad nor anyone else really knew their biological function.[11] As Judge Bryson wrote in his dissent at the Federal Circuit, Myriad's location of the *BRCA* genes was simply not "an act of invention."

At most, Myriad discovered (invented) a method for assessing a person's risk of cancer based on the presence of certain mutations in the *BRCA* genes. Justice Thomas, echoing Judge Bryson, did not rule out patents claiming such "new *applications* of knowledge about the *BRCA1* and *BRCA2* genes." Determining an individual's risk of breast cancer could be viewed as an application of the knowledge that Myriad gained about the *BRCA* genes.

The disease-correlated mutations in the *BRCA* genes were Myriad's actual contribution to science and, in my view, probably *should* have

---

11 All that Myriad knew at the time it identified the *BRCA* genes was that there was a *correlation* between a handful of mutations in the genes and an increased risk of breast or ovarian cancer. The mechanisms behind that correlation were completely unknown then. Today, we know that the proteins encoded by the *BRCA* genes are tumor suppressors. When the genes are damaged, they cannot code these proteins correctly, which allows tumors to grow unchecked. The mechanisms for these effects, however, are complex and still not wholly understood.

been eligible for patenting (more on this below). But by allowing Myriad to claim the entire *BRCA1* and *BRCA2* genes and all of their uses, both known and unknown, the Patent Office went too far. Myriad did not *invent* the genes themselves.

Allowing patents on naturally occurring genes as compositions of matter had wide-ranging effects and gave Myriad far more than it deserved based on its actual discovery. This is how composition of matter patents work. Suppose that I invent and patent a new rubbery elastic compound that I call "flubber." I have exclusive rights to every possible use of flubber—in bouncing balls, car bumpers, roof caulking, athletic shoes, and whatever else can be imagined.

Should the discoverer of a gene have the same broad rights as the inventor of a new substance? Myriad's composition of matter patents covering the *BRCA1* and *BRCA2* genes gave it exclusive rights not only to conduct diagnostic tests using those genes, but to do anything else imaginable with those genes. A *BRCA1* antibody to fight tumor cells? Patented. Using *BRCA2* to predict the occurrence of birth defects? Patented. An endless number of applications and uses for those genes, none of which were discovered by Myriad, are all covered by the broad patents claiming the genes as compositions of matter.[12] Which makes little sense, given that, unlike flubber, Myriad did not *invent* the BRCA genes. It merely discovered that they were useful in a particular application—predicting cancer risk.[13]

Patents arising from the discovery of gene-disease associations should always have been limited to the discovery actually made, and should

---

12 This broad scope was advantageous to Myriad, as it granted Eli Lilly an exclusive license to develop therapeutic drugs using the *BRCA* genes in exchange for significant up-front payments. Nevertheless, neither Lilly nor Myriad ever discovered any meaningful therapeutic use for the *BRCA* genes.

13 One popular economic theory developed by Edmund Kitch in the 1970s compares patent rights to land claims in mineral prospecting. In order to encourage efficient exploration for minerals, the government parceled out tracts of land to private parties, with the understanding that these parties would have strong incentives to prospect their own parcels, either individually or by subleasing portions to those better equipped to do so. Edmund W. Kitch (1977) "The Nature and Function of the Patent System," *Journal of Law & Economics* 20(2): 265–290, p. 266. This "prospect theory" has frequently been used to justify the grant of broad patent rights to single companies, though I and others have argued that it does not lead to optimal development of new technologies. See Contreras and Sherkow (2017).

never have been allowed to expand to cover human genes as compositions of matter. Human genes are quintessentially products of nature, not inventions.

What about Methods—"Hold the Mayo"

ABOVE, I ARGUE that those who discover a particular gene-disease association should not be permitted to claim the entire gene as a composition of matter, but *should* be permitted to obtain patent protection for their specific discovery—in the case of Myriad, the association of particular *BRCA* mutations with elevated risks of breast and ovarian cancer. This reasoning is supported by Justice Thomas, who makes it clear that his opinion does not relate to "new applications of knowledge about the *BRCA1* and *BRCA2* genes." The patentability of new *applications* of laws of nature finds its roots in Supreme Court decisions going back to the 1940s.

Yet all but one of Myriad's method claims were rejected, even by the patent-friendly Federal Circuit. Why? Because something changed with the Supreme Court's decision in *Mayo v. Prometheus*. In *Mayo*, as in the earlier cases, Justice Breyer acknowledges that the Court must "determine whether the claimed processes have transformed . . . unpatentable natural laws into patent-eligible applications of those laws." But in doing so, he seemingly raises the bar on what counts as a "new application" of a natural law. In particular, in order for a patent to issue, he requires that there be some "inventive concept" above and beyond the natural law and its application through "well-understood, routine, conventional activity." Thus, the Court in *Mayo* held that a patent claiming a method of adjusting a patient's drug dosage based on the level of metabolites in the patient's blood was merely a straightforward application of a natural correlation observed in the human body, and was therefore ineligible for patent protection.[14]

---

14 Recall that the Federal Circuit's prior test allowed the claims in *Mayo* because they resulted in a physical "transformation" in a patient's body. I would not advocate a return to the Federal Circuit's peculiar "machine or transformation" test either.

The result of *Mayo* and the many lower court decisions following it has been that the genetic diagnostics industry is largely left without patents. Thus, while *AMP v. Myriad* eliminated patents claiming human genes as compositions of matter, *Mayo* eliminated patents on the application of knowledge arising from the observation of human genes. But it is not only genetic diagnostics that have been affected. Recent judicial rulings have severely limited patents on other biomedical innovations such as new antibodies. Is this too much?

Critics of these cases argue that the unavailability of patents on human genes and related biotech discoveries will hinder the progress of science, that we need patents to encourage companies to take risks and spend large amounts on R&D. Without the carrot of exclusivity hanging just within reach, who would, or could, afford to spend a billion dollars to discover the next miracle cure?

This concern is a legitimate one. Clearly, the potential profits to be made by patenting *BRCA1* and *BRCA2* motivated Myriad to search for the genes. Without the promise of patents, Myriad never would have been formed or funded. Nevertheless, the *BRCA* genes would almost certainly have been discovered even if Myriad had never existed. In the race to discover these particular genes, competing academic labs were never far behind the company. Prominent researchers like Mary-Claire King and Francis Collins, who had already contributed substantially to the search, would have found them eventually, even without the extra capital and equipment brought to bear by Myriad. In the case of *BRCA2*, a lab in London was on Myriad's heels, losing the race to publish the genetic sequence by a single day. So denying Myriad a patent on these discoveries would hardly have impacted scientific progress. Academic labs had other incentives to find the *BRCA* genes, including the promise of greater governmental grant funding and the enhancement of their academic reputations (as Chris Hansen wryly noted during oral arguments—Nobel Prizes). But the facts of *AMP v. Myriad* are not typical. In most other scientific fields, groups of government-funded academic labs are *not* furiously racing to make the same discovery. In these areas, without companies that are incentivized to conduct R&D by the promise of exclusivity down the road, some lifesaving discoveries might *not* be made. And this is not ideal either.

So perhaps the Supreme Court's test for patent eligibility *is* too strict— not under *AMP v. Myriad* (where, as I have argued above, it is still too permissive), but under *Mayo*. That is, the requirement under *Mayo* that an application of a truly novel discovery cannot be patented unless the discoverer adds some *invention* above and beyond application of the discovery may ask too much. I believe that it is the intent of the patent laws, and of the Constitution, that useful and non-obvious applications of truly novel discoveries be eligible for patent protection,[15] provided that patents are limited to the specific application(s) of the discovery made by the inventor, and not expanded through artful claim drafting to cover anything else that a skilled patent lawyer can dream up.[16]

Under a liberalized *Mayo* standard for method claims, Myriad would have been entitled to patent a method of predicting an elevated risk of breast or ovarian cancer in a patient by detecting specified mutations in her *BRCA* genes. The claims would not have covered the entire gene, nor would they cover any mutations of the gene not specifically disclosed (i.e., discovered) by Myriad, nor would they cover any use or application of those genes other than the ones specifically spelled out by Myriad. But within those narrow confines, I would have allowed a patent to issue.[17]

This approach resonates with earlier Supreme Court precedent like *Diamond v. Diehr* (1981) in which the Court upheld a patent on a new method for curing rubber that was based on the well-known Arrhenius equation. The equation itself was not patentable, but its application in

---

15 This theory is best summed up by Professors Jeffrey Lefsin and Peter Menell in their *amicus* brief supporting the grant of *certiorari* in *Sequenom, Inc. v. Ariosa Diagnostics, Inc.* (Fed. Cir. 2015), a case in which a truly revolutionary breakthrough was denied patent protection because it embodied a "law of nature." Regrettably, the Supreme Court declined to hear the case.

16 I cannot emphasize enough the need for patent claims to be drawn narrowly to the actual and specific discoveries made. Over-claiming, as encouraged by the Kitch prospect theory (see note 13, above) tends to pre-empt too many uses of a natural law or relationship, thus dampening innovation that might otherwise occur.

17 For a variety of reasons, Myriad's European *BRCA* patents are generally limited to this narrow scope.

the narrow context of curing rubber was viewed as an invention eligible for patent protection. I think this must be what Justice Breyer intended when he described patents on "applications of natural laws" in *Mayo*.

So where would such a modified test for method claims leave us? Among other things, Myriad would still have patents covering *BRCA* testing. Even without broadly claiming *BRCA1* and *BRCA2* as compositions of matter, method claims like these would have enabled Myriad to lock up the *BRCA* testing market—at least for mutations discovered by Myriad—as it did for many years.[18] The end result for patients and competing labs would have been the same—limited access, no competition, high prices. And that's not ideal either.

But we shouldn't blame the patent system for all of the world's woes, or twist the patent system into unnatural configurations in order to avoid them. These problems arise not only from the existence of patents, but from other systemic issues that can and should be addressed. Two of the biggest culprits are the university licensing system and the healthcare payment system.

All Hail, Alma Mater

DURING AMERICA'S POST-WAR boom, thanks in part to the success of academic-government collaboration on the Manhattan Project, the Federal government began to pour money into university labs. The prevailing wisdom was that funding basic research would help not only national defense, but the U.S. economy. Although an increasing share of each year's Nobel prizes went to American scientists, relatively little academic research was finding its way into the commercial sector. Unlike Japan, where the government directly funded industrial research in fields like semiconductors and consumer electronics, U.S. academic research seemed to have few commercial applications. The problem, many felt, was patents. Specifically, *not enough* patents.

---

18 What Myriad would have lost is the ability to license other applications of the *BRCA* genes (e.g., therapeutics) to companies like Eli Lilly.

In response, Senators Birch Bayh, a Democrat from Indiana, and Bob Dole, a Republican from Kansas, introduced legislation to harness the power of the American research establishment to help private industry. The resulting Bayh–Dole Act of 1980 provides that when a university or other federally funded institution develops a new technology, the institution is entitled to the patent. Moreover, if the institution *fails* to patent the technology, it will lose those rights. In effect, universities are penalized for *not* patenting their inventions. The Bayh-Dole Act also provides that any institution earning revenue from one of these patents must *share* some of the profits with the individual inventors. The statute doesn't specify how much each inventor should get, but most universities have developed a rough three-way split of these profits: one-third to the inventors, one-third to their academic department, and one-third to the university itself.

This generous arrangement offered large potential rewards to entrepreneurial academics. A researcher could form a start-up company to commercialize the scientific discoveries made in his or her laboratory, the university would get the patents and then grant the start-up company exclusive rights to utilize those patents in return for a share of the revenue, and external investors—venture capitalists and, at a later stage, investors in public markets—would provide the capital necessary to fund the commercial growth of the company.

Spinout companies were particularly attractive in Utah, where the state university's geographical isolation worked against the kind of easy collaboration that institutions in Silicon Valley and the East Coast enjoyed with local companies. As a result, the University of Utah quickly began to rival much larger research universities in terms of company formation. During the decade after the Bayh-Dole Act was passed, the University of Utah created sixty-two spinout companies (including Myriad), second in number only to MIT.

Like many universities, the University of Utah established a technology transfer office (often referred to as a TTO) to obtain and manage patents on university research. By the mid-1980s, Utah's TTO was busy. The university was enjoying a boom in biomedical engineering, sparked by the first artificial heart—the Jarvik-7—which had been developed at the university and implanted in a living patient's chest in 1982. By the mid-eighties projects were under way in Utah to develop artificial arms, hands, kidneys, bladders, fallopian tubes, and sphincters. Some enterprising promoter dubbed the Salt Lake region "Bionic Valley," a moniker that, thankfully, never took hold. One of the most significant aspects of the university's artificial organ program was its early outreach to private industry, a move that some branded, not entirely kindly, "academic capitalism."

The TTO occupied a brand new office suite in the university's Research Park, a 320-acre development that now includes a Marriott hotel and dozens of low-rise labs and office buildings. By 1986, Utah's TTO was earning roughly a million dollars per year from research funding and patent royalties. Academic capitalism was alive and well at the U.

Then, in the summer of 1988, the U's research landscape was transformed again. The head of the chemistry department, a soft-spoken researcher named Stanley Pons, and Martin Fleischmann, a visiting researcher from England, thought they had stumbled onto the Holy Grail of energy research: sustained nuclear fusion at room temperature—*cold fusion*. If it was real, the process could generate more energy than it consumed, a world-changing proposition. Of course, the cold fusion honeymoon didn't last long, and by 1991, after wasting two years and more than five million dollars of taxpayer money, the University of Utah shut down its cold fusion institute and leased the state of the art facility in Research Park to the newly formed Myriad.

Nevertheless, Myriad followed the same pattern as earlier ventures at the U and other major research universities around the country. When Mark Skolnick and Pete Meldrum formed Myriad in 1991, their first step was to secure rights to all breast cancer genetic discoveries made by Skolnick's university lab. The U agreed to license any patents arising from those discoveries exclusively to Myriad, enabling Myriad to exercise nearly complete control over them. In return, the university received a small share of Myriad stock, plus a 2 percent royalty on Myriad's *BRCA* testing revenue.[19] Of course, with a successful product like BRACAnalysis, this royalty became significant. Court documents show that, over the lifetime of the *BRCA* patents, the University of Utah earned about $40 million in royalties.

When, in the mid-1990s, Myriad and the university made clear their intention to obtain patents on the *BRCA* genes, there was an outcry from other academic groups. Mary-Claire King lamented that a company had won the "race" to find *BRCA1*, and that a university had exclusively licensed its patents to that company. There was a strong suggestion that if King at Berkeley or Collins at Michigan had won the *BRCA* race, the controversy over the resulting patents would never have arisen.

There may be some truth to this suggestion, though less, perhaps, than its proponents may wish to believe. The liberal licensing models to which King, Collins, and others pointed were established in the 1980s, during the early days of university technology transfer. The *CFTR* gene associated with cystic fibrosis was discovered in 1989 by a team of researchers including Collins. The University of Michigan and Hospital for Sick Children applied for a patent, then licensed that patent on a non-exclusive, royalty-free basis to any lab that wanted to test the gene. A similar situation existed with the *HEXA* gene for Tay-Sachs disease, which was discovered at NIH in 1984. NIH obtained a patent on the gene, but chose to make licenses available without charge.

Not quite as magnanimously, Stanford University and the University of California, which held patents on recombinant DNA techniques, licensed those patents for a modest fee. Over the life of the patents, the universities granted 468 licenses earning them more than $250 million. A few years later, Columbia University adopted a similar approach with its patents covering DNA cotransformation, from which it earned nearly $800 million. While the recombinant DNA and cotransformation patents were hugely profitable, the universities that held them adopted a strategy of licensing them broadly on a non-exclusive basis to the entire field. This strategy worked because recombinant DNA and DNA

---

19 The university separately agreed in 1995 to split this royalty with NIH, which was also a co-owner of several of the foundational *BRCA* patents.

cotransformation are generally applicable research tools that they anticipated could be used by every genetics lab in the world.

Not so with gene patents. Unlike a generally applicable research tool,
there are far fewer potential licensees for a patent covering a particular
disease-related gene. Thus, unlike the 468 licensees that Stanford and
UC enrolled for their recombinant DNA patents, the owner of a disease-related gene patent might optimistically expect to find a dozen
commercial and university labs willing to pay for a non-exclusive license.
The financial return from such a licensing program is not great.

Prevailing wisdom in the technology licensing field holds that a single
exclusive licensee will often pay more than a smattering of non-exclusive
licensees. And while the institutions that held the *CFTR* and *HEXA*
patents chose to make them available to others without charge, that
altruistic approach was quickly supplanted in the 1990s by more profit-oriented strategies. Nevertheless, most gene patents were licensed
exclusively to companies seeking to earn both testing and licensing
revenue from them.

Thus, Myriad and hundreds of companies like it obtained exclusive
license rights to patented university technologies with very few strings
attached. In 2017, Professor Jacob Sherkow and I coined the term "surrogate
licensing" to describe these arrangements in which a research institution
effectively relinquishes all of its rights in a patented technology to a private
company, and abdicates its corresponding public responsibilities to that
company. Thus, tens of millions of dollars in federal research funding
flow to U.S. research institutions, which then patent them and subse-
quently license those patents to private companies to do what they wish.

Universities are, by and large, non-profit organizations chartered to
pursue social and educational goals. Companies are not. When universities divest their technology to private companies in exchange for
financial remuneration, they effectively cede discretion over the uses and
exploitation of that technology to the private sector. Company directors
have a duty to maximize the financial return to their shareholders, not
to provide affordable healthcare or medical technology to those in need.
Social goals like these are the province of the government and charitable
enterprises—enterprises like non-profit universities.

So if companies like Myriad take aggressive stances regarding pricing and access to their products, they are only acting as companies do. Don't blame the tiger for behaving like a tiger, blame the careless guards who allowed the tiger to roam free in the village. It is the public organizations operating in the background or, more precisely, failing to act, that are equally responsible for any inequities that result from the commercial exploitation of publicly funded R&D.

Take, for example, Myriad's commercialization of the *BRCA* genes. As described earlier, there was significant criticism of the company's exclusion of competitors, high pricing, and broad promotion of its test to the public. Much of this criticism came from government officials— especially NIH directors like Francis Collins and Harold Varmus. And recall that, though NIH was one of the earliest applicants for patents on human DNA segments, it reversed that position in 1994, soon becoming one of the leading opponents of DNA patenting.

Against this backdrop, few remember that NIH was part-owner of several of Myriad's *BRCA1* patents due to work that two NIH scientists did in collaboration with Myriad during the race to discover the gene. Varmus himself gave control of those patents to Myriad in 1995 when he settled a dispute over "inventorship" of the gene. To Varmus, apparently, securing credit for his inventors was more important than the eventual economic consequences of the resulting patents.

In the 1980s, NIH even made an effort to rein in drug pricing by requiring a "fair pricing" clause in all of its cooperative R&D agreements with private industry. This mandatory contractual language, widely reviled by the pharmaceutical industry, was adopted by NIH in response to a controversy surrounding the AIDS drug AZT. The drug, which was released in 1987 by Burroughs Wellcome, bore the then-stratospheric price tag of $8,000 per year.[20] Yet, as AIDS activists were quick to point out, Burroughs Wellcome discovered neither the drug nor its effectiveness against AIDS. A failed cancer treatment, AZT's potential use against

---

20 Compared to today's astronomical prices for the latest gene therapy treatments (Novartis's Zolgensma costs $2.1 million per treatment), the $8,000 price tag for AZT seems quaint. Yet, at the time, the *New York Times* called AZT "the most expensive prescription drug in history." ("AZT's Inhuman Cost," *NY Times*, Aug. 28, 1989).

AIDS was first suspected by scientists at NIH's National Cancer Institute. To encourage Burroughs to bring AZT to market, NIH allowed the company to retain full ownership of the resulting patent. But once that happened, there was no way to constrain Burroughs's pricing of the drug, and it charged what it felt the market would bear.

To prevent further instances of price gouging, in 1989 NIH inserted a new fair pricing clause into all of its cooperative research and development agreements with private industry, requiring that there be a "reasonable relationship between the pricing of a licensed product, the public investment in that product, and the health and safety needs of the public." But in Varmus's view, despite its worthy aims, NIH's fair pricing clause had had little impact on drug pricing. Instead, it seemed to make companies reluctant to cooperate with government labs, or at least to sign agreements with them. Which would be preferable, he asked, a high-minded pricing policy that resulted in little or no collaboration, or more collaboration without the fair pricing policy? Ever the pragmatist, in 1995 Varmus decided to eliminate the fair pricing clause from NIH's standard research agreement, reasoning that this would better "promote research that can enhance the health of the American people."

If the federal government had truly been committed to the fair pricing of biomedical innovations, it could have held its ground in early negotiations with the pharmaceutical industry and Myriad. It is particularly difficult to understand how a multi-billion dollar government agency could have capitulated so completely to the demands of a tiny Utah-based start-up that had not yet released a product or earned a single penny of revenue. There is no good reason why NIH's license agreement with Myriad contained no safeguards regarding pricing of or access to any BRCA-related inventions.

The same sins of omission can be attributed to the University of Utah. In multiple agreements with Myriad, the university gave Myriad complete discretion over exploiting the BRCA patent rights—largely funded by federal grants—that emerged from the university. So long as the university received its royalties, it reserved no right to control or approve any aspect of Myriad's commercial program.

Perhaps this lack of oversight was forgivable, or at least

understandable, in 1991, when Myriad was formed and its first license agreements with the university were signed. Those were still the early, heady days of university technology transfer. Institutions across the country were salivating at the thought of blockbuster profits from the next Genentech or Biogen. The University of Utah, in particular, had aspirations born of the success of the Jarvik artificial heart program, coupled with a desperate need to redeem itself after the humiliating and costly cold fusion fiasco. Utah's TTO officials didn't know in 1991 whether Myriad, then merely a gleam in Mark Skolnick's eye, would ever amount to anything, but at least they had avoided the worst excesses of the sweetheart deal offered to Pons and Fleischmann.

But a dozen years later, in 2003, the university could have done better. By then, Myriad had patented both the *BRCA1* and *BRCA2* genes and eliminated all competition in the field of *BRCA* testing, news exposés about the company had already appeared in outlets ranging from *Nature* to the *Boston Globe Magazine,* a growing chorus of researchers, genetic counselors, and bioethicists had criticized the company's policies, and a major government task force had been convened to study the impact of gene patenting on research and patient care. In that year, the University of Utah fought Myriad over unpaid royalties and Myriad's failure to assign patents under the terms of their agreements. A formal arbitration was commenced, and in settlement Myriad paid the university $3.5 million and assigned it rights to 23 patents. The university immediately licensed those patents back to Myriad under the terms of their original 1991 exclusive agreement.

Had the university focused on promoting public health in addition to its own financial rewards, it could have imposed additional requirements on Myriad when it settled the dispute in 2003. Those requirements could have included pricing constraints and access guarantees on Myriad's tests, a requirement that Myriad permit second opinions or confirmatory testing, and a requirement that Myriad share *BRCA* data with the research community.[21]

---

21 While it was predictable that it would do so, Myriad did not actually discontinue participation in the global academic consortium that shared breast cancer genetic information, the Breast Cancer Information Core (BIC), until 2004.

Lest you think that such contractual commitments are radical or unusual, they are not, and Myriad itself was subject to such commitments. Back in 1998 when Myriad acquired its only commercial competitor, Oncormed, it also acquired Oncormed's *BRCA* patent license from the UK-based Cancer Research Centre, which had raced with Myriad to find the *BRCA2* gene. In the license, CRC insisted that others be given broad access to any *BRCA* diagnostic tests that Oncormed offered. Whether Myriad complied with those conditions is less clear.

Nevertheless, the impetus for research universities to exert greater control over their surrogate licensees' commercial behavior has begun to pick up steam. In 2007, the heads of eleven major U.S. research universities including Harvard, Yale, MIT, and the University of California, committed to abide by a set of core intellectual property licensing values known as the "Nine Points." These include commitments by the universities to require that their licensees make research tools as broadly available as possible, to grant exclusive licenses sparingly, to facilitate access to health-related technologies in the developing world, and the like. To date, over one hundred universities and other research institutions around the world, including the University of Utah, have committed to abide by the Nine Points.

So-called "ethical licensing" by research institutions has attracted even greater attention recently with the development (and patenting) of CRISPR gene-editing technology. The foundational CRISPR patents are held by Berkeley and the Broad Institute of Harvard and MIT. In its patent licensing agreement with Monsanto, the Broad has prohibited the use of CRISPR technology in three controversial areas: the creation of gene drives that have the potential to wipe out entire species in a single stroke, sterile "terminator" seeds that prevent plants from reproducing naturally, and genetic modifications intended to "improve" tobacco products. Numerous other examples of voluntary commitments by patent holders have been documented in the literature—including agreements not to transfer patents to "trolls" that will only use them to sue others, to refrain from enforcing patents against small businesses, and to charge only "fair, reasonable and nondiscriminatory" royalties.

Likewise, it was not preordained that the University of Utah had to grant Myriad an *exclusive* license to its *BRCA*-related discoveries. As Sherkow and I acknowledge, exclusivity may indeed be required to incentivize expensive R&D efforts such as drug development, testing, and approval. But this is not always so. Sometimes, as in the case of the *BRCA* genes, somebody would have discovered them even without the exclusivity promised by patents. Several competing groups almost did. So when a university is considering whether to grant a company exclusive or nonexclusive rights to its discoveries, it should consider two things: whether exclusivity is needed to ensure that the discoveries find their way into useful products, and whether the products themselves might achieve a greater social impact if they are available from multiple sources (e.g., a research tool like a gene sequencing technique that can be used by every lab in the world). I am not alone in arguing that if either of these factors weighs in favor of granting non-exclusive licenses, then universities, with their public missions in mind, should not grant exclusive rights to their discoveries, even if doing so might yield them greater short-term profits.

And when universities lose sight of their public missions, perhaps NIH and other funding agencies could take a more active role in nudging them back toward their public charters. A large percentage of academic research in the United States is funded by the taxpayers—in fiscal year 2020, more than $39 billion through NIH alone, and about $134 billion across all state and federal research funding agencies. In 2019, a team of researchers estimated that nearly a third of U.S. patents arose from federally funded research. Given this significant level of support, commentators over the years have wondered why the benefits of this research should flow primarily to private companies, rather than to the taxpayers. Couldn't funding agencies require that research universities abide by the types of fair licensing practices discussed above?

As Rai and Eisenberg noted years ago, an amendment to the Bayh-Dole Act might be required to give funding agencies this authority. But why not? Such an amendment might be warranted, particularly in the case of federally funded research that is likely to have a significant effect on human health and welfare—like searching for the *BRCA* genes.

Thus, if there is blame to be assigned for limited access, high prices, and a lack of competition in the area of genetic diagnostic testing—not just for *BRCA*, but across the board—there is plenty to go around. Certainly, companies like Myriad could have focused more on patients and the human cost of their lengthy coverage battles with the insurance industry and their unwillingness to compromise corporate profits. But fault can also be found with the Patent Office, which adopted a particularly myopic interpretation of patent eligibility in allowing the patenting of human genes as compositions of matter, as well as the government agencies and universities that neglected to fulfill their public missions when granting a private company unrestricted exclusive rights to a valuable federally funded discovery.

Paying It Forward

ANOTHER SIGNIFICANT FACTOR that limited access to Myriad's *BRCA* test was pricing. Admittedly, Myriad's list price for BRACAnalysis, which rose from $2,400 in 1996 to about $4,000 today (including $700 for the BART component), is not pocket change for most people. Yet these sums pale in comparison to the staggering figures that are being charged for the latest gene therapies, including the record-breaking $2.1 million price tag that Novartis recently announced for Zolgensma, a gene-based treatment for spinal muscular atrophy (SMA).

The key, of course, is insurance coverage. Few families caring for children with SMA will pay the $2.1 million price for Zolgensma, as most insurance policies in the United States, as well as Medicare and Medicaid, cover such treatments. But genetic testing for otherwise healthy individuals remains more difficult to get covered. As discussed in chapter 10, it took Myriad nearly a decade to convince large insurance carriers to cover *BRCA* testing, and then only with the imposition of strict eligibility requirements (recall the story of Kathleen Maxian and her sister Eileen). It was clearly in Myriad's self-interest to persuade insurance carriers to cover *BRCA* testing, and it did its best to do so. Eventually, this effort succeeded, and many of the tragic stories that emerged from the *BRCA* testing debates will not likely recur.

This being said, pricing for medical services in the United States is in a state of crisis. Americans pay far more for healthcare than the citizens of any other nation in the world in exchange for health outcomes that are middling at best. It is therefore incumbent upon federal and state legislatures, not the courts, to fix our overpriced healthcare system and ensure that every American has access to decent healthcare. This issue is far bigger than diagnostic genetic testing—it affects every American and should be fixed. The precise nature in which that will happen is a topic for another book.

Myriad and the Making of Law

SETTING ASIDE PATENTS and genetics, *AMP v. Myriad* offers some lessons about the law and lawmaking, more broadly. This remarkable case sheds light on the rough and tumble manner in which the common law is made. As then future Justice Ginsburg's husband says in the film *On the Basis of Sex*, "The law is never finished. It is a work in progress. And ever will be." This ever-evolving law is not, generally speaking, designed from the ground up, like a skyscraper or a car or a computer chip. Rather, it accretes, case by case, judge by judge, like a gigantic coral formation far beneath the waves. Yes, there is a structure, a skeletal framework upon which individual cases must build, but the randomness and sheer volume of human events effectively guarantee that no central planner can predict every set of circumstances that the law will need to resolve.[22]

This is especially true of scientific and technological development.

---

22 The importance of serendipity in forming legal doctrine cannot be understated, and there is no better example of this phenomenon than *AMP v. Myriad*. In this case, the absence or alteration of any number of chance events might have changed the outcome, and the law, dramatically. Most important may be the chance appearance of a number of trained scientists in positions critical to the outcome of the case at exactly the right moments to tip the scales—particularly Tania Simoncelli as the ACLU's first and only science advisor, Herman Yue as Judge Sweet's law clerk, and Jessica Palmer as an intern on the DOJ Civil Division's appellate staff. These individuals brought the perspective of the scientific community, which largely opposed gene patents, to bear from within organizational structures that ordinarily might have given a greater degree of deference to the expertise and policy preferences of the Patent Office.

The legal system offers a framework grounded in notions of fair play, justice, and public welfare that is, we hope, flexible enough to adapt to the breathtaking pace of scientific advancement. Yet that adaptation does not happen overnight. Genetic diagnostics is just one example of a new technology that "broke" the legal system of the day, but the past is littered with examples of similar legal inflection points caused by technological advances, including the telegraph, the railroad, the video recorder, the internet, social media, and GPS tracking. And there are more on the horizon—gene editing, artificial intelligence, augmented reality, autonomous vehicles, interstellar travel, and advances that we haven't even begun to think about yet.

For better or worse, courts—judges and juries—are required to decide cases involving these new technologies. They don't always get it right. In some cases, as exemplified by Justice Scalia's short concurrence in *AMP v. Myriad*, they might not understand all of the nuances of the particular technologies before them. Nevertheless, it is a hallmark of our democratic system that legal decisions, even concerning the most complex technologies, are presented to lay juries and judges. They do their best, and, on balance, I think that our hybrid system is better than governance by specialists and technocrats alone. After all, it took non-experts like Chris Hansen, who had never tried a patent case, and Tania Simoncelli, a non-lawyer, to recognize the legal absurdity of treating human genes as new compositions of matter. Excessive familiarity with a particular technical domain can sometimes blind one to otherwise obvious inequities.

This being said, judges and juries don't set out to design the law for future generations. Rather, they are charged with deciding the particular cases before them. They respond to the facts of these cases—the human stories that motivate them, that brought parties into collision and dispute. There are narratives behind every case. These narratives shape not only the outcome as to the parties—who wins, who loses, who pays, who goes home empty handed—they shape the law as well. In *AMP v. Myriad*, the different narrative strains developed by scientists, the biotech industry, the patent establishment, the medical community and cancer patients themselves each shaped the reasoning used by the courts to decide the case.

AMP v. Myriad also demonstrates the often under-appreciated impact of politics on our legal system. Some have criticized the significant role that NIH, the White House, the Department of Justice and other governmental bodies played in the case. But I view this as a natural feature of our legal system. Our three branches of government will continue to collide, collaborate, and collude to create new law and to change old law as it becomes obsolete. The law will inevitably adapt, though sometimes in fits and starts. And in many cases the end result will not be perfect. This is not necessarily a bad thing, but it does lend a measure of unpredictability to the law and to all of us who must live under it. But this unpredictability, despite the hardships it inflicts on some, also leaves openings for the smart, the determined, the idealistic, and the lucky to seek change when change is needed most. The vast, undersea formation that is the common law accretes and evolves bit by bit, unpredictably and seemingly at random, until, when viewed from afar, it assumes an air of majesty, larger than its individual components, like the human genome itself.

PRINCIPAL CHARACTERS

LORI ANDREWS—Law professor, Illinois Institute of Technology Chicago-Kent School of Law; member of ACLU Patent Committee and advisor to ACLU legal team

WILLIAM BRYSON—Acting Solicitor General of the United States (1989, 1993); Circuit Judge, Court of Appeals for the Federal Circuit (1994–present); author of dissenting opinions in *AMP v. Myriad*

DAN BURK—Law professor, University of Minnesota, then University of California Irvine; member of ACLU Patent Committee

GREG CASTANIAS—Partner, Jones Day, lead appellate counsel to Myriad Genetics

LISBETH CERIANI—Breast cancer patient and *BRCA* mutation carrier; Plaintiff, *AMP v. Myriad*

RAY CHEN—Solicitor, U.S. Patent and Trademark Office

WENDY CHUNG—Researcher, Columbia University; plaintiff in *AMP v. Myriad*

FRANCIS COLLINS—Researcher, University of Michigan; Director of the Human Genome Project/NIH Genome Institute (1993–2008); Director of U.S. National Institutes of Health (2009–present)

LAURA CORUZZI—Partner, Jones Day

REBECCA EISENBERG—Law professor, University of Michigan

MARK FREEMAN—Attorney, U.S. Department of Justice Civil Division, Appellate Staff

ARUPA GANGULY—Co-Director, University of Pennsylvania Genetic Diagnostics Laboratory; plaintiff in *AMP v. Myriad*

GENAE GIRARD—Breast cancer patient and *BRCA* mutation carrier; plaintiff in *AMP v. Myriad*

CHRISTOPHER HANSEN—ACLU National Legal Staff, lead ACLU counsel in *AMP v. Myriad*

KATHY HUDSON—Director, Genetics and Public Policy Center, Johns Hopkins University (2002–09); Chief of Staff, National Institutes of Health (2009–10)

BEN JACKSON—Intellectual Property Counsel, Myriad Genetics (2007–2019), General Counsel (2019–present)

ELENA KAGAN—Solicitor General of the United States (2009–10); Associate Justice, U.S. Supreme Court (2010–present)

DAVID KAPPOS—Director, U.S. Patent and Trademark Office (2009–13)

NEAL KATYAL—Acting Solicitor General of the United States (2010–11)

HAIG KAZAZIAN—Director, University of Pennsylvania Genetic Diagnostics Laboratory (1994–2010); Plaintiff in *AMP v. Myriad*

MARY-CLAIRE KING—Researcher, University of California Berkeley (1976–95); University of Washington (1995–present)

JIM KOHLENBERGER—Chief of Staff, Office of Science and Technology Policy (OSTP) (2009–11)

ERIC LANDER—President and Founding Director, Broad Institute of MIT and Harvard (2004–21); Co-chair, President's Council of Advisors on Science and Technology (PCAST) (2009–17); Director, Office of Science Technology Policy and Science Advisor to the President (2021–); author of influential *amicus* brief in *AMP v. Myriad*

LENORA LAPIDUS—Director, ACLU Women's Rights Project (2001–19)

JENNIFER LIEB—Washington, DC biotechnology consultant

RUNI LIMARY—cancer patient and *BRCA* mutation carrier; plaintiff in *AMP v. Myriad*

ALAN LOURIE—Circuit Judge, U.S. Court of Appeals for the Federal Circuit (1990–present); author of Federal Circuit's plurality opinions in *AMP v. Myriad*

RICHARD (RICK) MARSH—General Counsel, Myriad Genetics (2002–19)

CHRISTOPHER MASON—Researcher, Cornell-Weill Medical Center; advisor to ACLU litigation team

ELLEN MATLOFF—Director of Cancer Genetic Counseling, Yale University (1995–2014); plaintiff in *AMP v. Myriad*

KATHLEEN MAXIAN—Ovarian cancer patient and *BRCA* mutation carrier; spokesperson for ACLU

PETER MELDRUM   CEO, Myriad Genetics (1991–2015)

KIMBERLY MOORE—Circuit Judge (2006-2021) and Chief Judge (2021–   ), U.S. Court of Appeals for the Federal Circuit; author of concurring opinions in *AMP v. Myriad*

KEVIN NOONAN—Partner, McDonnell Boehnen Hulbert & Berghoff LLP; co-founder and contributor, *Patent Docs* blog

HARRY OSTRER—Director, Molecular Genetics Laboratory of NYU Langone Medical Center (1990–2011); Director, Genetic and Genomic Testing, Montefiore Hospital at Albert Einstein Medical College of Yeshiva University (2011–present)

JESSICA PALMER—Student intern, U.S. Department of Justice Civil Division, Appellate Staff (2010)

SANDRA PARK—Attorney, ACLU Women's Rights Project; plaintiffs' counsel in *AMP v. Myriad*

BRIAN POISSANT—Partner, Jones Day; lead trial counsel for Myriad Genetics

RANDALL RADER— Circuit Judge (1990-2010) and Chief Judge (2010-2014), U.S. Court of Appeals for the Federal Circuit

DAN RAVICHER—Director, Public Patent Foundation; plaintiffs' counsel in *AMP v. Myriad*

ANTHONY ROMERO—Executive Director, ACLU (2001–present)

STEVEN SHAPIRO—Legal Director, ACLU (1987–2016)

TANIA SIMONCELLI—Science Advisor, ACLU (2003–10)

MARK SKOLNICK—Founder and Chief Scientific Officer, Myriad
Genetics (1991–2010)

LAWRENCE SUMMERS—Director, National Economic Council
(2009–10)

ROBERT SWEET—Deputy Mayor of New York City (1966–69); District
Judge (1978–91) and Senior District Judge (1991–2019), U.S. District
Court for the Southern District of New York; author of opinion in
AMP v. Myriad

CLARENCE THOMAS—Associate Justice, United States Supreme Court
(1991–present); author of Supreme Court's opinion in *AMP v. Myriad*

HAROLD VARMUS—Director, National Institutes of Health (1993–99);
Nobel Prize in Physiology or Medicine (1989)

DONALD VERRILLI JR.—Solicitor General of the United States
(2011–16)

JAMES WATSON—Director, Human Genome Project (1990–92); Nobel
Prize in Physiology or Medicine (1962); co-discoverer of the chemical
structure of DNA

HERMAN YUE—Law clerk, Judge Robert Sweet (2009–10)

JAY ZHANG—Vice President and Chief Patent Counsel, Myriad
Genetics (2000–10)

A NOTE ABOUT SOURCES

AMP v. Myriad is a lawsuit, and as such the voluminous record of its official proceedings is public. I have relied extensively on the litigation record in the case, including court opinions, party briefs and pleadings, witness declarations, *amicus* briefs, and the transcripts of hearings and oral arguments. I have not directly cited this body of material except in a few key instances, though I have tried to make my discussion clear enough that an interested reader should have little difficulty finding it.

In addition to the litigation record, I reviewed a wide variety of other primary documents, most of which are publicly available. These include press releases, financial statements, securities filings, and product literature released by Myriad Genetics, Patent Office records; public statements and policies of governmental agencies and professional associations; state and federal legislation and legislative histories; and reports and studies by government agencies and academics. I also obtained a wealth of useful information via Freedom of Information Act (FOIA) requests to a host of federal and state governmental agencies.[1]

There are numerous firsthand accounts of the events described in this book. Many of the attorneys and parties involved in *AMP v. Myriad* have published articles and books; were interviewed on television, radio, and online sites; and appeared in TED talks and panels across the country. These personal accounts were invaluable in my research, and I cite them in the notes when I have used material from them.

I have also relied on the extensive secondary literature surrounding gene discovery, and particularly the race to sequence the *BRCA* genes, as well as an extensive literature analyzing and critiquing Myriad Genetics itself, which appears in books, magazines, newspaper articles, and

---

1 This being said, not all agencies were responsive to my FOIA requests, and I am still waiting, after several years, to receive a response from NIH, which blames its unconscionable delay, among other things, on a "hiring freeze."

academic journals. I owe a debt of gratitude to the many legal scholars who have thought long and hard about the issues of patent eligibility raised by *AMP v. Myriad* and other cases. The bibliography that appears on page 367 lists some of the more significant literature in this area, both directly relevant to *AMP v. Myriad* and more generally applicable to the areas and issues covered by the book. The notes make extensive references to this literature.

One particularly valuable source of information regarding the progress of *AMP v. Myriad* through the courts was the *Patent Docs* blog co-founded by Kevin Noonan, which offers a multi-year, blow-by-blow account of the litigation—sometimes impassioned, often opinionated, but always accurate. Other blogs that offer useful information about the case include *IP Watchdog, Patently-O,* and the *Genomics Law Report* (now the *Privacy Report*). The community owes a debt to these public-spirited commentators, who have created a historical record that would otherwise be largely inaccessible to the public.

But the most useful and valuable sources of information for this book came from more than one hundred personal interviews[2] that I conducted with the attorneys, advocates, parties, judges, government officials, and patients who played a role in the complex story of *AMP v. Myriad*. I am grateful to each and every one of them. They shared their experiences, opinions, recollections, and convictions with me. Out of respect for the privacy of these individuals, I do not name them here. They know who they are, and I hope that I have done their stories justice.

2　The interviews conducted for this book were deemed exempt from human subjects research requirements by the University of Utah Institutional Review Board (Nov. 20, 2014).

BIBLIOGRAPHY

Abramson, Bruce (2007) *The Secret Circuit: The Little-Known Court Where the Rules of the Information Age Unfold* (Rowman & Littlefield).

Andrews, Lori B. (2002) "Genes and patent policy: rethinking intellectual property rights" *Nature Reviews—Genetics* 3: 803–808.

Andrews, Lori and Dorothy Nelkin (2001) *Body Bazaar: The Market for Human Tissue in the Biotechnology Age* (Crown).

Annas, George and Sherman Elias (2015) *Genomic Messages: How the Evolving Science of Genetics Affects our Health, Families, and Future* (HarperOne).

Association of University Technology Managers (AUTM) (2007) "In the Public Interest: Nine Points to Consider in Licensing University Technology."

Bailey, Ronald (1999) "Warning: Bioethics May Be Hazardous to Your Health" *Reason*, Aug/Sep. 1999.

Baldwin, A. Lane and Robert Cook-Deegan (2013) "Constructing narratives of heroism and villainy: case study of Myriad's BRACAnalysis® compared to Genentech's Herceptin®" *Genome Medicine* 5(8).

Blanton, Kimberly (2002) "Corporate Takeover: Exploiting the U.S. Patent System" *Boston Globe Magazine*, Feb. 24, 2002.

Boldrin, Michele and David K. Levine (2008) *Against Intellectual Monopoly* (Cambridge University Press).

Burk, Dan L. (2014) "The Curious Incident of the Supreme Court in *Myriad Genetics*" *Notre Dame Law Review* 90(2): 505–542.

Burk, Dan L. and Mark A. Lemley (2009) *The Patent Crisis and How the Courts Can Solve It* (Univ. Chicago Press).

Caulfield, Timothy (2008) "Human Gene Patents: Proof of Problems? *Chicago-Kent Law Review* 84(1): 133-145.

Chien, Colleen V. (2011) "Patent Amicus Briefs: What the Courts' Friends Can Teach Us about the Patent System" *UC Irvine Law Review* 1: 397–433.

Cho, Mildred K., et al. (2003) "Effects of Patents and Licenses on the Provision of Clinical Genetic Testing Services" *Journal of Molecular Diagnostics* 5(1): 3–8.

Colaianni, Alessandra, Subhashini Chandrasekharan, and Robert Cook-Deegan (2010), "Impact of Gene Patents and Licensing Practices on Access to Genetic Testing and Carrier Screening for Tay-Sachs and Canavan Disease" *Genetics in Medicine* 12(4):S5–S14.

Collins, Francis S. (2011) *The Language of Life: DNA and the Revolution in Personalized Medicine* (Harper).

Contreras, Jorge L. (2011) "Bermuda's Legacy: Patents, Policy and the Design of the Genome Commons" *Minnesota Journal of Law, Science and Technology* 12: 61–125.

—(2016) "Narratives of Gene Patenting" *Florida State University Law Review* 43: 1133–1199.

—(2018) "The Anticommons at 20: Concerns for research continue" *Science* 361: 335–337.

—(2021) "Association for Molecular Pathology v. Myriad Genetics: A Critical Reassessment" *Michigan Technology Law Review* 27(1):1-54.

Contreras, Jorge L. and Jacob S. Sherkow (2017) "CRISPR, surrogate licensing, and scientific discovery" *Science* 355(6326): 698–700.

Cook-Deegan, Robert (1994) *The Gene Wars—Science, Politics, and the Human Genome* (Norton).

Cook-Deegan, Robert, et al. (2010) "Impact of gene patents and licensing practices on access to genetic testing for inherited susceptibility to cancer: Comparing breast and ovarian cancers with colon cancers" *Genetics in Medicine* 12(4): S15–S38.

Cook-Deegan, Robert and Christopher Heaney (2010) "Patents in Genomics and Human Genetics" *Annual Review of Genomics and Human Genetics* 11: 383–425.

Cook-Deegan, Robert, John M. Conley, James P. Evans, and Daniel Vorhaus (2013) "The next controversy in genetic testing: clinical data as trade secrets?" *European Journal of Human Genetics* 21: 585–588.

Cotter, Thomas F. (2018) *Patent Wars: How Patents Impact Our Daily Lives* (Oxford University Press).

Crichton, Michael (1990) *Jurassic Park* (Knopf).

—(2006) *Next* (HarperCollins).

—(2007) "Patenting Life" *NY Times*, Feb. 13, 2007.

Critchfield, Gregory C. (2009) "Declaration," *Assn. Molecular Pathology v. Myriad Genetics* (SDNY, filed Dec. 23, 2009).

Davies, Kevin and Michael White (1995) *Breakthrough: The Race to Find the Breast Cancer Gene* (Wiley).

Doll, John J. (1998) "The Patenting of DNA" *Science* 280: 689–690.

Dreyfuss, Rochelle C., Jane Nielsen, and Dianne Nicol (2018) "Patenting nature—a comparative perspective" *Journal of Law and the Biosciences* 5(3): 550–589.

Duan, Charles (2021) "Gene Patents, Drug Prices, and Scientific Research: Unexpected Effects of Recently Proposed Patent Eligibility Legislation" *Marquette Intellectual Property Law Review* 24(2): 139-168.

Eisenberg, Rebecca S. (1990) "Patenting the Human Genome" *Emory Law Journal* 39: 721–745.

—(1996) "Public Research and Private Development: Patents and Technology Transfer in Government Sponsored Research" *Virginia Law Review* 82: 1663–1727.

—(2002) "Why the Gene Patenting Controversy Persists" *Academic Medicine* 77:1380–1387.

—(2015) "Diagnostics Need Not Apply." *Boston University Journal of Science & Technology Law* 21(2): 256–286.

Eisenberg, Rebecca S. and Robert Cook-Deegan (2018) "Universities: The Fallen Angels of Bayh-Dole?" *Daedelus* 147(4): 76–89.

Ghosh, Shubha (2020) "Myriad post-Myriad" *Science and Public Policy*, 2020, 1-9.

Gitschier, Jane (2013) "Evidence Is Evidence: An Interview with Mary-Claire King," *PloS Genetics* 9(9).

Gold, E. Richard and Julia Carbone (2010) "Myriad Genetics: In the eye of the policy storm" *Genetics in Medicine* 12(4 Suppl): S39—S70.

Heaney, Christopher, Julia Carbone, Richard Gold, Tania Bubela, Christopher M. Holman, Alessandra Colaianni, Tracy Lewis, Robert Cook-Deegan (2009) "The Perils of Taking Property Too Far" *Stanford Journal of Law, Science & Policy* 1: 46–64.

Heller, Michael A. and Rebecca S. Eisenberg (1998) "Can Patents Deter Innovation? The Anticommons in Biomedical Research" *Science* 280(5364): 698–701.

Holman, Christopher M. (2008) "Trends in Human Gene Patent Litigation" *Science* 322(5899): 198–199.

—(2012). "Will Gene Patents Derail the Next Generation of Genetic Technologies?: A Reassessment of The Evidence Suggests Not," *UMKC Law Review* 80:563, 582–83.

Hull, Sara Chandros and Kiran Prasad (2001) "Reading between the Lines: Direct-to-Consumer Advertising of Genetic Testing" *Hastings Cent Rep.* 31(3): 33–35.

Institute of Medicine (IOM) (1994) Assessing Genetic Risks: Implications for Health and Social Policy (National Academies Press).

—(2006) Cancer-Related Genetic Testing and Counseling: Workshop Proceedings (National Academies Press).

Jensen, Kyle and Fiona Murray (2005) "Intellectual Property Landscape of the Human Genome" *Science* 310:239.

Jost, Kenneth (2013) "Patenting Human Genes" *CQ Researcher* 23(20): 473–496.

Kepler, Thomas B., Colin Crossman, and Robert Cook-Deegan (2010) "Metastasizing patent claims on *BRCA1*" *Genomics* 95(5): 312–314.

Kevles, Daniel J. (2002) "Of Mice and Money: The Story of the World's First Animal Patent" *Daedalus* 131(2): 78–88.

Kevles, Daniel J. and Leroy Hood, eds. (1992) *The Code of Codes: Scientific and Social Issues in the Human Genome Project* (Harvard University Press).

King, Mary-Claire (2014) "'The Race' to Clone *BRCA1*" *Science* 343: 1462–1465.

Knowles, Sherry and Anthony Prosser (2018) "Unconstitutional Application of 35 U.S.C.§101 by the U.S. Supreme Court" *John Marshall Review of Intellectual Property Law* 18: 144–168.

Koepsell, David (2009) *Who Owns You? The Corporate Gold-Rush to Patent Your Genes* (Wiley).

Krimsky, Sheldon and Tania Simoncelli (2011) Genetic Justice—DNA Data Banks, Criminal Investigations and Civil Liberties (Columbia Univ. Press).

Lazarus, Richard J. (2019) The Rule of Five: Making Climate History at the Supreme Court (Belknap).

Leonard, Debra (2002) "Medical Practice and Gene Patents: A Personal Perspective" *Academic Medicine* 77(12): 1388–1391.

Lewin Group, The (2005) "The Value of Diagnostics Innovation, Adoption and Diffusion into Health Care," Jul. 2005.

Magnus, David, Arthur Caplan and Glenn McGee, eds. (2002) *Who Owns Life?* (Prometheus Books).

Matloff, Ellen and Arthur Caplan (2008) "Direct to Confusion: Lessons Learned from Marketing BRCA Testing," *American Journal of Bioethics* 8(6): 5–8.

Merrill, Stephen A., Anne-Marie Mazza, eds. (2006) *Reaping the Benefits of Genomic and Proteomic Research: Intellectual Property Rights, Innovation, and Public Health* (National Research Council).

McElheny, Victor K. (2010) *Drawing the Map of Life: Inside the Human* Genome Project (Basic Books).

Mukherjee, Siddhartha (2010) *The Emperor of all Maladies—A Biography of Cancer* (Scribner).

—(2016) *The Gene: An Intimate History* (Scribner).

National Institutes of Health (NIH) (1999) "Principles and Guidelines for Recipients of NIH Research Grants and Contracts on Obtaining and Disseminating Biomedical Research Resources: Final Notice" *Federal Register* 64: 72,090, 72,093, Dec. 23, 1999.

—(2005) "Best Practices for the Licensing of Genomic Inventions: Final Notice" *Federal Register* 70: 18,413, Apr. 11, 2005.

National Research Council (1997) *Intellectual Property Rights and Research Tools in Molecular Biology* (National Academy Press).

Nicol, Dianne, Rochelle Cooper Dreyfuss, Richard E. Gold, Wei Li, Johnathon Liddicoat, and Geertrui Van Overwalle "International Divergence in Gene Patenting" *Annual Review of Genomics and Human Genetics* 20: 519–541.

Noonan, Kevin E. (2009a) "Falsehoods, Distortions and Outright Lies in the Gene Patenting Debate," *Patent Docs*, Jun. 15, 2009.

Ouellette, Lisa Larrimore (2015) "Patentable Subject Matter and Nonpatent Innovation Incentives" 5 *UC Irvine Law Review* 5(5): 1115–1145.

Ouellette, Lisa Larrimore, and Rebecca Weires (2019) "University Patenting: Is Private Law Serving Public Values?" *Michigan State Law Review* 2019: 1329-1387.

Palombi, Luigi (2009) *Gene Cartels: Biotech in the Age of Free Trade* (Edward Elgar).

Parthasarathy, Shobita (2007) *Building Genetic Medicine: Breast Cancer, Technology, and the Comparative Politics of Health Care* (MIT Press).

—(2017) *Patent Politics: Life Forms, Markets, and the Public Interest in the United States and Europe* (University of Chicago Press).

Park, Sandra S. (2014) "Gene Patents and the Public Interest: Litigating *Association for Molecular Pathology v. Myriad Genetics* and Lessons Moving Forward" *N.C. Journal of Law and Technology* 15: 519–536.

—(2018) "The Challenge to Gene Patents as Feminist Patent Litigation" *Technology and Innovation* 19: 659–670.

Pierce, Brandon L., Christopher S. Carlson, Patricia C. Kuszler, Janet L. Stanford, and Melissa A. Austin (2009) "The impact of patents on the development of genome-based clinical diagnostics: an analysis of case studies" *Genetics in Medicine* 11: 202–209.

Poste, George (1995) "The Case for Genomic Patenting" *Nature* 378: 534–536.

Rai, Arti K. (2012) "Patent Validity across the Executive Branch: Ex Ante Foundations for Policy Development" *Duke Law Journal* 61:1237–1281.

—(2014) "Diagnostic Patents at the Supreme Court" *Marquette Intellectual Property Law Review* 18(1): 1–9.

Rai, Arti K. and Rebecca S. Eisenberg (2003) "Bayh—Dole Reform and the Progress of Biomedicine" *Law and Contemporary Problems* 66: 289–314.

Rai, Arti K. and Robert Cook-Deegan (2013) "Moving beyond 'Isolated' Gene Patents" *Science* 341: 137–138.

Rimmer, Matthew (2008) *Intellectual Property and Biotechnology: Biological Inventions* (Edward Elgar).

—(2013) "The Empire of Cancer: Gene Patents and Cancer Voices" *Journal of Law, Information and Science* 22(2).

Rinehart, Amelia Smith (2015) "Myriad Lessons Learned" *UC Irvine Law Review* 5: 1147–1191.

Rogers, Eric J. (2010) "Can You Patent Genes? Yes and No" *Journal of the Patent and Trademark Office Society* 93: 19–56.

Rooksby, Jacob (2016) *The Branding of the American Mind: How Universities Capture, Manage, and Monetize Intellectual Property and Why It Matters* (Johns Hopkins University Press).

Rosenfeld, Jeffrey and Christopher E. Mason (2013) "Pervasive sequence patents cover the entire human genome" *Genome Medicine* 5(27): 1–7.

Rosenthal, Elizabeth (2017) *An American Sickness: How Healthcare Became Big Business and How You Can Take It Back* (Penguin).

Rutgers Oral History Archives (2015), Interview with Chris Hansen conducted by Shaun Illingworth, Mar. 3, 2015.

Sarnoff, Joshua D. (2011) "Patent-Eligible Inventions after Bilski: History and Theory" *Hastings Law Journal* 63: 53-126.

Secretary's Advisory Committee on Genetics, Health, and Society (SACGHS) (2010) "Gene Patents and Licensing Practices and Their Impact on Patient Access to Genetic Tests."

Sherkow, Jacob S. and Henry T. Greely (2015) "The History of Patenting Genetic Material" *Annual Review of Genetics*, 49:161–182.

Shreeve, James (2004) *The Genome War: How Craig Venter Tried to Capture the Code of Life and Save the World* (Knopf).

Skloot, Rebecca (2010) *The Immortal Life of Henrietta Lacks* (Broadway Books).

Simoncelli, Tania and Sandra S. Park (2015) "Making the Case against Gene Patents" *Perspectives on Science* 23(1): 106–145.

Strandburg, Katherine J. (2012) "Much Ado about Preemption", *Houston Law Review* 50: 563-622.

Task Force on Genetic Testing (1997) *Final Report: Promoting Safe and Effective Genetic Testing in the United States* (Sept. 1997).

Tu, Shine, et al. (2014) "Response to 'pervasive sequence patents cover the entire human genome'" *Genome Medicine* 6(14): 1–3.

United States Senate (1992) Hearing before the Subcommittee on Patents, Copyrights and Trademarks of the Committee on the Judiciary, The Genome Project: The Ethical Issues of Gene Patenting, U.S. Senate, 102nd Cong., 2nd Sess., Sep. 22, 1992.

Van Zimmeren, Esther, et al. (2014) "The BRCA Patent Controversies—An international review of patent disputes" in *Breast Cancer Gene Research and Medical Practices—Transnational Perspectives in the Time of BRCA* 151–174 (Sahra Gibbon, et al., eds., Routledge).

Venter, Craig (2007) *A Life Decoded: My Genome, My Life* (Penguin).

Vishnubhakat, Saurabh (2020) "The Normative Molecule: Patent Rights and DNA," *HYLE—International Journal for Philosophy and Chemistry* 26: 55–78.

Waldholz, Michael (1997) *Curing Cancer: The Story of the Men and Women Unlocking the Secrets of Our Deadliest Illness* (Touchstone).

Watson, James D. (2013) Brief of James D. Watson, Ph.D. as *Amicus Curiae* "In Support of Neither Party, *Assn. Molecular Pathology v. Myriad Genetics*" (U.S. 2013).

Young, Alexandra (2014) *Prelude to "Pigs Fly:" The Early History of the Myriad Case* (undergraduate thesis, Duke University, submitted Apr. 20, 2014).

NOTES

Preface

14 *"Though I can't take credit for it . . ."*—The SNP Consortium succeeded in large part due to the early support of individuals including Alan Williamson at Merck, Brian Spear at Abbott Laboratories, Allen Roses at Duke University, Michael Morgan at the Wellcome Trust, its Chairman, Arthur Holden and many more. For general discussions of the SNP Consortium, see Arthur L. Holden (2002) "The SNP Consortium: Summary of a Private Consortium Effort to Develop an Applied Map of the Human Genome" *BioTechniques* 32:S22–S26, Contreras (2011, pp. 95–97).

Chapter 1: Who Can We Sue?

4 *"the Reverend Jerry Falwell . . ."*—Gustav Niebuhr "AFTER THE ATTACKS: FINDING FAULT; U.S. 'Secular' Groups Set Tone For Terror Attacks, Falwell Says, *New York Times*, Sept. 14, 2001.

4 *"The ACLU's campaigns . . ."*—For a history of the ACLU, see Samuel Walker, *In Defense of American Liberties: A History of the ACLU* (2nd ed., Carbondale and Edwardsville, IL: Southern Ill. Univ. Press, 1999) and Judy Kutulas, *The American Civil Liberties Union & the Making of Modern Liberalism 1930–1960* (Chapel Hill, NC: Univ. of NC Press, 2006).

9 *"Some topics, like DNA fingerprinting . . ."*—With Simoncelli's help, the ACLU eventually brought a case challenging the legality of retaining DNA evidence after the exoneration of a criminal suspect. She writes about this issue in her book *Genetic Justice* (Krimsky and Simoncelli (2011)).

10 *"As Hansen explained . . ."*—Rutgers interview (2015, p. 23).

11 *"a disorienting surrealist print"*—"The Hunted Sky," Yves Tanguy (1951).

12 *"could a genetic . . ."*—See *Sutton v. United Airlines, Inc.*, 527 U.S. 471 (1999) (narrowly construing Americans with Disabilities Act to exclude individuals with a genetic predisposition to an illness that has not yet manifested itself).

12 *"cadres of the genetically unemployable . . ."*—Nachana Wilker, executive director of the Council for Responsible Genetics (quoted in *CQ Researcher*, Oct. 18, 1991, p. 790).

12 *"the ACLU was already . . ."*—See Statement of Legislative Consultant Ron Weich on Genetic Privacy And Non-Discrimination for the Senate Health, Education, Labor, and Pensions Committee on Behalf of the American Civil Liberties Union, for inclusion in the record of the hearing of the Senate Committee on Health, Education, Labor, and Pensions on Genetic Privacy and Non-Discrimination, July 25, 2001. This legislation was eventually passed in the United States as the 2008 Genetic Information Non-Discrimination Act (GINA).

Chapter 2: The World in the Helix

57 *"Today he is probably best known . . ."*—Afshinnekoo et al., Geospatial Resolution of Human and Bacterial Diversity with City-Scale Metagenomics, CELS 1:1–15, Jul. 29, 2015. Among the fascinating findings of this study were that nearly half of DNA samples collected were

of unknown origin, hundreds of different species of bacteria exist within the subway system, and the bacterial profile of the closed South Ferry subway station, which was flooded during Hurricane Sandy, still resembles that of a marine ecosystem.

15 *"explain why the average Dutchman . . ."*—eLife, A century of trends in adult human height, Jul. 26, 2016.

15 *"We share about 99 percent of our DNA code with chimpanzees . . ."*—This fact was first elucidated by Mary-Claire King as part of her PhD work. Mary-Claire King and A. C. Wilson (1975) "Evolution at Two Levels in Humans and Chimpanzees" *Science* 188: 107–116.

16 *"the famous 'double helix' structure . . ."*—James Watson and Francis Crick (together with independent investigator Maurice Wilkins) won the Nobel Prize in 1962 for identifying the double-helical structure of DNA. Their groundbreaking accomplishment was made possible with the help of one of the unsung heroines of the scientific world, Rosalind Franklin, whose x-ray diffraction photographs of DNA molecules led Watson and Crick to theorize that the structure of DNA was a double helix. Watson's firsthand account of the discovery, *The Double Helix* (1968) remains a classic, and its later editions include an introduction recognizing Franklin's contribution.

17 *"The HGP was one of the largest . . ."*—The dramatic story of the Human Genome Project and the race to sequence the human genome has been chronicled many times. See, e.g., Kevles and Hood (1992); Cook-Deegan (1994); Shreeve (2004); Venter (2007); McElheny (2010).

18 *"U.S. President Bill Clinton, who proclaimed that . . ."*—"Reading the Book of Life: White House Remarks on Decoding of Genome," *New York Times*, June 27, 2000, at F8.

18 *"refinements continue to be made today"*—See Michael Eisenstein (2021) "Closing in on a Complete Human Genome" *Nature* 590: 679-681 ("Advances in sequencing technology mean that scientists are on the verge of finally finishing an end-to-end human genome map.")

Chapter 3: The Gene Queen

20 *"In profiling Andrews . . ."*—MORE Magazine, Nov. 2001, pp. 68–72.

20 *"Watson insisted that Congress earmark . . ."*—Kathi E. Hanna, "The Ethical, Legal, and Social Implications Program of the National Center for Human Genome Research. A Missed Opportunity?" in Institute of Medicine (1995) *Society's Choices: Social and Ethical Decision Making in Biomedicine.* Washington, DC: The National Academies Press.

21 *"In 1995, Lori Andrews became its chair . . ."*—Andrews resigned her position as chair of the ELSI Working Group in 1996 as a result of a conflict with Dr. Francis Collins, then director of the HGP.

21 *"The Greenbergs' two children . . ."*—Lucinda Hahn, "Owning a Piece of Jonathan," Chicago Magazine, May 2003. Canavan disease is an inherited condition known as a leukodystrophy, which causes damage to nerve fibers in the brain. It most frequently occurs in individuals of Ashkenazi (eastern European) Jewish descent. Symptoms typically appear by age 3–5 months, with death usually occurring prior to the age of twenty. There is still no cure. See National Center for Advancing Translational Science, Genetic and Rare Disease Information Center (GARD), Canavan disease, https://rarediseases.info.nih.gov/diseases/5984/canavan-disease (reviewed Oct. 22, 2017).

21 *"contribute their children's blood . . ."*—Greenberg v. Miami Children's Hosp., 264 F.Supp.2d 1064, 1067 (S.D. Fla. 2003).

21 *"Matalon's new employer . . ."*—When the Greenbergs first approached Dr. Matalon in 1987, he was affiliated with University of Illinois at Chicago. He became director of research at Miami Children's Hospital in 1989.

22 *"a clinical instructor . . ."*—Laurie Ellen Leader and Edward M. Kraus on the Chicago-Kent faculty were the named attorneys in the case.

22 *"prime time TV news show . . ."*—Whose Body Is It Anyway? (60 Minutes, Feb. 23, 2001).

Chapter 4: Mr. Lincoln's Boat

25 *"working model . . ."*—from 1790 to 1880, applicants for U.S. patents were required to submit a miniature working model of their invention, with a base no larger than twelve inches square. This requirement was intended to make clear the nature and purpose of the invention, particularly given the poor writing skills of many inventors during the period. Approximately 200,000 models were submitted to the Patent Office over its history.

25 *"U.S. Patent No. 6,469 . . ."*—"Manner of Buoying Vessels." Unfortunately for river barge captains across the country, the greater sweep of history overtook the young Lincoln, and his labor-saving device was never produced commercially.

25 *"Nobody wishes more than I . . ."*—Letter to Oliver Evans (May 1807), *V Writings of Thomas Jefferson*, at 75–76 (Washington ed.). The Supreme Court recognized Jefferson's contributions to the shaping of American patent law in *Graham v. John Deere Co.*, 383 U.S. 1 (1966).

25 *"Thomas Edison received more than a thousand patents . . ."*—For a fascinating look at Edison's extensive patent estate, see the Thomas A. Edison Papers maintained by Rutgers University, http://edison.rutgers.edu/miscpats.htm.

26 *"They are enshrined in the Constitution . . ."*—U.S. Const., Art. I, Sec. 8, Cl. 8.

26 *"Lincoln said of patents . . ."*—Abraham Lincoln (1859) "Second Lecture on Discoveries and Inventions [February 11, 1859]" *Collected Works of Abraham Lincoln*, vol. 3, p. 363.

27 *"science and technology will grind to a halt . . ."*—There is increasing evidence that the incentive of patent exclusivity may not be necessary for all inventive output. Open source software developers, for example, give their code away for free, yet more of it continues to be developed. For a discussion of these phenomena, see, e.g., Boldrin and Levine (2009), Burk and Lemley (2009).

27 *"Since 1930, horticulturalists . . ."*—Plant Patent Act of 1930, 46 Stat. 703, codified as 35 U.S.C. Ch. 15; see Daniel J. Kevles (2008), "Protections, Privileges, and Patents: Intellectual Property in American Horticulture The Late Nineteenth Century to 1930," 152 *Proceedings of the American Philosophical Society* 207–213.

27 *"This landmark case . . ."*—*Diamond v. Chakrabarty*, 447 U. S. 303, 309 (1980). Numerous articles have been written about the background and history of the Chakrabarty case. See Daniel J. Kevles, "Ananda Chakrabarty Wins a Patent: Biotechnology, Law, and Society, 1972–1980," *Historical Studies in the Physical and Biological Sciences*, Vol. 25, No. 1 (1994), pp. 111–135; Rebecca S. Eisenberg, "The Story of *Diamond v. Chakrabarty*: Technological Change and the Subject Matter Boundaries of the Patent System" in *Intellectual Property Stories*, J. C. Ginsburg and R. C. Dreyfuss, eds. New York: Foundation Press, 2006; pp. 327–357; Christopher B. Seaman and Sheena X. Wang, *An Inside History of the Burger Court's Patent Eligibility Jurisprudence*, 53 Akron Law Review 915, 944–54 (2020).

27 *"In 1988, Harvard scientists . . ."*—See Kevles (2002). U.S. Patent No. 4,736,866 for "Transgenic Non–Human Mammals"

27 *"patents on newly discovered sequences of human DNA . . ."*—Though there are debates about the "first" patent on human DNA, one early contender is U.S. Patent No. 4,363,877 (Dec. 14, 1982) issued to investigators at the University of California at San Francisco on the *CSH1*

gene. For the early history of human gene patenting, see Eisenberg (1990) and Sherkow and Greely (2015).

27 *"ethical and religious concerns . . ."*—See Paul J. Heald (2019) "Christian Libertarianism and the Curious Lack of Religious Objections to the Patenting of Life Forms in the United States," in *Patents on Life: Religious, Moral, and Social Justice Aspects of Biotechnology and Intellectual Property* (T. Berg, R. Cholij and S. Ravenscroft (eds.), pp. 152-164), and Sarnoff (2011, pp. 84-90).

28 *"a pair of researchers at MIT . . ."*—Jensen and Murray (2005).

28 *"One commentator called it . . ."*—Koepsell (2009).

28 *"This popular imagery . . ."*—See, e.g., Francis Fukuyama, *Our Posthuman Future: Consequences of the Biotechnology Revolution* 148–50 (2002); Gordon Graham, *Genes: A Philosophical Inquiry* 113–42 (2002). Baruch A. Brody, "Protecting Human Dignity and the Patenting of Human Genes," in *Perspectives on Genetic Patenting: Religion, Science, and Industry in Dialogue* 111, 122 (Audrey R. Chapman ed., 1999) (noting objections to the patenting of human genes rooted in a desire to avert the development of a future eugenics movement).

28 *"viewed by some as encouraging . . ."*—See Jeremy Rifkin, *Who Should Play God?* (1978). In 1995, a coalition of 185 religious leaders issued a *Joint Appeal against Human and Animal Patenting*, in which they proclaimed that "the genetic blueprints of life are the province of God and cannot be owned as 'patented inventions' by any human being or institution."

29 *"They argued that GE's patent . . ."*—Brief on Behalf of the Peoples Business Commission, Amicus Curiae at *3, *Parker v. Chakrabarty* 447 U.S. 303 (1980) (No. 79–136) at 22.

29 *"the ill-fated OncoMouse . . ."*—See generally Kevles (2002).

29 *"Patents on human DNA . . ."*—For an excellent discussion of the early history of gene patenting, see Eisenberg (1990) and Greely and Sherkow (2015), dating the first human gene patent to 1982 (U.S. Pat. No. 4,363,877).

29 *"albeit a complex one . . ."*—*Amgen, Inc. v. Chugai Pharma. Co.*, 927 F.2d 1200, 2016 (1991).

30 *"NIH researcher recalls that . . ."*—Colaianni et al. (2010).

30 *"In 1990, J. Craig Venter . . ."*—Venter is best known for founding the company Celera Genomics, which, in 1998, announced that it would sequence the entire human genome a full four years before the publicly funded Human Genome Project, an announcement that sparked one of the most heated scientific races in modern history. See Shreeve (2004), Venter (2007).

30 *"by 1991 he was using automated gene sequencing machines . . ."*—Leslie Roberts, "Genome Patent Fight Erupts," *Science*, Oct. 11, 1991, p. 254.

30 *"called Venter's plan 'sheer lunacy' . . ."*—Roberts, ibid.

30 *"other influential voices . . ."*—See Eisenberg (2002, n.13), collecting sources.

30 *"Even NIH's own internal advisory committees . . ."*—Cook-Deegan (1994, p. 394).

30 *"effectively abandoning its EST patents . . ."*—Christopher Anderson, "NIH Drops Bid for Gene Patents," 263 *Science* 909, 909 (1994).

31 *"and continue, even today, to define . . ."*—See, e.g., Contreras (2011), Mason et al. (2018).

31 *"Another important purpose of the Bermuda Principles . . ."*—Contreras (2011).

31 "Nature *reported that more than 350 . . ."*—Claire O'Brien, "U.S. Decision Will Not Limit Gene Patents," 385 *Nature* 755, 755 (1997).

31 *"gene-of-the-week syndrome . . ."*—Andrews (2001, p. 46).

Chapter 5: The ACLU Way

32   "*Rebecca Eisenberg at University of Michigan . . .*"—Eisenberg (1990).

33   "*a documentary that premiered at Sundance . . .*"—*Frozen Angels* (Umbrella Films, et al. 2005).

33   "*a semi-historical yarn . . .*"—*The Pirate Latitudes* (HarperCollins, 2009); published posthumously following Crichton's death in 2008.

33   "Next, *a tale of corporate biotechnology . . .*"—*Next* (HarperCollins, 2006) was not Crichton's finest literary production. See Jorge L. Contreras, *Book Review: NEXT and Michael Crichton's Five-Step Program for Biotechnology Law Reform*, 48 JURIMETRICS 337 (2008).

33   "*her own genetics-based mystery novel . . .*"—Andrews published *Sequence* (St. Martin's Press, 2006) in the same year that Crichton released *Next.*

34   "*published an op-ed . . .*"—Crichton (2007).

34   "*Colorado jury sided with Metabolite . . .*"—*Metabolite Labs, Inc. v. Lab. Corp.*, No. 99-Z-870 (D.Colo. Dec. 3, 2001); *Metabolite Labs., Inc. v. Lab. Corp.*, No. 99-Z-870 (D.Colo. Nov. 19, 2001).

34   "*The Supreme Court grants . . .*"—See SCOTUSblog, Supreme Court Procedure, http://www.scotusblog.com/reference/educational-resources/supreme-court-procedure/. See further discussion at chapter 23, "Doing the Math."

36   "*Back in 2000...*"—Alex Berenson and Nicholas Wade (2000) "Clinton-Blair Statement on Genome Leads to Big Sell-Off" *NY Times*, Mar. 15, 2000.

37   "*Even the U.S. solicitor general . . .*"—In about twenty-five cases per year, and almost all patent cases, the Supreme Court asks the solicitor general of the United States for his or her views regarding a case before the Court. This is called a "Call for Views of the Solicitor General" or CVSG. See Am. Bar Assn., *Calls for the views of the solicitor general: An obscure but important part of Supreme Court practice*, Jun. 26, 2017, https://www.americanbar.org/publications/trends/2016-2017/july-august-2017/calls-for-the-views-of-the-solicitor-general/

38   "*Christopher Alan Hansen . . .*"—Material on Chris Hansen's upbringing and background is taken from Rutgers (2015) and the author's numerous personal interviews with Hansen.

39   "*Ronald Coase . . .*"—Ronald H. Coase (1910–2013) won the 1991 Nobel Memorial Prize in 38 for his groundbreaking work on transaction cost economics.

39   "*involuntary hepatitis experiments . . .*"—David Rothman, "Were Tuskegee and Willowbrook 'Studies in Nature'?" *The Hastings Center Report* 12:5–7 (1982).

39   "*a further expose . . .*"—Geraldo Rivera, *Willowbrook: The Last Great Disgrace*, WABC-TV (1972).

39   "*That case . . .*"—*NYSARC & Parisi v. Carey*, 393 F. Supp. 716 (EDNY, 1975). For an account of the massive Willowbrook litigation campaign, see David Rothman and Sheila Rothman, *The Willowbrook Wars* (New York: Columbia Univ. Press, 1984) and Chris Hansen's personal account, "Willowbrook," *Social Policy*, vol. 41 (Nov.–Dec. 1979).

40   "*multiple statutes . . .*"—Protection and Advocacy System of the Developmental Disabilities Assistance and Bill of Rights Act (1975), Education for All Handicapped Children Act (1975), Civil Rights of Institutionalized Persons Act of 1980.

40   "*Brown v. Board of Education . . .*"—In 1954, the U.S. Supreme Court ruled in *Brown v. Board of Education of Topeka* (347 U.S. 483) that "separate but equal" educational facilities violate the Equal Protection Clause of the Fourteenth Amendment of the U.S. Constitution. In a follow-on 1955 decision, the Court ordered states to desegregate their public school systems "with all deliberate speed." (349 U.S. 294). In Kansas, the Topeka school board abolished its

mandatory segregation regulations, but did little else to ensure school desegregation. Thus, in 1978, the ACLU, together with a group of African American attorneys, re-opened *Brown* to enforce the Supreme Court's desegregation mandate.

41 *"inside-the-Beltway"*—Refers to Washington, DC, insiders—named for the I-95/495 interstate freeway that encircles the District of Columbia and parts of Maryland and Virginia.

41 *"the Court dismissed . . ."*—548 U.S. 124 (Jun. 22, 2006).

42 *"lengthy dissenting opinion . . ."*—Among other things, Justice Breyer has been influential in shaping the Supreme Court's jurisprudence regarding the admissibility of scientific evidence in court proceedings.

42 *"'law-of-nature' question . . ."*—Metabolite, 548 U.S. at 128.

Chapter 6: Product of Nature

44 *"There are policies . . ."*—Interestingly, the ACLU's policy documents are not published or made generally available to the public, largely to forestall public criticism. The ACLU is, however, open to sharing its policies with those who are interested.

44 *"striking down parts of the federal statute . . ."*—Reno v. Am. Civil Liberties Union, 521 U.S. 844 (1997) (striking down portions of the Communications Decency Act).

44 *"the 1925 'Scopes Monkey Trial' . . ."*—The ACLU defended John Scopes, the biology teacher who was indicted for teaching the theory of evolution in a Tennessee public high school. The fascinating story of the Scopes "Monkey" trial can be found in numerous books as well as the award-winning 1960 film *Inherit the Wind* starring Spencer Tracy as a fictionalized version of Clarence Darrow, who was recruited by the ACLU to lead the defense.

44 *"Bert Foer, the head of . . ."*—In 1998, Albert "Bert" Foer founded the American Antitrust Institute (AAI), generally regarded as a pro-plaintiff antitrust advocacy group. I am currently a member of AAI's advisory board.

44 *"Hamid Kashani, an Indianapolis-based . . ."*—Though Kashani was listed as a member of the committee, none of the other members recalled his ever attending a meeting.

45 *"Burk was widely acknowledged . . ."*—Dan Burk is consistently ranked as one of the five most-cited intellectual property scholars in the U.S. https://leiterlawschool.typepad.com/leiter/2018/09/20-most-cited-intellectual-property-cyberlaw-scholars-in-the-us-for-the-period-2013-2017.html

45 *"According to one delegate . . ."*—Rebecca Rand (2007) "National Board Meeting" ACLU of Minnesota—Civil Liberties News 36(3): 5.

46 *"Lawrence Sung at the University of Maryland . . ."*—Sung 2007 testimony at 7 (cited in De Guilio, 2010, at 300).

48 *"Today, patented machines include . . ."*—See Manual of Patent Examination Procedure, 2106.l.

50 *"does not make them into patentable compositions of matter. . ."*—See *The Wood Paper Patent*, 90 U.S. 566, 594 (1874), holding, that while purified wood pulp is not patent-eligible, the novel process for extracting pulp as well as new boiler designs were eligible for patent protection. Following *Wood-Paper*, the Supreme Court and a number of lower courts denied patent protection to a range of different purified substances including tungsten (*General Electric Co. v. De Forest Radio Co.*, 28 F.2d 641 [3d Cir. 1928]), alizarine (a natural red dye) (*Cochrane v. Badische Anilin & Soda Fabrik*, 111 U.S. 293 (1884), vanadium (In re: Marden, 47 F.2d 958, 1059 [C.C.P.A. 1931]) and pine needle fibers (Ex Parte Latimer, 1889 Dec. Comm'r Pat. 123, 125 [1889]).

50 *"lawyers for the biotech firm Genentech . . ."*—Mukherjee (2016, p. 245), Palombi (2009, pp. 264–65). For an interesting discussion of early twentieth century debates over the patenting of isolated and purified insulin, see Palombi (2009, pp. 252–59).

51 *"the greatest jurist of his time"*—"Judge Learned Hand Dies; On U.S. Bench 52 Years," *New York Times*, Aug. 19, 1961, at 1 (quoted in Harkness, Jon M. (2011) "Dicta on Adrenaline!: Myriad Problems with Learned Hand's Product-of-Nature: Pronouncements in *Parke-Davis v. Mulford*," *Journal of the Patent and Trademark Office Soc'y* 93(4): 363–99.

51 *"a complex case brought by pharmaceutical giant . . ."*—Historical accounts of the *Parke-Davis* case can be found in Harkness (2011), Katherine J. Strandburg (2013) "Derogatory to Professional Character? Physician Innovation and Patents as Boundary-Spanning Mechanisms" *New York University Law and Economics Working Papers*, No. 357 (pp. 19–21), Cotter (2018, pp. 75–78), and Christopher Beauchamp (2013) "Patenting Nature: A Problem of History" *Stanford Technology Law Review* 16(2): 257–311.

52 *"a reputation as an anti-patent advocate . . ."*—Strandburg (2013, p. 19).

52 *"I cannot stop without calling attention . . ."*—189 F. at 115.

52 *"Hand upheld the patent . . ."*—189 F. at 103.

52 *"was affirmed in glowing terms . . ."*—196 F. 496 (2d Cir. 1912).

53 *"As Rebecca Eisenberg has pointed out . . ."*—Eisenberg (2002). These early patents did not seek to claim the DNA sequence of entire genes, but a sequence of coding DNA inserted into a bacterium or other organism for replication.

53 *"In 1999 it published . . ."*—U.S. Patent and Trademark Office (1999) "Revised Interim Utility Examination Guidelines" *Federal Register* 64: 71440.

53 *"comments were submitted . . ."*—U.S. Patent and Trademark Office (2000) "Public Comments on the United States Patent and Trademark Office "Revised Interim Utility Examination Guidelines""

Chapter 7: On the Hill

55 *"a different front: legislation . . ."*—Few areas of law better illustrate the co-dependent operation of the three branches of federal government than patent law. Congress, expressly authorized by the Constitution, has enacted the Patent Act, which establishes the basic machinery of the patent system. Within that system, the Patent and Trademark Office, part of the executive branch's Department of Commerce, considers patent applications and issues new patents. The judiciary, in turn, both interprets and fills gaps in the Patent Act, and evaluates whether individual patents are valid. When the courts interpret the Patent Act in a manner that displeases Congress, Congress may amend the act to "correct" the problem. All three branches of government thus work in tandem to adjust and adapt the law as technology develops and times change.

55 *"three-year moratorium on patenting genetic material . . ."*— Proposed amendment to H.R. 2507, 102nd Cong., introduced by Sen. Mark O. Hatfield, reproduced in U.S. Senate (1992, pp. 185–92).

55 *"Orrin Hatch, the long-serving . . ."*—Prepared Statement of Sen. Orrin Hatch, U.S. Senate (1992, p. 3).

55 *"legislation that would have immunized physicians . . ."*—Genomic Research and Diagnostic Accessibility Act of 2002, H.R. 3967, 107th Cong. (introduced Mar. 14, 2002).

56 *"similar immunity that had been extended . . ."*—Pub. L. No. 104–208, 110 Stat. 3009 (1996), codified at 35 U.S.C. § 287I.

56   *"Genomic Research and Accessibility Act"*—H.R. 977, 110th Cong. (2007).

56   "New York Times *op-ed* . . ."—Crichton (2007).

58   *"In a series of posts* . . ."—Kevin E. Noonan (2008) "Proponent of Gene Patent Ban to Leave
      Congress" *Patent Docs*, Dec. 10, 2008, https://www.patentdocs.org/2008/12/proponent-of-
      gene-patent-ban-to-leave-congress.html (responding to an incorrect rumor that Rep. Becerra
      was poised to leave Congress to accept a position as U.S. Trade Representative); Kevin E.
      Noonan (2007) "The Continuing Threat to Human Gene Patenting" *Patent Docs*, Oct. 16,
      2007, https://www.patentdocs.org/2007/10/the-continuing-.html;Noonan (2009a).

59   *"As reported by* . . ."—Shreeve (2004, p. 14).

59   *"to chat with his staff* . . ."—Congressional staff or aides play a critical role in the legislative
      process. They brief their senator or representative on all legislation that he or she is asked
      to consider, sponsor, support, or oppose, and often take a leading role in determining the
      office's policy position on many issues.

60   *"who had just released* . . ."—*A Patent System for the 21st Century* (2004), *Patents in the
      Knowledge-Based Economy* (2003).

61   *"his own genetics bill* . . ."—Senator Obama introduced the Genomics and Personalized
      Medicine Act of 2006, S.3822 (109th Cong.) in the Senate on August 3, 2006. The bill was
      reintroduced on Mar. 23, 2007, as the Genomics and Personalized Medicine Act of 2007,
      S.976 (110th Cong.).

62   *"Berman's Patent Reform Act of 2007* . . ."—On April 18, 2007, Rep. Howard Berman (D-Cal)
      introduced the Patent Reform Act of 2007 110th Cong. (H.R. 1908, S. 1145). The bill sought
      to overhaul the patent system in a number of ways, along the lines of an earlier bill introduced
      in 2005.

62   *"the real patent reform bill* . . ."—The Senate version of the Patent Reform Act was reported
      out of committee on January 24, 2008. The bill stalled and was not put to a vote. Though
      the Patent Reform Act failed in the 110th Congress, a subsequent bill was introduced in
      2009, and was finally enacted as the Leahy-Smith America Invents Act three years later, in
      2011 (Pub. L. 112–29).

62   *"a single day of hearings* . . ."—Stifling or Stimulating—The Role of Gene Patents in Research
      and Genetic Testing: Hearing before the Subcomm. on Courts, the Internet and Intellectual
      Property of the H. Comm. on the Judiciary, 110th Cong. 61 (Oct. 30, 2007).

62   *"No further action was taken* . . ."—Congressman Becerra is reported to have considered
      reintroducing the Genomic Research and Accessibility Act in 2010, but appears not to have
      done so. See Kevin E Noonan (2010) "He's Baaack! Congressman Becerra Once Again Tries
      to Ban Gene Patenting by Statute," *PatentDocs*, Apr. 8, 2010.

62   *"the House finally passed* . . ."—The Genetic Information Nondiscrimination Act, H.R. 493,
      was approved by the House on April 25, 2007, by a vote of 420–9–3. A companion bill was
      approved by the Senate in April 2008, went back to the House and was approved on May
      1. GINA was signed by President Bush on May 21, 2008.

62   *"Senator Obama's bill also died in committee* . . ."—Though Senator Obama's personalized
      medicine bill gained little traction in Congress, its primary author, Ed Ramos, moved to
      NIH as a science policy analyst in 2008 and, from there, helped to implement many of its
      proposals within the agency.

Chapter 8: Speaking of Patents

63   *"Voltaire's apocryphal proclamation . . ."*—The quote by Voltaire is believed to have originated
     with historian Evelyn Beatrice Hall in her book S. G. Tallentyre, *The Friends of Voltaire* pp.
     198–99 (John Murray: London, 1906).

63   *"he read a student newspaper story . . ."*—Rutgers (2015, p. 4).

64   *"In the same way, . . ."*—See *Buckley v. Valeo*, 435 U.S. 765 (1978) (political contributions);
     Branzburg v. Hayes, 408 U.S. 665 (1972) (investigative journalism).

64   *"but also thought . . ."*—The clearest statement of this theory, cited by both Andrews in her
     Metabolite brief and Hansen in his *Bilski* brief, is by Professor Laurence Tribe of Harvard
     Law School, who wrote, "The guarantee of free expression . . . is inextricably linked to the
     protection and preservation of open and unfettered mental activity." Laurence Tribe,
     American Constitutional Law § 15–7, at 1322 (2d ed. 1988).

64   *"Justice Benjamin Cardozo famously wrote . . ."*—*Palko v. Connecticut*, 302 U.S. 319, 326–327
     (1937). George Orwell's seminal dystopian novel *1984*, which introduced the term "thought
     crime" to the English lexicon, appeared in 1948.

64   *"In the late 1990s, Dan Burk . . ."*—Burk, Dan L. (2000) "Patenting Speech" *Texas Law Review*
     79:99. See also Thomas (2002, p. 589) (arguing that "the patent law allows private actors
     to impose more significant restraints on speech than has ever been possible through
     copyright").

65   *"a provocative article . . ."*—Lori B. Andrews (2006) "The Patent Office as Thought Police"
     Chronicle of Higher Education, Feb. 17, 2006.

65   *"his stinging dissent . . ."*—Justice Breyer did, however, suggest that researchers' freedom to
     explore the world of science was of concern, writing that the "presence [of patents] can
     discourage research by impeding the free exchange of information, for example by forcing
     researchers to avoid the use of potentially patented ideas, by leading them to conduct costly
     and time-consuming searches of existing or pending patents, by requiring complex licensing
     arrangements, and by raising the costs of using the patented information, sometimes
     prohibitively so." *LabCorp v. Metabolite*, Breyer, J., dissenting, at *3.

66   *"Bilski and Warsaw developed a formula . . ."*—Bilski and Warsaw described the background
     of their "invention" as follows:

> [The] patent application, entitled "Energy Risk Management Method," describes a
> method in which energy consumers, such as businesses and homeowners, are offered
> a fixed energy bill, for example, for the winter so they can avoid the risk of high
> heating bills due to abnormally cold weather. An intermediary or "commodity provider"
> sells natural gas, in this example, to a consumer at a fixed price based upon its risk
> position for a given period of time, thus isolating the consumer from an unusual spike
> in demand caused by a cold winter. Regardless of how much gas the consumer uses
> consistent with the method, the heating bill will remain fixed.
>
> Having assumed the risk of a very cold winter, the same commodity provider hedges
> against that risk by buying the energy commodity at a second fixed price from energy
> suppliers called "market participants." These market participants or suppliers have a
> risk position counter to the consumers, that is, they want to avoid the risk of a high
> drop in demand due to an unusually warm winter. A market participant could be, for
> example, someone who holds a large inventory of gas and wants to guarantee the sale
> of a portion of it by entering into a contract now. The risk assumed in the transactions
> with the market participants at the second fixed rate balances the risk of the consumer
> transactions at the first rate.

According to the patent application, setting the fixed price is not a simple process. The application discloses a complicated mathematical formula for calculating the price . . ." Petition for a Writ of Certiorari, *Bilski v. Doll* (2008), at 5–6.

66   *"one can't patent a mathematical formula . . ."*—Ex parte Bilski, 2006 WL 4080055 (B.P.A.I. Sept. 26, 2006).

66   *"The Federal Circuit allowed . . ."*—The Federal Circuit's 1998 decision in *State Street Bank & Trust Co. v. Signature Financial Group, Inc.*, 149 F.3d 1368 (Fed.Cir.1998) allegedly unleasheda "legal tsunami" of applications for patents on common business methods. In re: *Bilski* (Mayer, J., dissenting).

66   *"a waste of time and a waste of money . . ."*—Perry Cooper, "Full Court Patent Review Bids Often 'Waste of Time,' Judge Says," *Bloomberg Law*, Feb. 23. 2021 (quoting Judge Todd M. Hughes).

66   *"but without the time and expense . . ."*—One commentator estimates that the typical cost of preparing an *amicus* brief is between $10,000 and $20,000, far less than the cost of initiating litigation. Chien (2011, p. 399). This was part of Lori Andrews's pitch to the ACLU just two years earlier when unsuccessfully seeking the ACLU's support in filing an *amicus* brief in *Metabolite*.

67   *"Crichton, who had recently been diagnosed . . ."*—Crichton died on November 4, 2008.

67   *"reminding the Federal Circuit . . ."*—The dynamic between the federal appellate courts of the United States and the Supreme Court is a fraught and fascinating one. Courts hate to be reversed on appeal, and the Supreme Court, beginning in the late 2000s, seems to have begun a directed campaign of "correcting" the Federal Circuit's patent jurisprudence.

68   *"Judge Mayer dissented . . ."*—In re: *Bilski*, 545 F.3d 943 (Fed. Cir. 2008) (en banc) (Mayer, J., dissenting). Judge Mayer's interest in the intersection of patent law and free speech continued, and in 2016 he wrote a controversial concurring opinion in *Intellectual Ventures v. Symantec*, 838 F.3d 1307 (Fed. Cir. 2016), in which he argued that patents constricting the essential channels of online communication run afoul of the First Amendment. Unlike his dissenting opinion in *Bilski*, which largely went unnoticed, Judge Mayer's concurrence in *Intellectual Ventures* spawned a flurry of criticism and commentary, including another article by Prof. Dan Burk ("Patents and the First Amendment" *Washington Univ. Law Review* 96:197 (2018)), as well as vocal calls for his resignation from the bench (Eugene Quinn, *IPWatchdog, It Is Time for Judge Mayer to Step Down From the Federal Circuit*, (Oct. 6, 2016). True to his training as an army paratrooper and Vietnam combat ranger, Judge Mayer has not yet blinked.

68   "Bilski *would be appealed . . ."*—The Supreme Court heard the *Bilski* case in 2010 and, like each of the lower courts that heard the case, denied a patent to Bilski and Warsaw.

Chapter 9: The Power of Pink

72   *"Congress had passed a technical amendment . . ."*—Patent and Trademark Office Authorization Act of 2002, enacted as part of Public Law 107-273 on November 2, 2002. The 2002 legislation amended the Optional Inter Partes Reexamination Act, Public Law 106-113 . The act was part of the American Inventors Protection Act, passed on November 29, 1999, as S.R. 1948 (106th Cong. First Sess., 1999). Today, these procedures have largely been superseded by the 2011 America Invents Act.

74   *"wryly asking . . ."*—Eli Kintisch (2005) "A 'Robin Hood' Declares War on Lucrative U.S. Patents" *Science* 309: 1319–1320 (quoting patent lawyer and blogger Stephen Albainy-Jenei).

74   *"Wegner dubbed . . ."*—Kintisch (2005, p. 1319) (quoting Hal Wegner).

74   *"a leading trade magazine . . ."*—"Politics, power and passion—this year's MIP 50," *Managing Intellectual Property*, Jul. 1, 2008.

75   *"nearby Cardozo School of Law . . ."*—Around 2005, PubPat became affiliated with the Benjamin N. Cardozo School of Law of Yeshiva University in Greenwich Village. Cardozo, also the institutional home of attorney Barry Scheck's Innocence Project, was well-known for its emphasis on civil rights.

75   *"Courts exist to resolve actual disputes . . ."*—This is known as the "cases and controversies" requirement, and is enshrined in Article III of the Constitution.

75   *"The number in 2008 . . ."*—By early 2013, Mason and a colleague presented evidence to argue that at least some segments of every human gene were claimed by a U.S. patent. Jeffrey Rosenfeld and Christopher E Mason (2013) "Pervasive sequence patents cover the entire human genome" *Genome Medicine* 5(27). While this analysis has not been contested, others (including the author) have questioned some of the broad conclusions drawn in this paper. Shine Tu et al. (2014) "Letter to the Editor—Response to 'pervasive sequence patents cover the entire human genome' *Genome Medicine* 6(14).

76   *"Miracle drugs like Genentech's Herceptin . . ."*—For an excellent discussion of the development of Herceptin and Gleevec, and the frustrating search for other targeted cancer therapies, see Mukherjee (2010, pp. 412–444). See also Baldwin and Cook-Deegan (2013, pp. 3–4) (discussing development of Herceptin).

78   Table: SACHGS (2010), Cook-Deegan et al (2010), Pierce et al (2009, pp. 203–04), Merrill and Mazza (2006, pp. 65–67), compiled by the author.

79   *"the CFTR gene . . ."*—For an in-depth discussion of the patenting and licensing of the *CFTR* gene, see Subhashini Chandrasekharan, Christopher Heaney, Tamara James, Chris Conover, and Robert Cook-Deegan (2010), "Impact of gene patents and licensing practices on access to genetic testing for cystic fibrosis" *Genetics in Medicine* 12(4):S194–S211; Mollie A. Minear, Cristina Kapustij, Kaeleen Boden, Subhashini Chandrasekharan, and Robert Cook-Deegan (2013) "Cystic Fibrosis Patents: A Case Study Of Successful Licensing," *Les Nouvelles*, Mar. 2013: 21–30.

79   *"the gene for Tay-Sachs disease . . ."*—For an in-depth discussion of the patenting and licensing of the Tay-Sachs gene *HEXA*, see Colaianni et al (2010).

80   *"the genes associated with a condition called . . ."*—For an in-depth discussion of the patenting and licensing of the Long QT Syndrome gene, see Misha Angrist, Subhashini Chandrasekharan, Christopher Heaney, and Robert Cook-Deegan (2010), "Impact of gene patents and licensing practices on access to genetic testing for long QT syndrome" *Genetics in Medicine* 12(4):S111–S154.

80   *"Chung described the case . . ."*—Marc Grodman (2007) Testimony before Subcommittee on Courts, the Internet, and Intellectual Property of the Committee on the Judiciary, House Of Representatives, One Hundred Tenth Congress, First Session, Oct. 30, 2007, p. 35.

81   *"They're certainly the poster child . . ."*—See Gold and Carbone (2010), Baldwin and Cook-Deegan (2013),

82   *"Lori Andrews and Dorothy Nelkin's popular book . . ."*—Andrews (2001, pp. 43–44).

83   *"the largest study of diagnostic gene patents . . ."*—Cook-Deegan's project, which began in 2006, resulted in the SACHGS (2010) report, as well as a number of separately published academic papers in the journal *Genetics in Medicine*. The data was released throughout the progress of the study in a series of draft reports and papers.

Chapter 10: We've Got You Covered

86   *"under Governor Mitt Romney . . ."*—Romney served as Governor of Massachusetts from 2003 to 2007.

89   *"fifty-seven different claims payment carriers . . ."*—Under the Medicare Prescription Drug, Improvement and Modernization Act (MMA) of 2003, the fifty-seven part B claims payment carriers were reorganized as fifteen regional Medicare Administrative Contractors. This reorganization was completed in 2008.

89   *"about nine hundred smaller insurance carriers . . ."*—https://www.iii.org/fact-statistic/facts-statistics-industry-overview (current as of 2017)

89   *"nearly five hundred HMOs . . ."*—https://www.kff.org/other/state-indicator/number-of-hmos/?currentTimeframe=0&sortModel=%7B%22colId%22:%22Location%22,%22sort%22:%22asc%22%7D (current as of Jan. 2016)

91   *"the payer's cost would have been less . . ."*—These costs are estimated, in 2006 dollars, in Margaret L. Holland, Alissa Huston, and Katia Noyes (2009) "Cost-Effectiveness of Testing for Breast Cancer Susceptibility Genes," *Value in Health* 12(2): 207–216, p. 211.

91   *"someone else's responsibility . . ."*—As explained by one governmental panel, "Because plan members change plans at fairly frequent intervals, private health insurance coverage for preventive services can be difficult to rationalize from an economic standpoint since the cost savings may not accrue to the plan that paid for the service." Secretary's Advisory Committee on Genetics, Health, and Society (SACGHS) (2006) "Coverage and Reimbursement of Genetic Tests and Services" (Feb. 2006, p. 21).

92   *"the Social Security Act . . ."*—Social Security Act of 1935, § 1862(a)(1)(A).

93   *"would have to demonstrate . . ."*—In most cases, this involved the submission of a "medical necessity letter" or "insurance justification letter" on behalf of the patient. Heather L. Shappell and Ellen T. Matloff (2001) "Writing Effective Insurance Justification Letters for Cancer Genetic Testing: A Streamlined Approach," *Journal of Genetic Counseling* 10(4): 331–341. Interestingly, the cost of genetic counseling was seldom covered by insurance.

93   *"The company warned investors . . ."*—Myriad Genetics, Inc. (1995) *Prospectus*, Aug. 17, 1995, pp. 6–7.

94   *"claimed that only 5.7 percent of reimbursement claims . . ."*—Shappell and Matloff (2001, p. 332).

94   *"Aetna would cover BRACAnalysis . . ."*—Aetna's reimbursement criteria are reported by Shobita Parthasarathy: "(1) two or more first-degree (e.g., mother) or second-degree (e.g., aunt) relatives on the same side of the family with breast or ovarian cancer, regardless of age of diagnosis, (2) two relatives with early-onset breast or ovarian cancer, (3) a family member with a *BRCA* mutation, (4) breast cancer in a male patient or relative, (5) ovarian cancer at any age and breast cancer at any age, both on the same side of the family, (6) of Ashkenazi Jewish descent and a relative with breast or ovarian cancer at any age, (7) other circumstances with authorization of Aetna's medical director." Parthasarathy (2007, p. 131 (citing Aetna U.S. Healthcare [1999] "Prior Authorization Request Form for Breast and Ovarian [Cancer] Genetic Testing").

94   *"The head of its diagnostics business . . ."*—Critchfield (2009, p. 15).

94   *"some level of coverage from private insurers . . ."*—The Patient Protection and Affordable Care Act of 2010 (sometimes referred to as "Obamacare"), and a set of 2013 guidelines relating to that statute, required that "*BRCA* testing and counseling, if determined appropriate by a woman's health care provider," must be covered without cost sharing. These later developments appear to have increased patient access to *BRCA* testing. See U.S. Centers for Disease Control and Prevention (2014) "*BRCA* Genetic Testing and Receipt of Preventive

Interventions among Women Aged 18–64 Years with Employer-Sponsored Health Insurance in Nonmetropolitan and Metropolitan Areas — United States, 2009–2014", pp. 6–7.

95   *"Researchers found that . . ."*—See Ellen T. Matloff, Heather Shappell, Karina Brierley, Barbara A. Bernhardt, Wendy McKinnon, and Beth N. Peshkin (2000) "What Would You Do? Specialists' Perspectives on Cancer Genetic Testing, Prophylactic Surgery, and Insurance Discrimination" *Journal of Clinical Oncology* 18(12): 2484–2492; Soo-Chin Lee, Barbara A. Bernhardt, and Kathy J. Helzlsouer (2002) "Utilization of *BRCA1/2* genetic testing in the clinical setting: report from a single institution" *Cancer* 94:1876–1885; Emily A. Peterson, Kara J. Milliron, Karen E. Lewis, Susan D. Goold, and Sofia D. Merajver (2002) "Health insurance and discrimination concerns and *BRCA1/2* testing in a clinic population. *Cancer Epidemiology, Biomarkers & Prevention* 11:79–87.

95   *"fears of genetic discrimination . . ."*—See, e.g., Lawrence O. Gostin (1991) "Genetic discrimination: The use of genetically based diagnostic and prognostic tests by employers and insurers" *American Journal of Law and Medicine* 17(1&2):109–144; Paul R. Bilings, Mel A. Kohn, Margaret de Cuevas, Jonathan Beckwith, Joseph S. Alper, and Marvin R. Natowicz (1992) "Discrimination as a Consequence of Genetic Testing" *American Journal of Human Genetics* 50:476–482.

95   *"a few anemic legislative Band-Aids . . ."*—Limited progress was made in 1996 with the passage of the Health Insurance Portability and Accountability Act ("HIPAA"), which, among other things, prohibits group health insurers from using genetic information to determine eligibility or premiums, and from treating genetic information as a preexisting condition barring coverage. Health Insurance Portability and Accountability Act of 1996, codified at 29 U.S.C. §§ 1181–82 and 42 U.S.C. §§ 300gg-41. Some protection against genetic discrimination also existed in the employment area. For example, in 2000 President Clinton issued an executive order prohibiting discrimination against federal employees on the basis of genetic information. Executive Order No. 13,145, *Federal Register* 65: 6877 (Feb. 8, 2000).

95   *"in the early 1990s . . . "*—Collins (2010, pp. 115–117).

95   *"in 2008, President Bush signed . . ."*—Genetic Information Nondiscrimination Act of 2008, Pub. L. No. 110-233, 122 Stat. 881 (2008). For a detailed account of the legislative history of GINA, see Roberts (2010, pp. 447–451).

95   *"Senator Ted Kennedy . . ."*—Quoted in Kathy L. Hudson, M. K. Holohan, and Francis S. Collins (2008) "Keeping Pace with the Times—The Genetic Information Nondiscrimination Act of 2008" *New England Journal of Medicine* 358(25): 2661–2663, Jun. 19, 2008.

95   *"while critics claim . . ."*—For two contrasting tenth anniversary retrospectives on GINA, see Barbara J. Evans (2019) "The Genetic Information Nondiscrimination Act at Age 10: GINA's Controversial Assertion that Data Transparency Protects Privacy and Civil Rights" *William & Mary Law Review* 60(6): 2017–2109; and Bradley A. Areheart and Jessica L. Roberts (2019) "GINA, Big Data, and the Future of Employee Privacy" *Yale Law Journal* 128: 710–790.

Chapter 11: BART

106   *"Might they even contemplate . . ."*—Patients diagnosed with Huntington's disease, an incurable and fatal genetic disorder, have been found to have a higher than average incidence of suicide. When the gene for Huntington's disease was discovered, researchers hypothesized that carriers of the gene might also be at higher risk for suicidal thoughts. T. B. Robins Wahlin et al. (2000) "High suicidal ideation in persons testing for Huntington's disease," *Acta Neurologica Scandinavica* 102:150–161, p. 151. By extension, some researchers asked whether similar suicidal tendencies might occur in individuals with identified *BRCA* mutations. Biesecker et al. (1993) "Genetic Counseling for Families With Inherited Susceptibility to Breast and Ovarian Cancer" *Journal of the American Medical Assn.* 269(15): 1970–74, p. 1973.

106 *"one influential set of guidelines . . ."*—See Bailey (1999) (quoting bioethics report presented at 1996 meeting and Collins 1996 Senate testimony).

107 *"Myriad wouldn't offer the BART test . . ."*—Perhaps in response to situations such as the one faced by Kathleen Maxian, by 2009 Myriad claims that it performed BART testing at no additional charge for all high-risk patients, and made it available at an additional charge for low-risk patients. Critchfield (2009, p. 25). However, the controversy over the charges for BART testing continued at least through 2011. See Ellen T. Matloff et al. (2011) "An Open Letter to Myriad Genetics Laboratories re: BART Testing," Jul. 11, 2011.

Chapter 12: Patents and Plaintiffs

108 *"settled on Myraid Genetics . . ."*—Technically, cases like *Bilski*, in which an applicant is challenging the PTO's rejection of a patent application, are brought against the director of the PTO. Thus, at the Federal Circuit, the case was styled *Bilski v. Doll*, after the then-acting director of the PTO, John Doll (or simply *In re: Bilski*—In the matter of Bilski), and at the Supreme Court, the case became known as *Bilski v. Kappos*, after David Kappos, who was appointed director of the PTO in 2009.

111 *"fifteen base pairs . . ."*—These claims were aggressive even within the biotech industry. Later studies found that 80 percent of the genes in the human genome contain DNA sequences that would have infringed these claims (Kepler et al. [2010]).

111 *"fifteen claims from"*—For a thorough but understandable discussion of the details of the Utah/Myriad patents that were challenged, see Cotter (2018, p. 68–70). See also Kepler et al. (2010) (analyzing some of the claims).

112 *"in his book* The Common Thread*"*—John Sulston and Georgina Ferry (2002) *The Common Thread: The Story of Science, Politics, Ethics, and the Human Genome* (Joseph Henry Press).

114 *"jumping genes"*—See Leslie A. Pray (2008) "Transposons: The Jumping Genes" *Nature Education* 1(1): 204.

114 *"new technique . . ."*—Arupa Ganguly, Matthew J. Rock, Darwin J. Prockop (1993) "Conformation-sensitive gel electrophoresis for rapid detection of single-base differences in double-stranded PCR products and DNA fragments: evidence for solvent-induced bends in DNA heteroduplexes" *Proceedings of the National Academy of Sciences* 90:10325–10329.

115 *"He told Skolnick . . ."*—This episode is recounted less colorfully in court papers filed later by Myriad: "On or about September 29, 1998, Dr. Kazazian informed a Myriad executive that, although he acknowledged the existence of Myriad's patents and the legal rights flowing therefrom, he would continue with the activities which Myriad considered to infringe its patents." *Myriad Genetics, Inc. v. University of Pennsylvania*, Case No. 2:98CV 0829S, Complaint at p. 6 (D. Utah, filed Nov. 19, 1998).

115 *"Based on that finding . . ."*—*Julie Anderson v. The Trustees of the University of Pennsylvania*, Case No. 1:09-cv-02345 (D.D.C. filed Dec. 10, 2009).

116 *"a misdiagnosis by Myriad's one-time competitor . . ."*—Rick Weiss (1999) "A Defective Side to Genetic Testing," *Wash. Post*, Jul. 26, 1999.

116 *"Myriad asked the court for an injunction . . ."*—*Myriad Genetics, Inc. v. University of Pennsylvania*, Case No. 2:98CV 0829S (D. Utah, filed Nov. 19, 1998).

116 *"who would eventually leave the university . . ."*—Barbara Weber left the University of Pennsylvania in 2005 to become a vice president at GlaxoSmithKline, after which she moved to Novartis as senior vice president for oncology and translational medicine. In 2015 she moved to Third Rock Ventures as a partner.

116 *"the path forward was clear . . ."*—Myriad's lawsuit against the University of Pennsylvania did not advance far. Possibly, it was filed simply to get the attention of Penn's legal department, which it succeeded in doing. The action was dismissed without prejudice for failure to serve process on the defendant on April 20, 1999.

117 *"He was particularly interested in . . ."*—In addition to numerous scientific papers, Ostrer is the author of the book *Legacy: A Genetic History of the Jewish People* (Oxford University Press, 2012).

119 *"scientific frenzy . . ."*—The dramatic and history-making race to sequence the *BRCA* genes has been the subject of two books (Davies and White [1995] and Waldholz [1997]), several television documentaries, at least one feature film (*Decoding Annie Parker* [Ozymandias Productions et al., 2013]), numerous news stories, and countless scholarly articles.

119 *"in 1989 his lab . . ."*—John R. Riordan et al. (1989). "Identification of the Cystic Fibrosis Gene: Cloning and Characterization of Complementary DNA," *Science* 245:1066–73.

119 *"a significant discovery . . ."*—Cook-Deegan (1994, p. 45), Helen Pearson (2009). "One Gene, Twenty Years," *Nature* 460:165–69.

119 *"was able to assemble . . ."*—Myriad's aggregation of all of the patents relating to *BRCA1/2* is a fascinating story in itself. Some of this history can be found in Contreras (2011).

122 *"a suit involving blood analysis patents . . ."*—Brief for Amici Curiae the American College of Medical Genetics, the American Medical Association, the American Society of Human Genetics, the Association of American Medical Colleges, The Association of Professors of Human and Medical Genetics, the Association for Molecular Pathology, and the College of American Pathologists in Support of Defendants-Appellees, *Prometheus Laboratories, Inc. v. Mayo Collaborative Services* (Fed. Cir., April 6, 2009).

123 *"She was particularly incensed . . ."*—Matloff and Caplan (2008).

123 *"an ad for BRACAnalysis . . ."*—Myriad's advertisement *Stagebill* for the Kennedy Center's sold-out run of *Wit* starring Judith Light has generated its own cottage industry of criticism. See, e.g., Parthasarathy (2017); Rimmer (2008, p. 187); Blanton (2002); Sara Chandros Hull and Kiran Prasad (2001) "Reading between the Lines: Direct-to-Consumer Advertising of Genetic Testing" *The Hastings Center Report*, 31(3): 33–35, May–Jun., 2001.

124 *"the university was still smarting . . ."*—Donald G. McNeil Jr. (2001) "Yale Pressed to Help Cut Drug Costs in Africa" *New York Times*, Mar. 12, 2001.

125 *"The program, which was run . . ."*—Alexandra Minna Stein (2009) "A Quiet Revolution: The Birth of the Genetic Counselor at Sarah Lawrence College, 1969," *Journal of Genetic Counseling* 18: 1–11.

125 *"A half dozen other colleges . . ."*—From 1971 to 1973, genetic counseling degree programs were started at the University of Pittsburgh, University of Denver, Rutgers University, University of California Berkeley and University of California Irvine. Stein (2009, p. 8).

127 *"NSGC was not interested . . ."*—Following the ACLU's favorable ruling at the District Court in 2010, NSGC issued a statement encouraging patent holders to "provide sensible license agreements that will improve future health care quality," but still not opposing the practice of gene patenting itself. Natl. Soc'y of Genetic Counselors (2010) "Statement from National Society of Genetic Counselors regarding Gene Patenting," Mar. 30, 2010.

127 *"Boston Women's Health Book Collective . . ."*—Young (2014, p. 21) (reporting on interview with Judy Norsigian, executive director and a founder of the collective).

127 *"Our Bodies, Ourselves . . ."*—According to ourbodiesourselves.org, "In 1970, a group of women in the Boston area self-published "Women and Their Bodies," a 193-page booklet that dared

to address sexuality and reproductive health, including abortion. They distributed it for 75 cents. A year later, they changed the title to "Our Bodies, Ourselves"—and changed the women's health movement around the world. The U.S. book was updated in print through 2011 and online until 2018, while global adaptions are still in development today." (last visited Jul. 26, 2019).

127 *"BCA first made national headlines . . ."*—See Mukherjee (2010, pp. 425–26). Even more vocal and radical breast cancer activist groups existed. See Davies and White (1995, pp. 257–58) (describing groups such as One in Nine, Cancer Patients Action Alliance (CANACT) and Women's Health and Mobilization (WHAM)).

128 *"NBCC's website . . ."*—National Breast Cancer Coalition (1995) "Presymptomatic genetic testing for heritable breast cancer risk," Press Release, Sep. 28, 1995.

128 *"a position that NBCC wasn't ready . . ."*—Breast Cancer Action had a similar policy, but changed it around the time they decided to join the ACLU's lawsuit as plaintiffs.

130 *"Many of its members participated . . ."*—For a glimpse of the types of questions and issues of concern to women carrying *BRCA* mutations, see Dina Roth Port, *Previvors: Facing the Breast Cancer Gene and Making Life-Changing Decisions* (Avery, 2010), Jonathan Herman and Teri Smieja, *Letters to Doctors: Patients Educating Medical Professionals through Practical True-Life Experiences* (Twin Trinity Media, 2013).

131 *"To some extent, Ceriani's participation . . ."*—Ceriani-Park StoryCorps interview, Mar. 2016.

Chapter 13: Pulling the Trigger

133 *"in any district . . ."*—The rules for choosing the district in which to bring a patent suit were tightened in 2017 following the Supreme Court's decision in *TC Heartland v. Kraft Food* (U.S. 2017).

134 *"One of the first . . ."*—Blanton (2002).

138 *"What the* New York Times *called . . ."*—Natalie Angier, "Scientists Identify a Mutant Gene Tied to Hereditary Breast Cancer," *New York Times*, Sep. 15, 1994.

140 *"particularly the loss of . . ."*—Dealbook (2009) "A.C.L.U. Loses Its $20 Million Man" *New York Times*, Dec. 9, 2009.

141 *"ACLU's PR team . . ."*—See Parthasarathy (2017, p. 174) (referring to the "ACLU's public relations machine").

PART II: LITIGATION

143 *"Could you patent the sun . . ."*—For an in-depth discussion of Jonas Salk's discovery of the polio vaccine and failure to patent it, see Jane Smith, *Patenting the Sun: Polio and the Salk Vaccine* (New York: William Morrow, 1990).

Chapter 14: The Big Guns

148 *"all of the other lawsuits . . ."*—For a discussion of gene patent litigation during and before this period, see Holman (2008) and Palombi (2009, pp. 276–302).

148 *"challenges that Myriad . . ."* For a discussion of Myriad's international patent disputes, see Gold and Carbone (2010, pp.S49–S57), Rimmer (2013), and Van Zimmeren et al. (2014).

148 *"Myriad had never enforced its patents against researchers . . ."*—Gold and Carbone (2010, p. S44), Caulfield (2008).

149 *"Thousands of them."*—According to court papers filed by Myriad, between 2005 and 2009, Myriad provided more than 3,000 low-income, uninsured patients with free *BRCA* testing. Critchfield (2009, p. 16).

149 *"he could work on expanding . . ."*—Perhaps in response to situations such as those alleged by the ACLU, by December 2009 Myriad claims that it performed BART testing at no additional charge for all high-risk patients, and made it available at an additional charge for low-risk patients. Critchfield (2009, p. 25).

150 *"their tests actually saved people's lives . . ."*—Jackson summarized his perspective in a 2013 article. Benjamin Jackson (2013) "A Patient-Centric Look at Gene Patents" *IP Watchdog*, May 9, 2013.

153 *"first two centuries . . ."*—An engaging early history of the Patent Office can be found in Kenneth W. Dobyns (1994) *The Patent Office Pony: A History of the Early Patent Office* (Docent Press).

155 *"Biotechnology Industry Association (BIO) . . ."*—Though BIO seeks to represent the entire biotechnology industry and has over 1,000 member companies, companies in the molecular diagnostics sector have generally not joined BIO. Thus, neither Myriad Genetics nor other companies offering genetic diagnostic testing (LabCorp, Quest, Athena) have traditionally been BIO members. Nevertheless, BIO emerged as one of the most active advocates in the gene patenting debate.

155 "Kojo Nnamdi Show . . ."—The segment aired on Jun. 4, 2009. For commentary, see Donald Zuhn (2009a) "Gene Patenting Debate Continues" *Patent Docs blog*, Jun. 9, 2009.

156 "Science Friday . . ."—The segment aired on Dec. 11, 2009. For commentary, see Donald Zuhn (2009b) "Gene Patenting Debate Continues—Round Three" *Patent Docs blog*, Dec. 17, 2009.

156 *"blog post . . ."*—Noonan (2009a).

156 *"John Conley asked whether . . ."*—John Conley (2009) "The *ACLU v. Myriad Genetics* Suit: Legitimate Challenge or Publicity Stunt?" *Genomics Law Report*, Jun. 4, 2009.

157 *"wrote in his . . ."*—Gene Quinn (2009) "ACLU Files Frivolous Lawsuit Challenging Patents" *IP Watchdog blog*, May 14, 2009.

157 *"calls for judicial sanctions . . ."*—See Dale B. Halling (2009) "ACLU Should Be Hit with Rule 11 Sanctions" *IP Watchdog blog*, Nov. 20, 2009.

157 *"obtaining patents in . . ."*—Pedram Sameni (2020) "What Patent Prosecution Market Size Means for Cos., Counsel" *Law360*, Oct. 14, 2020.

Chapter 15: S.D.N.Y.

159 *"They review the briefs . . ."*—As explained by Jeffrey Toobin, the typical duties of a judicial clerk involve "reading the briefs in cases, summarizing the arguments in memos, writing first drafts of opinions, and picking up the judge's wife at the train." Jeffrey Toobin (1991) *Opening Arguments: A Young Lawyer's First Case: United States v. Oliver North* (Anchor Books).

160 *"cases assigned at random . . ."*—Each case is assigned to a judge shortly after the complaint is filed with the court. However, unless one stands in the court clerk's office and takes note of each case being assigned, there is no convenient way to keep track of assignments to a particular judge's chambers until the weekly distribution of complaints on Tuesday mornings.

160 *"Muse and his shipmates . . ."*—This episode became the subject of a feature-length film, *Captain Phillips* (2013), starring Tom Hanks in the title role as the captain of the hijacked ship.

161 *"he had read a story . . ."*—John Schwartz (2009) "Cancer Patients Challenge the Patenting of a Gene" *New York Times*, May 12, 2009.

161 *"a patent application . . ."*   U.S. Patent No. 7,892,539B2 "TRAIL-R as a negative regulator of innate immune cell responses" (Feb. 22, 2011).

Chapter 16: Chicken and Egg

167 *"handful of patent . . ."*—see, e.g., *Weddingchannel.com, Inc. v. The Knot, Inc.*, 2005 U.S. Dist. LEXIS 991 (S.D.N.Y. 2005).

167 *"Sweet ruled for McDonald's . . ."*—*Pelman v. McDonald's Corp.*, 237 F. Supp. 2d 512, 532–33 (S.D.N.Y. 2003).

168 *"PXE . . ."*—The fascinating story of the founding of PXEI can be found in Terry SF, Terry PF, Rauen KA, Uitto J, Bercovitch LG (2007). Advocacy groups as research organizations: the PXE international example. *Nat. Rev. Genet.* 8:157—64.

168 *"Walking through New York's Times Square . . ."*—*Loper v. New York City Police Dept.*, 802 F.Supp. 1029 (S.D.N.Y. 1992).

169 *"The judge agreed with Ravicher . . ."*—669 F. Supp. 2d 365 (S.D.N.Y., Nov. 2, 2009).

169 *"Judge J. Edward Lumbard . . ."*—Judge Lumbard, though not well-known today, was a significant force in his day. Among his many prominent law clerks was Steve Shapiro, the ACLU's legal director.

179 *"his translator and his guide . . ."*—Robert W. Sweet, comments made to American Soc'y of Human Genetics, Oct. 14, 2011.

180 *"Ceriani, Limary, and Girard were eager! . . ."*—Girard had released a book about her personal experiences with cancer, *Off the Rack: Chronicles of a thirty-something, single, breast cancer survivor* (Strong Books, 2010).

181 *"was featured in the popular SELF Magazine . . ."*—Ginny Graves "Who Owns Your Genes?" *SELF*, Oct. 16, 2010.

181 *"issued his decision . . ."*—Judge Sweet's opinion, including subsequent corrections and amendments, is reported, among other places, at 702 F. Supp. 2d 181 (S.D.N.Y. 2010).

181 *"the contrary precedent in Davis v. Mulford . . ."*—To some degree, Judge Sweet's beef with the larger-than-life Hand is personal. Sweet puckishly notes, at the end of a lengthy footnote, that Hand, who was known for his dominating courtroom presence, once turned his back on Sweet, then a young attorney arguing in his courtroom. 702 F. Supp. 2d at 225 n.46. Sweet then notes that while the *Parke-Davis* decision "deserves careful review," one is reminded of the adage "Quote Learned, but follow Gus" (referring to Hand's brother, Augustus Noble Hand, who also served on the Second Circuit). 702 F. Supp. 2d at 225 n.46 (citing James Oakes [1995] "Personal Reflections on Learned Hand and the Second Circuit" *Stanford Law Review* 47: 387–394, n. 175). And, in the next footnote, Sweet further cuts Hand down to size, observing that "notwithstanding Judge Hand's reputation, his opinion in *Parke-Davis* was one of a district court judge and does not supersede contrary statements of the law by the [court of appeals] or the Supreme Court." 702 F. Supp. 2d at 226 n.47.

182 *"did not need to address the constitutional issues . . ."*—Under the doctrine of "constitutional avoidance," courts should not decide constitutional questions unnecessarily. 702 F. Supp. 2d at 237–38.

183 *"Pigs Fly . . ."*—John Conley and Dan Vorhaus (2010) "Pigs Fly: Federal Court Invalidates Myriad's Patent Claims" *Genomic Law Report*, Mar. 30, 2010.

183 *"Another blogger . . ."*—Eric Guttag (2010) "Foaming at the Mouth: The Inane Ruling in the Gene Patents Case" *IPWatchdog*, Mar. 30, 2010.

183 *"But Gene Quinn topped them all . . ."*—Gene Quinn (2010) "Hakuna Matada, the ACLU Gene Patent Victory Will Be Short Lived" *IPWatchdog*, Mar. 31, 2010.

184 *"war over human nature . . ."*—Jim Dwyer (2010) "In Patent Fight, Nature, 1; Company, 0" *New York Times*, Mar. 30, 2010.

184 *"the two Nobel laureates . . ."*—Joseph Stiglitz and John Sulston (2010) "The Case against Gene Patents," *Wall St. Journal*, Apr. 16, 2010.

184 *"The editorial board of* Nature Biotechnology *. . ."*—Editorial (2010) "Sitting up and taking notice" *Nature Biotechnology* 28(5): 381.

184 *"his evangelical Christian credentials . . ."*—Francis Collins discusses his Christian beliefs and their compatibility with science in his book *The Language of God: A Scientist Presents Evidence for Belief* (Free Press, 2006).

185 *"playing with 'verve' . . ."*—Deborah Jowitt (2006) "Making Music Dance" *Village Voice*, Jun. 13, 2006.

Chapter 17: We're from the Government

187 *"her top deputy . . ."*—Neal Katyal served as principal deputy solicitor general of the United States from February 3, 2009 to May 17, 2010.

188 *"a landmark Supreme Court victory . . ."*—*Hamdan v. Rumsfeld*, U.S. (2006).

188 *"Paris Hilton of the Legal Elite . . ."*—David Lat, "Neal Katyal: The Paris Hilton of the Legal Elite?" *Above the Law*, Aug. 9, 2006.

188 *"the Thurgood Marshall of his era . . . "*—National Public Radio, *Morning Edition* (2006) "Law Professor Beats the Odds in Detainee Case" Sep. 5, 2006.

190 *"one of the President's rallying cries . . ."*—In his October, 2008, debate with John McCain, Obama said "For my mother to die of cancer at the age of 53 and have to spend the last months of her life in the hospital room arguing with insurance companies because they're saying that this may be a pre-existing condition and they don't have to pay her treatment, there's something fundamentally wrong about that." In 2011, *New York Times* reporter Janny Scott challenged this statement in her book, *A Singular Woman: The Untold Story of Barack Obama's Mother* (Riverhead Books, 2011). Scott claims that Obama's mother was not denied health insurance coverage for her cancer, and that her dispute was with her employer's disability insurance carrier.

190 *"hundreds of different agencies, bureaus . . ."*—these entities are generally referred to as "components."

192 *"officials from the Patent Office . . ."*—The evolving role of the solicitor general in patent cases has been explored in a growing body of literature including Rai (2012), Ben Picozzi (2014) "The Government's Fire Dispatcher: The Solicitor General in Patent Law" *Yale Law and Policy Review* 33(2): 427–454, and Tejas Narechania (2020) "Defective Patent Deference" *Washington Law Review* 95: 869–945.

193 *"up the chain of command . . ."*—the individuals holding these offices at the time were Scott McIntosh (supervising attorney, appellate staff), Beth Brinkman (deputy assistant AG in charge of the appellate staff), and Tony West (assistant AG in charge of the Civil Division).

195 *"NIH was quietly assembling . . ."*—It was something of an embarrassment that NIH, as a result of the 1995 settlement of an inventorship dispute over the *BRCA1* patents, was actually

a co-owner of four of the *BRCA1* patents being challenged in the lawsuit, and that NIH had exclusively licensed its ownership interest in the patents to Myriad. By way of explanation (apology?) the DOJ notes in its *amicus* brief that "that result is anomalous: NIH ordinarily does not grant exclusive licenses under DNA patents for diagnostic applications." (Brief for the United States, Oct. 29, 2010, p. 6 n.2).

196 *"a self-declared science nerd . . ."*—Davey Alba (2016) "Obama Geeks Out over a Brain-Controlled Robotic Arm That 'Feels'" *Wired*, Oct. 14, 2016 (quoting President Obama: "I confess I'm a science geek. I'm a nerd. I won't make any apologies for it.")

196 *"older brother to . . ."*—The third Emanuel brother, Ari, is a Hollywood talent agent.

196 *"a dysfunctional mess . . ."*—Ezekiel J. Emanuel (2007) "What Cannot Be Said on Television about Health Care" *Journal of the American Medical Association* 297(19), May 16, 2007.

197 *"which has been referred to as . . ."*—Brian Palmer (2010) "All the President's Money Men," *Explainer*, Sep. 22, 2010.

197 *"outspoken but brilliant . . ."*—Journalist Stephen Dubner, interviewing Summers for the popular Freakonomics radio show and podcast, called Summers "one of the most brilliant economists of his generation (and perhaps the most irascible)." Stephen J. Dubner (2017) "Why Larry Summers Is the Economist Everyone Hates to Love" *Freakonomics*, Episode 303, Sep. 27, 2017.

197 *"Treasury Secretary Timothy Geithner . . ."*—Lori Montgomery (2010) "Lawrence Summers to leave economic council, return to Harvard" *Washington Post*, Sep. 21, 2010.

197 *"Summers and his staff . . ."*—NEC officials primarily involved in *AMP v. Myriad* included Diana Farrell, a Harvard MBA who had previously run the international think tank at McKinsey, and Phil Weiser, a telecommunications lawyer and law professor from Colorado whom Summers recruited from the Department of Justice Antitrust Division. Weiser is reputed to have played a particularly important role in coordinating the different White House organizations that weighed in on the gene patenting debate.

197 *"Nobody was accusing Myriad Genetics . . ."*—It is well-accepted that aggressive and even ruthless commercial behavior does not violate the antitrust laws unless it has the effect of harming competition. As the Court of Appeals for the Seventh Circuit has written, the "drive to succeed lies at the core of a rivalrous economy. Firms need not like their competitors; they need not cheer them on to success; a desire to extinguish one's rivals is entirely consistent with, often is the motive behind, competition." *A.A. Poultry Farms, Inc. v. Rose Acre Farms, Inc.*, 881 F.2d 1396, 1402 (7th Cir. 1989).

198 *"that orthodoxy was increasingly being questioned . . ."*—In 2008 and 2009, a trio of influential books appeared, each arguing that intellectual property, and patents in particular, were hindering scientific, technological, or economic progress. These included Boldrin and Levine (2008), Burk and Lemley (2009), and James Bessen and Michael J. Meurer (2009) *Patent Failure—How Judges, Bureaucrats, and Lawyers Put Innovators at Risk* (Princeton University Press).

198 *"district court explained that . . ."*—Declaration of Joseph E. Stiglitz, PhD., *AMP v. Myriad* at 12, 25 (S.D.N.Y., filed Jan. 19, 2010).

198 *"head of the Antitrust Division . . ."*—Christine Varney served as AAG in charge of the Antitrust Division at this time.

201 *"which the* Washington Post *ranked . . ."*—Alice Graeme (1936) "12 Murals Completed for Justice Department," *Washington Post*, Sep. 27, 1936, p. AA7.

Chapter 18: Splitting the Baby

203 *"even if it had the same 8,000 bases . . ."*—The science of genomics is continually advancing. In this book, I have described the *BRCA* genes based on data from NIH's RefSeq database as of May 2020. The data in Myriad's patents and the various judicial opinions differ slightly, reflecting the information available at the time.

203 *"cDNAs are useful . . ."*—See Cook-Deegan and Heaney (2010, p. 395).

204 *"synthesized cDNA molecules . . ."*—See U.S. Pat. No. 4,322,499 (1982).

204 *"This distinction between . . . "*—The idea that man-made cDNA should be eligible for patent protection, whereas naturally occurring genomic DNA should not, has been discussed in the literature for some time. It first appears to have been noted by Rebecca Eisenberg, one of the first legal scholars to consider gene patenting (1990, p. 727 n. 25) and was recognized by Leroy Hood, one of the founders of Applied Biosystems, and Dan Kevles, the historian of science, a couple of years later (Kevles and Hood [1992, pp. 313–14]. This being said, Professor Christopher Holman has argued that the process for cloning gDNA, as it is used in the diagnostics industry, is virtually identical to the process for creating cDNA. See "Brief of Amicus Curiae Law Professor Christopher M. Holman In Support of Neither Party" at 9 (Fed. Cir. Jun. 11, 2012) ("The methodology for producing cDNA is entirely analogous to the methodology for isolating genomic DNA.") While Professor Holman is undoubtedly correct as a matter of science, it is not clear that the method of production of gDNA and cDNA is really at issue. Thus, while some gDNA may be fabricated in the laboratory, it also occurs naturally in the human cell, and is thus different than cDNA in that regard.

208 *"sometimes permitted an agency . . ."*—For example, in the controversial "snail darter" endangered species case (*Tennessee Valley Authority v. Hill*, 437 U.S. 153 [1978]), the Department of Interior was permitted to add its views in support of the Endangered Species Act to the brief submitted by the government, which largely supported the position of the Tennessee Valley Authority, opposing application of the act. See Zygmunt J. B. Plater, *The Snail Darter and the Dam* 215–16, 237 (Yale Univ. Press, 2013).

211 *"His expert testimony in a Bronx courtroom . . ."*—*People v. Castro*, 545 N.Y.S.2d 985 (Sup.Ct. 1989). The evidence and Lander's role in the case are discussed in Harold M. Schmeck Jr. (1989) "DNA Findings Are Disputed by Scientists," *New York Times*, Mary 25, 1989.

211 *"The experience led the lawyers . . ."*—Eric Lander discusses his experience in the *Castro* case and its aftermath in a lively interview on themoth.org (recorded Jun. 4, 2011).

212 *"once reputed to be the most powerful woman in New York . . ."*—Gail Sheehy (1973) "The Life of the Most Powerful Woman in New York," *New York Magazine*, December 10, 1973.

213 *"Hal Wegner gleefully criticized . . ."*—Harold C. Wegner (2010) "Myriad DNA Patent-Eligibility Case (cont'd): Eric Holder Hijacks the Patent System, Flunks Patents 101," email distributed Oct. 30, 2010.

213 *"The authors of the* Genomic Law Report *. . ."*—Dan Vorhaus and John Conley (2010) "Swine Soar Higher in Myriad Thanks to U.S. Government's Amicus Brief" *Genomics Law Report*, Nov. 1, 2010.

213 *"It may be that . . ."*—Kevin E. Noonan (2010) "DOJ Tries to Be All Things to All Constituencies in Myriad Amicus Brief" *Patent Docs*, Oct. 31, 2010.

Chapter 19: The Patent Court

214 *"its well-deserved nickname . . ."*—See Abramson (2007, p. 1).

214 *"professor Rochelle Dreyfuss . . ."*—Rochelle Cooper Dreyfuss (1989) "The Federal Circuit: A Case Study in Specialized Courts" *NYU Law Review* 64: 1–77, p. 7.

214 *"The federal circuit was created . . ."*—For a discussion of the political maneuvering and debate that led to the creation of the Federal Circuit, see Dreyfuss (1989), *supra*, and Abramson (2007).

215 *"in decision after decision . . ."*—The pro-patent tendencies of the Federal Circuit, in comparison to the pre-1982 regional circuit, the district courts, and the Supreme Court, have been debated, measured, and analyzed in dozens of law review articles, beginning almost as soon as the court was created. For a sample of this debate, see Abramson (2007), William M. Landes and Richard A. Posner (2004) "An Empirical Analysis of the Patent Court" *University of Chicago Law Review*, 71(1): 111–128, and a special symposium and issue of the *Loyola of Los Angeles Law Review* entitled "The Federal Circuit as an Institution" 43(3) (symposium held on Oct. 30, 2009).

215 *"the de facto supreme court of patents . . ."*—Mark D. Janis (2001) "Patent Law in the Age of the Invisible Supreme Court, *University of Illinois Law Review* 2001: 387–419, p. 387.

218 *"wrote the annual survey . . ."*—See, e.g., Castanias, Gregory A. et al. (2007) "Survey of the Federal Circuit's Patent Law Decisions in 2006: A New Chapter in the Ongoing Dialogue with the Supreme Court" *American University Law Review* 56(4): 793–985.

218 *"a complex international case . . ."*—*Sinochem Intl. Co. v. Malaysia Intl. Shipping Corp.* 549 U.S. 422 (2007).

220 *"Rader was scheduled to speak . . ."*—Coverage of this panel can be found in John T. Aquino (2010) "Finding Gene Patents Unpatentable Too Blunt an Approach, Panelists Say" *BNA Patent, Trademark & Copyright Journal*, May 14, 2010.

222 *"In a recent* Science *article . . ."*—Sherry M. Knowles (2010) "Fixing the Legal Framework for Pharmaceutical Research" *Science* 327: 1083–1084.

222 *"federal law required . . ."*—28 U.S.C. § 455(a).

224 *"One waggish author . . ."*—Craig Anderson (2011) "Chief Judge Randall Rader Shoots from the Lip" *Daily Journal*, Apr. 6, 2011.

227 *"was submitted by the attorney general . . ."*—See Richard Kluger, *Simple Justice: The History of* Brown v. Board of Education *and Black America's Struggle for Equality*, 745 (Vintage Books, 1975).

Chapter 20: Magic Microscope

229 *"one court watcher . . ."* Tom Goldstein (2010) "Anticipating the Next Solicitor General" *SCOTUS Blog*, Aug. 13, 2010.

235 *"What did 'covalently bonded' mean?"*—A covalent bond is formed between two atoms when they share electron pairs. This type of bond is relatively strong, and results in the formation of stable compounds such as water (H_2O). When covalent bonds are broken, the constituent elements no longer form the compound (e.g., hydrogen and oxygen gas are different from their combined existence as water). Judge Lourie's point was that breaking a gene away from the rest of a DNA molecule (by breaking covalent chemical bonds) would similarly alter its nature, thus forming a new compound that was "made" by man.

Chapter 21: Last Man Standing

240 *"some viewed* Mayo *as a stalking horse . . ."*—Kevin E. Noonan (2012) "*Prometheus Laboratories, Inc. v. Mayo Collaborative Services* (Fed. Cir. 2010)" *Patent Docs*, Dec. 20, 2010.

240 *"actively interviewing . . ."*—Lyle Denniston (2011) "Verrilli begins, Katyal to depart" *SCOTUSblog*, Jun. 10, 2011.

241 *"on the basis of"*—For an excellent discussion of the standing issues in *Myriad*, see Megan

M. La Belle (2011) "Standing to Sue in the Myriad Genetics Case" *California Law Review Circuit* 2: 68–94.

242 *"Under applicable Supreme Court precedent . . ."*—The Supreme Court established the applicable standard for standing in declaratory judgment cases in *MedImmune, Inc. v. Genentech, Inc.*, 549 U.S. 118 (2007), which overruled the Federal Circuit's prior, more restrictive, test.

242 *"the concept of standing . . ."*—Under the Supreme Court's leading 1992 precedent, *Lujan v. Defenders of Wildlife*, the Court denied standing to environmentalists who claimed that they were injured by a Department of the Interior regulation that removed non-U.S. actions from the scope of the Endangered Species Act. The Court found that the harm alleged by the environmentalists—that the regulation would reduce their opportunity to observe wildlife—was insufficiently imminent or certain to support a lawsuit challenging the regulation. The latest word on standing in patent cases was handed down by the Supreme Court in *Medtronic, Inc. v. Mirowski Family Ventures, LLC*, 134 S. Ct. 843 (Jan. 22, 2014).

242 *"At a press conference . . ."*—Meredith Wadman (2011) "Arguments heard in high-profile patent case against Myriad Genetics" *Nature News Blog*, Apr. 4, 2011.

244 *"Kazazian and Ganguly had no standing to sue . . ."*—The question of standing, and whether the Federal Circuit correctly applied relevant Supreme Court precedent, has occasioned disagreement among scholars. See, for example, Rinehart (2015, pp. 1171–73) and La Belle (2011).

244 *"a book of his own . . ."*—Harry Ostrer (2012) *Legacy: A Genetic History of the Jewish People* (Oxford Univ. Press).

244 *"A month after . . ."*—Even before he knew that Ostrer would be the only plaintiff left with standing, Castanias thought it would be worth notifying the court of Ostrer's pending move. After all, nineteen plaintiffs were better than twenty. So on July 27, two days before the court released its opinion, Castanias filed a letter informing the court that Ostrer had announced his intention to leave NYU effective September 1. Fortunately for the plaintiffs, the court's opinion was already complete and in the process of being formatted and duplicated for publication. When the opinion was released, there was no mention of Ostrer's departure from NYU.

249 *"we had segregated schools . . ."*—The issue of school segregation was well-known to Hansen, who worked on the continuation of the landmark desegregation case *Brown v. Board of Education of Topeka* early in his career.

249 *"The Society had invited . . ."*—A recording of this discussion is available at https://www.ashg.org/press/webcast/ICHG2011-GenePatenting.shtml.

250 *"quoting from Goethe . . ."*—This quotation is generally attributed to the German poet, statesman, and natural philosopher Johann Wolfgang von Goethe (1749–1832).

PART III: HIGHEST COURT IN THE LAND

251 *"Patents cannot issue . . ."*—*Funk Brothers Seed Co. v. Kalo Inoculant Co.*, 333 U.S. 127 (1948).

Chapter 22: Déjà Vu All Over Again

252 *"asked Myriad whether . . ."*—Declaration of Ellen Matloff in *AMP v. Myriad* (2010), p. 2.

253 *"the New York Times . . ."*—Andrew Pollack (2011) "Despite Gene Patent Victory, Myriad Genetics Faces Challenges" *New York Times*, Aug. 24, 2011.

253 "The Oncology Nurse"—Cristi Radford (2011) "How Comprehensive is Comprehensive BRACAnalysis" *The Oncology Nurse* 4(8), Dec. 27, 2011.

256 *"massive, multi-gene tests . . ."*—But see Holman (2012), arguing that gene patents will not impede next-generation genetic tests.

256 *"The Supreme Court took nearly four months . . ."*—*Mayo Collaborative Services v. Prometheus Laboratories, Inc.*, 566 U.S. 66 (2012).

256 *"sent shock waves through the patent community . . ."*—Blogger Gene Quinn expressed the community's shock and outrage at the *Mayo* decision, hyperventilating that "those in the biotech, medical diagnostics and pharmaceutical industries have just been taken out behind the woodshed and summarily executed by the Supreme Court this morning. An enormous number of patents will now have no enforceable claims. Hundreds of billions of dollars in corporate value has been erased." Gene Quinn (2012) "Killing Industry: The Supreme Court Blows *Mayo v. Prometheus*," *IP Watchdog*, Mar. 20, 2012.

258 *"BIO's outside counsel . . ."*—Seth Waxman, a former solicitor general who now works for the law firm WilmerHale.

260 *"he later joked with a reporter . . ."*—Jan Wolfe (2012) "Litigator of the Week: Gregory Castanias of Jones Day" *The American Lawyer (Online)*, August 23, 2012.

261 *"giving some commentators . . ."*—Courtenay C. Brinckerhoff (2012) "Myriad Oral Arguments: Deja Vu?" *PharmaPatents Blog*, Jul. 22, 2012.

Chapter 23: Air Force 1

265 *"conservative Justice Scalia . . ."*—For a discussion of Justice Scalia's use of standing issues to derail public environmental litigation, see Lazarus (2019, pp. 188–9).

267 *"there was one factor . . ."*—For an informative discussion of the Supreme Court's practices regarding the granting of cert. in patent cases, see Christa J. Laser (2020) "Certiorari in Patent Cases" *AIPLA Quarterly Journal* 48(4): 569–621.

270 *"challenging Monsanto's patents . . ."*—See Eli Kintisch (2011) "Patent Foe Sues Monsanto on Modified Crops," *Science* online, Mar. 30, 2011.

271 *"seemed to be itching to say more . . ."*—Justice Breyer's discomfort with intellectual property protection can be traced all the way back to an article he wrote while still a junior faculty member at Harvard. Stephen Breyer (1970) "The Uneasy Case for Copyright: A Study of Copyright in Books, Photocopies, and Computer Programs" *Harvard Law Review* 84(2): 281–355.

271 *"As one commentator . . ."*—Dennis Crouch (2019) "My Take after Oral Arguments: Supreme Court Likely to Affirm in *Peter v. NantKwest*" *Patently-O blog*, Oct. 8, 2019.

271 *"especially after . . ."*—One can only obtain a patent on an "invention" that is not already known to the public (in the "prior art"). Thus, the effect of the Human Genome Project, which publicly released vast quantities of human DNA sequence information from 1998 through about 2001, was to preempt further patenting of this information. See Contreras (2011, pp. 86–87). Of course, this had no effect on the patents obtained before the HGP released its data.

271 *"a trademark dispute over . . ."*—*Already, LLC v. Nike, Inc.*, 568 U.S. 85 (2013).

274 *"wrote a blog post titled . . ."*—Kevin E. Noonan (2013) "Is It Time for Myriad to Concede in *AMP v Myriad* for the Good of the Biotechnology Industry?" *Patent Docs*, Jan. 23, 2013.

274 *"there was precedent on the books . . ."*—*Lear v. Atkins*, 395 U.S. 653 (1969).

274 *"if an appeal is rendered moot . . ."*—*U.S. Bancorp Mortgage Co. v. Bonner Mall Partnership*, 513 U.S. 18 (1994).

274 *"the risks and uncertainty of the Nike gambit were just too great . . ."*—Despite the hesitation

on the part of Myriad's counsel, the *"Nike* gambit"—issuing an irrevocable license to a litigant to remove any case or controversy between the parties—has proven to be successful in both patent and copyright cases following the Supreme Court's ruling in the *Air Force 1* case. In *St.-Amour v. Richmond Org.* , 2020 BL 73895, S.D.N.Y., No. 16-4464, Opinion (Feb. 28, 2020), the owner of the copyright in the famous Woodie Guthrie song "This Land Is Your Land" granted a license to a band that sought to challenge the validity of its copyright. Citing the *Nike* case, the New York court dismissed the case on the basis that it had become moot. And in the massive patent lawsuit between Apple and Qualcomm, Qualcomm's promise not to assert certain patents against Apple similarly mooted Apple's challenge to Qualcomm's patents. *In re: Qualcomm Litigation*, No. 3:17-cv-108-GPC-MDD, Order Granting Qualcomm's Motion for Partial Dismissal of Apple's First Amended Complaint and the CMS' Counterclaims (S.D. Cal., Nov. 20, 2018).

Chapter 24: With Friends like These

276 *"even to the extent that . . ."*—Chien (2011, p. 400). At this stage, many of the "usual suspects" appeared on the scene. One study of *amicus* briefs in patent appeals found that of the top ten filers, five were bar associations, two were companies (Intel and Eli Lilly), two were U.S. government entities (the solicitor general and the Patent Office), one was a patent-focused trade organization (the Intellectual Property Owners Association or IPO) and one was BIO. *Ibid.*, p. 416.

276 *"the average of four . . ."*—Chien (2011, p. 418) (counting *amicus* briefs in patent cases heard by a three-judge Federal Circuit panel and filed between 1989 and 2009, a total of 2,366 cases).

276 *"Over the last fifty years . . ."*—Joseph D. Kearney and Thomas W. Merrill (2000) "The Influence of Amicus Curiae Briefs on the Supreme Court" *University of Pennsylvania Law Review* 148: 743–855.

276 *"Professor Allison Orr Larsen reports . . ."*—Allison Orr Larsen (2014) "The Trouble with Amicus Facts," *Virginia Law Review* 100: 1757–1818, p. 1762.

277 *"One pair of law professors . . ."*—Brief of Kali N. Murray and Erika R. George as Amici Curiae in Support of Petitioners, pp. 21–24. Interestingly, one of these professors, Erika George, was (and, along with the author, continues to be) on the law faculty of the University of Utah, the owner of several of the challenged *BRCA* patents.

277 *"the Southern Baptist Convention . . ."*—Brief for Amici Curiae of the Ethics & Religious Liberty Commission of the Southern Baptist Convention and Prof. D. Brian Scarnecchia in Support of Petitioners, p. 4.

277 *"Lynch Syndrome International . . ."*—Brief for Amicus Curiae Lynch Syndrome International in Support of Respondents, p. 18.

278 *"in the words of Professor Larsen . . ."*—Larsen (2014, p. 1757).

278 *"Larsen found that . . ."*—Larsen (2014, p. 1778).

279 *"In a 2009* Fortune *magazine profile . . ."*—David Stipp (1999) "Hatching a DNA Giant" *Fortune*, May 24, 1999.

281 *"some of the most important intellectual property . . ."*—Among Verrilli's significant Supreme Court cases were *MGM Studios, Inc. v. Grokster, Ltd.*, 545 U.S. 913 (2005) (finding peer-to-peer file sharing sites liable for inducing copyright infringement) and *Golan v. Holder* 565 U.S. 302 (2012) (extending copyright protection to foreign works previously in the public domain).

281 *"of a contact sport . . ."*—Stephen M. Shapiro (1984) "Oral Argument in the Supreme Court of the United States," *Catholic University Law Review* 33: 529–553.

281 *"constantly interrupt the speaker . . ."*—In October 2019, the Court amended its internal procedures to give advocates two minutes of (generally) uninterrupted time to present their arguments before questions would begin. Guide for Counsel in Cases to be Argued before the Supreme Court of the United States (Oct. Term 2019, updated Oct. 3, 2019), p. 7. The two-minute rule was broken almost as soon as it was implemented. See Kimberly Strawbridge Robinson (2019) "Sotomayor First—and Second—Justice to Break 'Quiet Time' Rule," *Bloomberg Law*, Oct. 16, 2019.

282 *"between 2009 and 2011 . . ."*—Cynthia K. Conlon and Julie M. Karaba (2012) "May It Please the Court: Questions about Policy at Oral Argument" *Northwestern Journal of Law and Social Policy* 8: 89–120.

284 *"The Court itself offers . . ."*—Guide for Counsel in Cases to Be Argued before the Supreme Court of the United States (Oct. Term 2015), pp. 5–6.

284 *"As one distinguished Supreme Court advocate . . ."*—Stephen M. Shapiro (1984) "Oral Argument in the Supreme Court of the United States" *Catholic University Law Review* 33: 529–553, p. 539.

285 *"Give your full time and attention . . ."*—Supreme Court Guide for Counsel (2015, p. 9).

285 *"perhaps the most useful . . ."*—Supreme Court Guide for Counsel (2015, p. 8).

285 *"its one-millionth . . . "*—Myriad Genetics, Inc. "Company Milestones."

Chapter 25: Oyez, Oyez, Oyez!

286 *"a grandiose marble edifice . . ."*—Architect Cass Gilbert, designer of the Woolworth Building in New York, designed the Supreme Court building as a "temple of justice" "on a scale in keeping with the importance and dignity of the Court and the Judiciary as a coequal, independent branch of the United States Government, and as a symbol of the national ideal of justice in the highest sphere of activity." Supreme Court of the United States, Building History, https://www.supremecourt.gov/about/buildinghistory.aspx (visited Jul. 7, 2020). Not all commentators have had equally glowing reactions to the Supreme Court building. Richard Kluger, in his history of the school desegregation cases, calls it "at once the most elegant and most preposterous building in Washington . . . It is as if America wished to proclaim itself the reincarnation of Greece of the Golden Age or Rome at its imperial height—a status attained, as lives of nations are measured, virtually overnight." Richard Kluger, *Simple Justice: The History of Brown v. Board of Education and Black America's Struggle for Equality* 563–64 (Vintage Books, 1975).

286 *"In important cases . . ."*—In *Obergefell v. Hodges*, 576 U.S. 644 (2015), the famous "gay marriage" case, would-be spectators began to queue in front of the Court building three days before the hearing.

287 *"gifts of the Japanese government . . ."*—In 1912, the Mayor of Tokyo made a gift of 3,000 Yoshino cherry trees to the city of Washington, DC. The cherry trees, which blossom every March or April, are viewed by millions and blanket the Capitol grounds in delicate pink and white petals.

289 *"confiscate all electronics . . ."*—This requirement has surprised more than a few would-be observers at the Court. But no exceptions are made to this strict policy, except for counsel actually arguing before the Court.

290 *"waited all night . . ."*—Professor Josh Blackman offers a detailed look at how to get a seat at a Supreme Court argument in "How to Attend Oral Arguments at the Supreme Court," *Reason*, Nov. 18, 2019.

290 *"The court's protocol manual explains . . ."*—Supreme Court Guide for Counsel (2015, p. 4).

293 *"Waston had filed an* amicus *brief . . ."*—Amicus brief of James Watson (2013, p. 6 n.3) (when asked, after his first presentation of the structure of the DNA molecule, whether he would patent it, Watson replied that patenting "was out of the question").

297 *"From this very podium . . ."*—For a discussion of prominent advocates before the Supreme Court, see Richard J. Lazarus (2008) "Advocacy Matters before and within the Supreme Court: Transforming the Court by Transforming the Bar" *Georgetown Law Journal* 96: 1487-1564.

297 *"pancreatic cancer . . ."*—The disease eventually took Justice Ginsburg's life in 2020.

299 *"a seemingly unrelated case about employment law . . ."*—*NASA v. Nelson*, 562 U.S. 134 (2011).

299 *"She asked him whether an employer could legally ask an employee . . ."*—In actuality, the genetics-related employment scenario posed by Justice Sotomayor would probably have been prohibited under the Genetic Information Nondiscrimination Act of 2008 (GINA), which was extended to cover federal employers in 2009.

299 *"she had been married to one . . ."*—Lisa Capretto (2013) "Justice Sotomayor Talks Marriage, Divorce and What It's like to Be on the Supreme Court" *Huffington Post*, Mar. 29, 2013 (discussing comments made by Justice Sotomayor on *Oprah's Next Chapter*).

301 *"Justice Breyer was famous for regularly deploying . . ."*—See Kimberly Robinson (2021) "Stephen Breyer, The Supreme Court's King of Legal Hypotheticals," *Bloomberg Law*, April. 5, 2021.

301 *"classic Sotomayor . . ."*—Justice Sotomayor discusses her upbringing and the debt she owes to her grandmother in her autobiography, *My Beloved World* (Knopf, 2013).

303 *"it was unheard of . . ."*—Jacqueline Bell and Cristina Violante (2016) "BigLaw's Amicus Business Is Booming, But Is the Court Listening?" *Law360*, October 2, 2016 (discussing emerging trend on Supreme Court to refer to *amicus* briefs during oral argument).

303 *"In a 1984 paper, Nussbaum . . ."*—Su, T. S., Nussbaum, R. L., Airhart, S, Ledbetter, D. H., Mohandas T., O'Brien, W. E., Beaudet, A. L. (1984) "Human chromosomal assignments for 14 argininosuccinate synthetase pseudogenes: cloned DNAs as reagents for cytogenetic analysis," *American Journal of Human Genetics* 36(5): 954-64.

308 *"In* Science . . ."—Eliot Marshall (2013) "In a Flurry of Metaphors, Justices Debate a Limit on Gene Patents" *Science* 340: 421.

310 "TIME *called Jolie's announcement . . ."*—Alice Park (2013) "The Angelina Effect" *TIME*, May 26, 2013.

310 *"testing surged across . . ."*—See, e.g., Evans, D.G.R., Barwell, J., Eccles, D.M. et al. (2014) "The Angelina Jolie effect: how high celebrity profile can have a major impact on provision of cancer related services" *Breast Cancer Research* 16, 442; Troiano, G., Nante, N., & Cozzolino, M. (2017) "The Angelina Jolie effect—Impact on breast and ovarian cancer prevention—A systematic review of effects after the public announcement in May 2013," *Health Education Journal* 76(6), 707–715.

311 *"Andrew Cohen, writing . . ."*—Andrew Cohen, "The Supreme Court Case Looming over Angelina Jolie's Breast-Cancer Column," *The Atlantic*, May 15, 2013.

Chapter 26: 9–0

312 *"Known for his sphinxlike silence . . ."*—Michael O'Donnell (2019) "Deconstructing Clarence Thomas," *The Atlantic*, Sept. 2019.

312 *"Over the past decade . . ."*—Patent-related opinions by Justice Thomas include *eBay v. MercExchange* (2006) (availability of injunctive relief), *Quanta v. LG Electronics* (2008) (doctrine

of patent exhaustion), *Alice v. CLS Bank* (2014) (patentability of software), *TC Heartland v. Kraft Food* (2017) (proper venue for infringement suits), *Oil States Energy v. Greene's Energy Group* (2018) (Patent Office review process)(a 7–2 decision), and *Helsinn Healthcare v. Teva Pharmaceuticals* (2019) (effect of confidential prior art). For a recent analysis of Justice Thomas's intellectual property jurisprudence, see Robert W. Gomulkiewicz (2022) " The Supreme Court's Chief Justice of Intellectual Property Law" Nevada Law Review vol. 22.

316 *"As she later told Park . . ."*—Ceriani, StoryCorps interview, Mar. 26, 2016.

318 *"closed down by 6 percent . . ."*—See Jose Pagliery, "Myriad Genetics whipsawed on Supreme Court gene ruling," CNN Business, Jun. 13, 2013.

318 *"the* New Republic *. . ."*—Jeff Guo (2013) "The Supreme Court Reveals Its Ignorance of Genetics," New Republic, Jun. 13, 2013.

318 *"the indomitable Gene Quinn . . ."*—Gene Quinn (2013) "Supremes Rule Isolated DNA and Some cDNA Patent Ineligible," *IP Watchdog,* Jun. 13, 2013.

Chapter 27: Aftermath

320 *"the scientific journal* Nature *. . ."*—"365 Days: Nature's 10," Nature, Dec. 18, 2013.

322 *"and has proposed that every woman . . ."*—Mary-Claire King, Ephrat Levy-Lahad, Amnon Lahad (2014) "Population-Based Screening for *BRCA1* and *BRCA2*" *Journal of the American Medical Association* 312(11): 1091–1092.

322 *"Yet, even now . . ."*—See, e.g., U.S. Preventive Services Task Force (2019) "Risk Assessment, Genetic Counseling, and Genetic Testing for *BRCA*-Related Cancer: U.S. Preventive Services Task Force Recommendation Statement" *Journal of the American Medical Association* 322(7):652–665; Susan M. Domchek (2019) "Risk-Reducing Mastectomy in *BRCA1* and *BRCA2* Mutation Carriers—A Complex Discussion" *Journal of the American Medical Association* 321(1): 27–28.

322 *"At a recent bar association event ..."* — Utah IP Summit – February 23, 2018 (Hilton Hotel, Salt Lake City).

323 "SAS Institute v. Iancu . . ."—584 U.S., 138 S. Ct. 1348, 200 L. Ed. 2d 695 (2018).

323 *"ranked Jones Day . . ."*—Acritas (2020) "U.S. Law Firm Brand Index 2020."

323 *"broke a new record . . ."*—Tony Mauro, "Jones Day Lands a Record 11 Supreme Court Law Clerks as Associates," *Natl. Law Journal,* Nov. 13, 2018.

324 *"now leads the non-profit . . ."*—See ovariancancerproject.org.

324 *"Less than thirty days . . ."*—In re: BRCA1- & BRCA2-Based Hereditary Cancer Test Patent Litig., 3 F. Supp. 3d 1213 (D. Utah 2014).

324 *"it counted 515 patent claims . . ."*—*University of Utah Research Foundation v. Ambry Genetics Corp.,* Case No. 2:13-cv-00640-RJS, Motion for Preliminary Injunctive Relief and Memorandum in Support (D. Utah, Jul. 9, 2013).

325 *"ruled against the company . . ."*—In re: BRCA1- and BRCA2- Based Hereditary Cancer Test Patent Litigation, MDL Case No. 2:14-MD-2510, Case No. 2:13-Cv-00640-RJS, Memorandum Decision and Order Denying Plaintiffs' Motion for Preliminary Injunction (D. Utah, Mar. 10, 2014).

325 *"Somewhat surprisingly, the Federal Circuit . . ."*—In re: BRCA1- & BRCA2-Based Hereditary Cancer Test Patent Litig. 774 F.3d 755 (Fed. Cir. 2014).

325 *"Myriad continues to capitalize on the valuable data . . ."*—See Ghosh (2020); Christi J. Guerrini et al. (2017) "Myriad take two: Can genomic databases remain secret? *Science* 356(6338): 586-7.

326 *"As explained by Senator Tillis . . ."*—Sen. Thom Tillis (2019) "Press Release: Sens. Tillis and Coons and Reps. Collins, Johnson, and Stivers Release Draft Bill Text to Reform Section 101 of the Patent Act," May 22, 2019.

326 *"drafting letters to Congress . . ."*—The ACLU assembled an impressive coalition of more than 150 advocacy, medical, and patient organizations to sign a letter opposing the Coons-Tillis legislation. Letter from ACLU et al. to Sens. Coons and Tillis, and Reps. Collins, Johnson, and Stivers dated Jun. 3, 2019.

329 *"Francis Collins and others had proposed . . ."*—See Francis S. Collins (2004) "The case for a U.S. prospective cohort study of genes and environment" *Nature* 429: 475-477.

329 *"one of the biggest opportunities . . ."*—Thomas M. Burton, et al. (2015) "Obama Announces $215 Million Precision-Medicine Genetic Plan" *Wall St. Journal*, Jan. 30, 2015 (quoting President Barack Obama).

Appendix: The (Legal) Meaning of Myriad

331 *"Bob Cook-Deegan and . . ."*—Baldwin and Cook-Deegan (2013).

332 *"there had been significant litigation . . ."*—See, e.g., Palombi (2009, pp. 276–78) (discussing erythropoietin litigation).

333 *"on more conventional grounds . . ."*— Holman (2012, pp. 582–83).

336 *"As Rebecca Eisenberg has pointed out . . ."*—See Eisenberg (2002).

337 *"In a blog post . . ."*—Gene Quinn (2013) "Supremes Rule Isolated DNA and Some cDNA Patent Ineligible" *IP Watchdog*, Jun. 13, 2013.

337 *"primers were synthetic . . ."*—*BRCA1-* & *BRCA2-*Based Hereditary Cancer Test Patent Litig. (*Univ. of Utah Research Found. v. Ambry Genetics Corp.*), 774 F.3d 756, 761 (2014) ("A DNA structure with a function similar to that found in nature can only be patent eligible as a composition of matter if it has a unique structure, different from anything found in nature. Primers do not have such a different structure and are patent ineligible.")

337 *"patents should be available for innovative methods . . ."*— See, e.g., *The Wood Paper Patent*, 90 U.S. 566, 594 (1874) (while purified wood pulp is not patent-eligible, the novel process for extracting pulp as well as new boiler designs were eligible for patent protection). For a recent overview of the law of method or process patents, see Timothy R. Holbrook (2017) "Method Patent Exceptionalism" *Iowa Law Review* 102(3): 1001–62.

338 *"We even have a law . . ."*—35 U.S.C. § 271(g).

339 *"some of the largest patent battles . . ."*—Mukherjee (2016, p. 245), Cook-Deegan and Heaney (2010, p. 396).

340 *"has attracted the ire . . ."*—See, e.g., Burk (2014, pp. 507–08) (calling the distinction "incoherent") and Rai and Cook-Deegan (2013, p. 137) ("The Court's analysis does not connect the dots as to why claims to information in the form of cDNA are less problematic than claims to information in the form of gDNA").

342 *"Did Myriad 'invent' the BRCA genes . . ."*—This argument is a simplified version of the one advanced by Joshua Sarnoff in an *amicus* brief to the Supreme Court, which was signed by fifteen law professors, including the author.

343 *"a new rubbery elastic compound . . ."*—Flubber, of course, is inspired by the admittedly mediocre 1997 Robin Williams movie of the same name (Walt Disney Pictures, 1997).

344 *"Supreme Court decisions going back to the 1940s . . ."*—See, e.g., *Funk Brothers Seed Co. v. Kalo Inoculant Co.*, 333 U. S. 127, 130 (1948) ("If there is to be invention from [a discovery

of a law of nature], it must come from the application of the law of nature to a new and useful end"), *Diamond v. Diehr*, 450 U. S. 175, 187 (1981) ("an application of a law of nature or mathematical formula to a known structure or process may well be deserving of patent protection").

345 *"the result of Mayo . . ."*—See Eisenberg (2015).

345 *"Recent judicial rulings have severely limited . . ."*—See Jacob S. Sherkow (2020) "Adaptive Intellectual Property," (working paper, p.39) ("Antibody patents in the U.S. have recently faced a reckoning from the U.S. Court of Appeals for the Federal Circuit. After decades of being able to patent antibodies with only loose—if not minimal—restrictions on sequencing information, the Federal Circuit significantly cracked down on the practice in 2017 in *Amgen v. Sanofi*").

346 *"non-obvious applications . . ."*—Though the debate over patentability of genetic tests has largely focused on the hurdle of patentable subject matter under Section 101 of the Patent Act, the requirement that inventions be non-obvious under Section 103 is not trivial either. See SACGHS (2010, pp. 68–71).

346 "Diamond v. Diehr . . ."—450 U.S. 175 (1981).

347 *"the Federal government began to pour money . . ."*—In 1953, federal non-defense R&D funding was $2.2 billion. By 1980, it had reached $41.5 billion. AAAS, Historical Trends in Federal R&D, By Function: Defense and Nondefense R&D, 1953–2017.

348 *"Bayh-Dole Act of 1980 . . ."*—For an early history and analysis of the Bayh-Dole Act, see Eisenberg (1996).

348 *"This generous arrangement . . ."*—For discussions of the incentives that universities and other non-profit institutions have to make discoveries, see, e.g., Ouellette and Weires (2019), Rooksby (2017), Ouellette (2015), Eisenberg (2015, p. 283–84), SACGHS (2010, pp. 35, 90).

348 *"During the decade after the Bayh-Dole Act was passed . . ."*—Rory P. O'Shea et al. (2005) "Entrepreneurial orientation, technology transfer and spinoff performance of U.S. universities," *Research Policy* 34: 994–1009, p. 1002 (spinout statistics 1980–1994).

348 *"By the mid 1980s . . ."*—Bill Curry, "Utah University Spurs Creation of 'Bionic Valley': Artificial Body Parts May Aid Thousands," *Los Angeles Times*, Nov. 10, 1985.

349 *"By 1986, Utah's TTO was earning . . ."*—Robert Miller, Interview No. 352—James J. Brophy interview 1991, pp. 373–74.

349 *"they had stumbled onto . . ."*—The saga of the cold fusion debacle at the University of Utah is chronicled in Gary Taubes (1993) *Bad Science: The Short Life and Weird Times of Cold Fusion* (Random House). A detailed chronology of events can be found in B. V. Lewenstein and W. Baur (1991) "A Cold Fusion Chronology," *J. Radioanal. Nucl. Chemistry* 152, 273.

350 *"$40 million . . ."*—In re: *BRCA1-* and *BRCA2-* Based Hereditary Cancer Test Patent Litigation, MDL Case No. 2:14-MD-2510, Memorandum Decision and Order Denying Plaintiffs' Motion for Preliminary Injunction 5 (D.Utah, Mar. 10, 2014).

350 *"Mary-Claire King lamented that . . ."*—Gitschier (2013, pp. 4–5).

350 *"The University of Michigan . . ."*—For an in-depth discussion of the patenting and licensing of the *CFTR* gene, see Misha Angrist, Subhashini Chandrasekharan, Christopher Heaney, and Robert Cook-Deegan (2010), "Impact of gene patents and licensing practices on access to genetic testing for long QT syndrome" *Genetics in Medicine* 12(4):S111–S154; Mollie A. Minear, Cristina Kapustij, Kaeleen Boden, Subhashini Chandrasekharan, and Robert Cook-Deegan (2013) "Cystic Fibrosis Patents: A Case Study of Successful Licensing," *Les Nouvelles*, Mar. 2013: 21–30.

350 *"NIH obtained a patent on the gene . . ."*—For an in-depth discussion of the patenting and licensing of the Tay-Sachs gene *HEXA*, see Alessandra Colaianni, Subhashini Chandrasekharan, and Robert Cook-Deegan (2010), "Impact of Gene Patents and Licensing Practices on Access to Genetic Testing and Carrier Screening for Tay-Sachs and Canavan Disease," *Genetics in Medicine* 12(4):S5–S14.

350 *"Over the life of the patents . . ."*—Cook-Deegan and Heaney (2010, p. 392).

350 *"Columbia University . . ."*—Alessandra Colaianni and Robert Cook-Deegan (2009) "Columbia University's Axel patents: technology transfer and implications for the Bayh—Dole Act" *Milbank Quarterly* 87(3): 683–715.

351 *"generally applicable research tools . . ."*—Academics and NIH have urged the non-exclusive licensing of research tools since the 1990s. See NIH (1999); Heller and Eisenberg (1998). This being said, some have criticized the licensing of these technologies on a royalty bearing basis at all, noting that royalties paid to universities seldom promote the dissemination of the technologies, and simply inure to the financial benefit of the universities. Eisenberg and Cook-Deegan (2018, p. 78) (comparing payments made on platform technologies to a tax on innovation).

351 *"Most gene patents . . ."*—Today, Marie-Claire's former institution, UC Berkeley, is regarded by many as one of the most aggressive and profit-oriented universities engaged in academic technology transfer. In 2017, Jacob Sherkow and I wrote about Berkeley's exclusive license of all of its rights in the revolutionary CRISPR-Cas9 gene editing technology to a privately held company called Caribou Therapeutics that was formed by Berkeley researchers. Contreras and Sherkow (2017).

351 *"Jacob Sherkow and I . . ."*—Contreras and Sherkow (2017, pp. 698–99). As we explain, "These surrogates control a large and lucrative field for the exploitation of the licensed technology, and have significant freedom both to exploit it themselves and to seek partners and sublicensees. The surrogates take on the role of the patent owner and retain a lion's share of the resulting profits."

352 *"This mandatory contractual language . . ."*—Rachel Nowak, NIH in danger of losing out on *BRCA1* patent, 266 SCIENCE 209, 209 (1994).

353 *"To prevent further instances of price gouging . . ."*—Marlene Cimons and Victor F. Zonana (1989) "Manufacturer Reduces Price of AZT by 20%," *L.A. Times*, Sept. 19, 1989.

353 *"Varmus decided to eliminate . . ."*—Lauren Neergaard (1995) "NIH Drops 'Reasonable Price' Clause for Drug Company Collaboration" *Associated Press*, Apr. 11, 1995. The issue of "fair" drug pricing has continued to be a top priority issue in the United States.

354 "Nature . . ."—Andrews (2002).

354 "Boston Globe Magazine . . ."—Blanton (2002).

354 *"a growing chorus . . ."*—See, e.g., Cho et al. (2003), Eisenberg (2002), Leonard (2002), Hull and Prasad (2001).

354 *"a major government task force . . ."*—The SACGHS task force was established in the fall of 2002 to "to formulate advice and recommendations on the range of complex and sensitive medical, ethical, legal, and social issues raised by new technological developments in human genetics, including the development and use of genetic tests," and to study gene patenting, in particular. SACGHS (2010, p. ix).

354 *"could have included . . ."*—The issue of pricing in healthcare markets is a complex and fraught one, and there is no single contractual mechanism that can solve it. Nevertheless, some contractual commitment to price products at "fair" levels could help. First, one could ask

whether Myriad's pricing for BRACAnalysis was fair. There is evidence to suggest that it was not. Both its impressive profit margins (approaching 90 percent for diagnostic testing) and Mark Skolnick's embarrassing admission in the 2007 documentary *In the Family* suggest that there was room for a price reduction.

355 *"a set of core intellectual property . . ."*—AUTM (2007).

355 *"Today, over one hundred . . ."*—A current list of signatories to the Nine Points document is maintained here: https://autm.net/about-tech-transfer/principles-and-guidelines/nine-points-to-consider-when-licensing-university.

355 *"So-called 'ethical licensing' . . ."*—Christi J. Guerrini, Margaret A. Curnutte, Jacob S. Sherkow, and Christopher T. Scott (2016), "The rise of the ethical license," *Nature Biotechnology* 37(1): 22–24.

355 *"Numerous other examples . . ."*—See, e.g., Jorge L. Contreras (2018) "The Evolving Patent Pledge Landscape" CIGI Papers No. 166, Apr. 3, 2018; Jorge L. Contreras and Meredith Jacob (2017) *Patent Pledges: Global Perspectives on Patent Law's Private Ordering Frontier* (Edward Elgar).

356 *"As Sherkow and I acknowledge . . ."*—Contreras and Sherkow (2017). See also Cook-Deegan and Eisenberg (2018, pp. 77–8), Rai and Eisenberg (2003, p. 300–03).

356 *"I am not alone in arguing that . . ."*—See, e.g., Ian Ayres and Lisa Larrimore Ouellette (2017) "A Market Test For Bayh—Dole Patents" *Cornell Law Review* 102: 271–331 (arguing that the potential market for university technology should dictate whether it is licensed exclusively or non-exclusively); Watson (2013, pp. 19–20) ("If, for some reason, patents on human genes are deemed necessary, the next best, albeit imperfect, solution is to require those patent holders to license the patents to other researchers so that scientific progress is not obstructed"); AUTM (2007) (universities commit to limiting exclusive licensing of research tools); NIH (2005) (urging NIH-funded researchers to grant non-exclusive licenses with respect to broadly applicable research tools); Rai and Eisenberg (2003, p. 310) ("Congress should recognize that patenting and exclusive licensing are not always the best way to maximize the social value of inventions and discoveries made with federal funds"); NIH (1999) ("Where the subject invention is useful primarily as a research tool, inappropriate licensing practices are likely to thwart rather than promote utilization, commercialization, and public availability"); Heller and Eisenberg (1998) (theorizing that patenting and exclusive licensing of broadly applicable research tools could lead to anticommons effects).

356 *"In fiscal year 2020 . . ."*—Congressional Research Service, Federal Research and Development (R&D) Funding: FY2020 at (March 18, 2020).

356 *"In 2019, a team of researchers . . ."*—L. Fleming, H. Greene, G. Li, M. Marx, D. Yao (2019) "Government-funded research increasingly fuels innovation" *Science* 364:1139–1141.

356 *"Given this significant level of support . . ."*—Michael Sweeny (2012) "Correcting Bayh-Dole's Inefficiencies for the Taxpayer" *Northwestern Journal of Technology and Intellectual Property* 10(3): 295–311, pp. 306–07, and Ana Santos Rutschman (2018) "Vaccine Licensure in the Public Interest: Lessons from the Development of the U.S. Army Zika Vaccine" *Yale Law Journal Forum* 127: 651–666.

356 *"As Rai and Eisenberg noted . . ."*—Rai and Eisenberg (2003, p. 310) ("The time is ripe to fine-tune the Bayh-Dole Act to give funding agencies more latitude in guiding the patenting and licensing activities of their grantees.")

356 *"federallly funded research that . . ."*—Other proposals regarding taxpayer-funded research have also been made, including the reimbursement of federal grant funding from the profits

of resulting discoveries (see Sweeny [2012, pp. 306–07] and the amendment of the Patent Act to prohibit exclusive licensing for vaccines and other critical technologies developed at public expense (Rutschman [2018]).

358 *"Americans pay far more . . ."*—McKinsey and Co. (2008) "Accounting for the cost of U.S. healthcare: A new look at why Americans spend more" (Dec. 2008).

358 *"Future Justice Ginsburg's husband . . ."*—*On the Basis of Sex* (Focus Features, 2018) (line spoken by Martin Ginsburg).

359 *"the different narrative strains . . ."*—Seasoned trial lawyers will explain that the most important thing about arguing a case is not marshaling a mass of legal precedent, but telling a convincing story. In *AMP v. Myriad*, many different stories were told: the scientific race to discover the *BRCA* genes, the high cost of testing once Myriad eliminated its competition, the honest efforts of patent examiners following long-established policy, the economic engine of the biotech industry, the explosion of patents covering everything from business models to software to human genes, and so on. In a 2016 article (Contreras [2016]), I traced the deployment of these different narratives by the parties, the *amici*, and the judges in *AMP v. Myriad*. Not surprisingly, the narrative that someone chose to tell about the case was highly determinative of the outcome that that person wished to achieve. Thus, scientists who opposed Myriad's patents talked about the routine nature of gene sequencing, while Myriad and its scientists discussed the amazing breakthrough they made. Judge Sweet wrote about the access and insurance issues caused by Myriad's dominance of the market, though those issues were not really before the Court, and the judges of the Federal Circuit talked about the Patent Office, the longstanding system that it upheld, and the commercial effect of upsetting settled expectations. Finally, Justice Thomas, who largely ruled for the ACLU, told a story of Myriad's opportunistic capture of the *BRCA* testing market. Again, irrelevant to the resolution of the narrow legal question before the Court, but persuasive to him, and the rest of the Court, nonetheless.

ACKNOWLEDGMENTS

THIS BOOK WOULD not have been possible without the assistance of many, many people. I interviewed more than a hundred individuals while researching the material, some of whom went above and beyond the call of duty, sending me photos and documents, responding to late-night emails, and pointing me toward unturned stones in this complex and multifaceted story. I can't name them all here, and some of them wish to remain anonymous. But I would like to recognize, in particular, Lisa Cannon-Albright, Bob Cook-Deegan, Chris Hansen, Chris Mason, Ellen Matloff, Kathleen Maxian, Sandra Park, and Tania Simoncelli for putting up with repeated requests from me and always responding with enthusiasm. I am especially grateful to Lisa Schlager, who introduced me to FORCE and some of its courageous members in 2013, at the very outset of this project. Thanks are also due to Kevin Noonan and Molly Silfen, who so carefully read the entire manuscript, and to Brunson Hoole, who put up with my endless revisions and corrections.

Over the years, I was ably assisted by numerous student research assistants at American University and the University of Utah. These included Vikrant Deshmukh, Hilary Gawrilow, Amy Biegelsen, Brian Flach, Michael Eixenberger, Steve Avena, Luke Hanks, and Brady Nash. I am also grateful to Ripple Weistling and Ross McPhail, my research librarians at American and Utah, respectively, who helped me to track down any number of obscure documents.

I received funding for this project from a variety of sources over the years, including a Leonardo da Vinci Fellowship Research Grant from the Center for the Protection of Intellectual Property at George Mason University Antonin Scalia Law School, project funding from the Program on Information Justice and Intellectual Property (PIJIP) at American University Washington College of Law, a University Research Committee Faculty Research and Creative Grant from the University of Utah, a pilot grant from the Utah Center for Excellence in ELSI Research (UCEER), and research funding from the Huntsman Cancer Institute at the University of Utah.

The writing and publication of this book were made possible by a number of people. Jordan Fisher Smith taught me how to tell a story that others would be interested in reading. My agents at Inkwell Management, Michael Carlisle and Michael Mungiello, believed in a novice to the field of narrative nonfiction. Amy Gash at Algonquin saw the potential of a story about real people engaged in a meaningful legal pursuit and shepherded the publication of this book through a global pandemic, and my editor, Margot Herrera, suffered with me through endless revisions of the manuscript, improving each one with skill, knowledge, and good humor. I was lucky to have Algonquin's Stephanie Mendoza as my publicist.

Finally, I owe an enormous debt of gratitude to my wife, Kimberly. Not only did she read many early drafts of the manuscript and offer countless suggestions for improvement, but over the years she has taught me much of the little I know about molecular biology, genomics, oncology, public health, and the virtue of patience.

PHOTO CREDITS

FRONT. COURTESY OF BREAST CANCER ACTION

8   COURTESY OF TANIA SIMONCELLI

10   COURTESY OF MATT CAHNAN, *SALT LAKE TRIBUNE*

22   COURTESY OF LORI ANDREWS

26   PUBLIC DOMAIN

51   PUBLIC DOMAIN

58   COURTESY OF ASHLEY JONES/WEILL CORNELL MEDICINE

73   © LUCA LUCARINI, CC BY SA 3.0

101   COURTESY OF THOMAS P. MCNULTY, HOST/PHOTOGRAPHER, WJYE-FM

118   COURTESY OF AMERICAN CIVIL LIBERTIES UNION

120   COURTESY OF MARY LEVIN/UNIVERSITY OF WASHINGTON

124   COURTESY OF AMERICAN CIVIL LIBERTIES UNION

135   COURTESY OF AMERICAN CIVIL LIBERTIES UNION

136   COURTESY OF JOEL ENGARDIO

139   PHOTO STILL FROM *IN THE FAMILY*, DIRECTED BY JOANNA RUDNICK (KARTEMQUIN FILMS)

166   COURTESY OF HERMAN YUE

173   COURTESY OF JOHN CURTIS/YALE SCHOOL OF MEDICINE

177   COURTESY OF AMERICAN CIVIL LIBERTIES UNION

184   U.S. NATIONAL INSTITUTES OF HEALTH

189   CBS PHOTO ARCHIVE VIA GETTY IMAGES

217   JONES DAY

259   U.S. DEPARTMENT OF JUSTICE

287   MLADEN ANTONOV VIA GETTY IMAGES

288   COURTESY OF BREAST CANCER ACTION

289   COURTESY OF BREAST CANCER ACTION

307   COURTESY OF AMERICAN CIVIL LIBERTIES UNION

309   COURTESY OF AMERICAN CIVIL LIBERTIES UNION

311   © 2013 TIME USA LLC. ALL RIGHTS RESERVED. USED UNDER LICENSE.

313   COURTESY OF ARTHUR LIEN/SCOTUSBLOG

INDEX

QUESTIONS FOR DISCUSSION

1. The ACLU's case challenging gene patenting was brought nearly twenty-five years after these patents started to issue on human DNA in the United States—so long that many in the industry, even some judges, felt that it was too late to reject the practice. Why is this relevant in a legal case and do you agree with Chris Hansen's response to those arguments?

2. Though the issues surrounding gene patenting were known to advocates like Lori Andrews for years before the ACLU became involved, they were not generally known to the public. When the ACLU decided to bring this suit, its team chose *BRCA* as the focal point knowing it would get the most traction in the public eye. What role did public perception ultimately play in the case?

3. Why did the patent bar react so strongly and negatively to the case? Did you find any of their objections, exemplified by blog posts by people like Kevin Noonan and Gene Quinn, to be persuasive? Can you think of any other arguments they could have made?

4. How large a role did individual patients play in the case? How did their stories enhance the legal arguments and issues? Did learning their stories influence the way you felt about the case?

5. Every legal case involves teamwork, and *AMP v. Myriad* was no exception. Was any single person indispensable in bringing and winning the case? What did you think about Dan Ravicher's role in it? Was he justified in feeling overshadowed by the ACLU?

6. Genetic counselors played an important part in the case, both as plaintiffs and advisors to some of the patients. Do you know anyone who has met with a genetic counselor? How was their experience? Under what circumstances would you want such a consultation?

7. In the beginning, Myriad seemed to have an advantage in the case: its patents were validly issued by the Patent Office and the company was represented by one of the largest and most powerful law firms in the country. Did you think there were any obvious missteps in Myriad's defense, or was the ACLU's legal case simply stronger?

8. One of the striking things about *AMP v. Myriad* is how differently various courts and judges reacted to the case. What do you think of a legal system that is so dependent on the personal views and backgrounds of individual judges? Is it possible for judges to be truly impartial?

9. Why was the issue of "standing" so important in the case? What was special about Harry Ostrer that gave him alone standing to pursue the case? Do you think the case might have turned out differently if any other single plaintiff had been the one to have standing?

10. Ultimately, the compromise position proposed by the U.S. solicitor general forms the basis for the Supreme Court's ruling. Were you surprised by the significant divisions within the federal administration over gene patenting? Can you think of other legal issues that might generate similar controversies within the government?

11. Was there a villain in this case? Did you find yourself agreeing with Myriad at any point in the book?

12. What tangible results emerged from the decision in *AMP v. Myriad*? What do you imagine might happen if, as proposed by some senators, the ruling is overruled by congressional legislation?

JORGE L. CONTRERAS received his law degree from Harvard University and teaches intellectual property, science policy, and genetics law at the University of Utah. He has served on government advisory committees and testified before Congress on patent law matters. His articles have appeared in *Science*, *Nature*, and leading law reviews; he has been featured on NPR, PRI, and BBC Radio; and his opinions are cited in the *New York Times*, the *Wall Street Journal*, the *Economist*, and the *Washington Post*. Visit genomedefense.org for more court documents and other supplemental information about the case.